The Latest Scientific Problems Related to the Implementation and Diagnostics of Construction Objects

The Latest Scientific Problems Related to the Implementation and Diagnostics of Construction Objects

Editors

Bożena Hoła

Anna Hoła

MDPI • Basel • Beijing • Wuhan • Barcelona • Belgrade • Manchester • Tokyo • Cluj • Tianjin

Editors

Bożena Hoła
Wroclaw University of
Science and Technology
Poland

Anna Hoła
Wroclaw University of
Science and Technology
Poland

Editorial Office
MDPI
St. Alban-Anlage 66
4052 Basel, Switzerland

This is a reprint of articles from the Special Issue published online in the open access journal *Applied Sciences* (ISSN 2076-3417) (available at: https://www.mdpi.com/journal/applsci/special_issues/Implementation_Diagnostics_Construction_Objects).

For citation purposes, cite each article independently as indicated on the article page online and as indicated below:

LastName, A.A.; LastName, B.B.; LastName, C.C. Article Title. *Journal Name* **Year**, *Volume Number*, Page Range.

ISBN 978-3-0365-5635-2 (Hbk)
ISBN 978-3-0365-5636-9 (PDF)

Contents

About the Editors

Bożena Hoła

Bożena Hoła is professor for Civil Engineering at the Faculty of Civil Engineering at the Wrocław University of Science and Technology. In 2008, she was appointed the head of the Department of Technology and Management in Construction. In 2019, the President of the Republic of Poland continued her academic title as professor. Scientific interests concern the issues of safety and health protection in construction processes, management of construction processes, the use of artificial intelligence methods in solving decision-making problems in construction, construction waste management and assessment of the quality of construction works. The scientific achievements include over 160 publications, including 3 monographs. She was the head of the research project "Knowledge management system in a construction company" funded by the National Science Center (2011–2012), head of the research team of the project funded by the National Center for Research and Development, entitled "Model of Risk assessment of construction disasters, accidents and hazardous events at workplaces using construction scaffolding" (2016–2018). Currently, she is a head of the grant entitled. "Modeling the impact of near misses on accidents in the construction industry" funded by the National Science Center. She was awarded the medal of the National Education Commission, the medal of the Polish Association of Construction Engineers and Technicians named after Aleksander Dyżewski for achievements in the field of scientific activity in the field of construction and the decision of the President of the Republic of Poland—the Gold Medal for Long Service.

Anna Hoła

Anna Hoła (Ph.D., Eng. arch.), is employed at the Faculty of Civil Engineering, Wroclaw University of Science and Technology. She completed her master's degree in architecture and urban planning in 2006, and defended her doctoral thesis with distinction before the Council of the Faculty of Architecture at Wrocław University of Science and Technology in 2013. Her scientific achievements include more than 80 works, including articles published in journals listed in the Journal Citation Reports (JCR) database and national research projects carried out within the framework of scientific and research cooperation with the environment. She has participated in research projects: "Hybrid tomograph for the study of dampness and building condition," carried out in 2016–2018 by NETRIX S.A., funded by the Operational Program Intelligent Development 2014–2020, "Model of risk assessment of construction disasters, accidents and hazardous events at workplaces using construction scaffolding," funded by the National Center for Research and Development. She was the main contractor in the project "Horticultural Exhibitions and Displays in Wroclaw", implemented in 2010–2013 and funded by the National Science Center. Scientific interests include: research of soggy and salted brick masonry in historical buildings, improvement of in situ moisture testing methodology, diagnostics of historical buildings using non-destructive methods, non-destructive testing, applications of artificial intelligence in construction, secondary moisture protection and methods of masonry drying.

applied sciences

MDPI

Editorial

The Latest Scientific Problems Related to the Implementation and Diagnostics of Construction Objects

Bożena Hoła * and Anna Hoła

Faculty of Civil Engineering, Wrocław University of Science and Technology, Wybrzeże Wyspiańskiego 27, 50-370 Wrocław, Poland; anna.hola@pwr.edu.pl
* Correspondence: bozena.hola@pwr.edu.pl

Citation: Hoła, B.; Hoła, A. The Latest Scientific Problems Related to the Implementation and Diagnostics of Construction Objects. *Appl. Sci.* **2021**, *11*, 6184. https://doi.org/10.3390/app11136184

Received: 7 June 2021
Accepted: 29 June 2021
Published: 3 July 2021

1. Introduction

The construction industry is a sector of the economy that is characterized by a large variety of building structures, as well as a large variability in the conditions of their implementation. Particularly in times of rapid economic development, this great variability and diversity generates many new scientific problems that must be solved in order to further improve the quality of construction production and reduce construction costs and time. Moreover, in the construction industry, as in other sectors of the economy, great importance is attached to all environmental issues, as well as to the broadly understood sustainable development strategies. This means that new building materials, modifications of commonly known and widely used materials, new research methods, and methods of implementing and controlling construction processes are still being sought. The diagnostics of existing facilities is also gaining importance, as it determines the operational safety and durability of buildings.

This Special Issue entitled "The Latest Scientific Problems Related to the Implementation and Diagnostics of Construction Objects" aims to present and discuss the results of the latest research in the broadly understood field of construction engineering, in particular regarding: the modification of the composition of building materials with various micro and nanomaterials, by-products, or waste; modern methods of controlling construction processes; methods of planning and effective management in the construction industry; and also methods of diagnosing building structures. Articles published in this issue cover theoretical, experimental, applied, and modelling research. They are organized into several representative topics, and the main content of each article is briefly discussed.

2. Research in the Field of Building Materials

The most popular material that is used in the construction industry is concrete. Its common use is primarily influenced by its high compressive strength, its relatively high durability and resistance to various factors, the ease of forming elements, and the availability of components and their low cost. Concrete is a composite, the basic components of which are a cement matrix and aggregate. In recent years there has been a growing interest in modifying cement composites with finer materials (e.g., various types of fibres or nanoparticles) in order to improve their parameters.

Article [1] presents an assessment of the creep of the cement matrix of self-compacting concrete modified with the addition of SiO_2, TiO_2, and Al_2O_3 nanoparticles using the cavity method. Depending on the type of nanoparticles used, an increase or a decrease in the creep coefficient CIT was found when compared to the reference series. It was found that the addition of SiO_2 and Al_2O_3 nanoparticles in the amount of 4.0% of the cement mass results in an unfavourably higher value of the creep coefficient (CIT) of the cement matrix. In turn, the use of TiO_2 nanoparticles in the amount of 4.0% of the cement mass results in a favourable reduction in the creep coefficient CIT. The statistical analysis of the obtained test results indicates, however, that the addition of nanoparticles does not significantly affect the creep of the cement matrix of self-compacting concrete.

Depending on the dimensions of concrete elements, aggregates of different granulation are used in building structures. The properties of the aggregate determine the strength and durability of the concrete. Taking this fact into account, the authors of article [2] examined the influence of the maximum graining of aggregate on the strength properties and modulus of elasticity of concrete. The research showed that the strength properties of the aggregate are not only proportional to the maximum size of the aggregate grain, but also to the aggregate's crushing strength. However, no analogous relationships were found in terms of the modulus of elasticity of the tested concrete.

As already mentioned, one of the main requirements for building structures is durability. In reinforced concrete structures, the direct factor that influences the durability of concrete is the corrosion of reinforcing steel. In paper [3], one of the non-destructive methods was used to assess the degree of corrosion of the reinforcement, namely the galvanostatic pulse method (GPM). It is an electrochemical method that uses the physicochemical properties of concrete and steel. Using this method, the influence of the temperature of the tested element on the results of such parameters as: corrosion current density, stationary potential of the reinforcement, and the resistivity of lagging, was investigated. The differences in the values of these parameters, which were measured on the same samples, but at different temperatures, amounted to several dozen percent in some cases. This means that measurements of actual structural elements conducted with the use of the GPM, e.g., at different times of the year, may lead to an incorrect estimation of the probability of corrosion of the reinforcement in the studied area, and also to an incorrect assessment of its corrosive activity over time. According to the authors, it is advisable to define appropriate temperature correction factors for measurements performed with the use of the GPM.

It is more and more common to use various fibrous composites that improve the technical parameters of concrete. The use of FRP (Fibre Reinforced Polymers) materials is of particular interest. Paper [4] proposes a new original mathematical formula for predicting the compressive strength of FRP-confined concrete cylinders. The formula was developed on the basis of the output data obtained from a neural network. The results of the study show that the mathematical formula proposed by the authors allows the compressive strength in concrete cylinders reinforced with FRP tapes to be estimated with greater accuracy when compared to other existing formulas. The authors emphasize that over 96% of the results obtained with the proposed formula are fully consistent with the results of experimental research. The proposed calculation method can be easily applied using a calculator, which is particularly important at the stage of preliminary engineering projects.

As a result of the growing environmental awareness of society, it has become very popular to use natural plant fibres as an addition to new thermo-insulating composite materials. Article [5] presents the results of research concerning the physical and thermomechanical properties of a new composite based on cement mortar reinforced with alpha fibres (AF) sourced from a species of grass growing in an area of the Mediterranean basin. It was shown that the addition of 5% of weight means that the composite material is lighter by about 15%, its thermal insulation properties improve by about 57%, and its heat diffusion damping coefficient increases by about 49%. Moreover, the mechanical bending and compressive strength of the composite increases by up to 10% with an AF content of 1%.

A significant proportion of building materials are capillary-porous materials that are characterized by a high degree of water absorption. In many technologies that are used in the construction industry, it is necessary to dry such materials. Paper [6] presents a mathematical model of drying a thin-layer capillary-porous material, which enables changes in the material's moisture and its drying time, depending on the drying temperature and the initial moisture content, to be forecasted. The results obtained from the model were confirmed to be in line with the experimental data known from the literature related to the drying of ceramic blocks used in the construction industry.

3. Methods of Controlling Construction Processes

The implementation of each construction project is related to the three following parameters: the scope of the project, implementation time, and budget. Changing one of these parameters causes changes to the others, and consequently affects the fourth parameter, i.e., the quality of the project. One of the key tasks of an investor and contractor at the stage of planning and implementing construction works is to measure the progress of construction works while taking into account the planned dates and costs. During the implementation of construction works, various types of disturbances often occur, which make the prepared implementation schedules obsolete. As a consequence, the original milestones identified in the project schedule are delayed. The authors of paper [7] undertook research that aimed to define the optimal set of actions of responding to schedule delays. They proposed a simulation method for selecting schedule compression measures, i.e., accelerating processes, and for determining the best moment to take such actions in the event of disruptions. The proposed method allows the costs of activities that cause schedule failure and the costs of delays to be minimized, as well as the resilience of the schedule to be increased by reducing the differences between the actual and planned start of the process. The developed model is meant to serve as a tool that supports decision-making by construction site managers in the event of finding disturbances in the course of construction works.

Appropriate cash flow planning is of key importance for investors and contractors. The S curve is a very helpful tool for planning, monitoring, and controlling construction projects in terms of time and costs. Knowledge of the planned and actual course of cumulative financial outlays over time, as well as the shape of the S curve and its deviations, allows rational actions to be taken in order to achieve success and the intended goal in the implementation of the investment. The aim of article [8] is to analyse the course of an exemplary construction project, to compare the costs of the planned works with the actual costs of the performed works, as well as to indicate the reasons of failure to meet the planned deadlines and the project's budget. The authors of the article analysed the financial expenditure for the implementation of a construction investment, which were incurred in 20-month cycles. On the basis of these results, charts and tables of the planned and actual cumulative costs of the completed investment were prepared, the detailed analysis of which allows interesting conclusions to be drawn. The development of a methodology for planning the cumulative cost curve in construction projects will enable better planning of financial outlays.

The image of every building is shaped by its facades. Currently, traditional concrete forms have been replaced with light casings in the form of aluminium-glass facades and ventilated facades. The authors of article [9] investigated the influence of various identified factors on the costs of implementing a building's facade system. On the basis of the collected quantitative and qualitative data, which were obtained as a result of research concerning the design documentation and cost estimates of public utility buildings, as well as on the basis of interviews conducted with experts, factors that have a real impact on the costs of aluminium-glass facades and ventilated facades were identified. The indicated factors were analysed and classified using the MICMAC structural analysis method. Finally, six groups of factors that influence the cost of facade systems were determined, including: regulatory factors that do not have a very strong impact on the cost level, but show a strong correlation with other factors; determinants that have a very strong impact on costs; and the group of external factors that have the least impact on the estimation of facade costs.

An important element in the implementation of construction investments is the quality of works. Defects affecting the quality of works are common in the construction industry in all countries. Previous studies of defects in residential buildings mainly focused on the defects that occur at the stage of works acceptance. The authors of article [10] examined damage in residential buildings that was reported during the warranty period. The statistical analysis of the research results showed that more than half of it was justified. Understanding the existence of defects in buildings is a fundamental prerequisite for their prevention and elimination. Research and analysis of defects that occur during the warranty period can significantly affect the development of defect management procedures and the creation of a knowledge map that concerns the frequency of defects in individual places in a building and a building's elements. There are costs associated with repairing damage. Knowledge about defects occurring in buildings can therefore be used for better planning of an investment budget.

It has recently become common practice to use small plots of land located in dense urban developments for construction purposes. In such places, new "infill buildings" are created. In the case of revitalizing existing historic buildings, their facade walls are often used to build new buildings. In both of these cases, the implementation of the facility includes deep excavations that can cause serious damage to existing buildings in the vicinity. Article [11] focuses on the problem of interaction mechanisms between soil and the structure of buildings located near deep excavations. The authors analysed various risk factors related to the construction of new infill buildings and the revitalization of historic buildings when using only their facade walls. The reaction of buildings to deformations caused by deep excavations is influenced by the accuracy of determining the deformations and stresses caused by these excavations. Examples of current solutions for securing the walls of existing buildings, as well as the method of monitoring vertical deformations with the use of the Hydrostatic Levelling Cell (HLC) system are presented.

The construction industry is one of the most dangerous branches on the labour market. Providing safe working conditions for construction workers is the basic task of every entrepreneur. Unfortunately, a significant proportion of accidents in the construction industry are caused by reasons attributable to an employee. One of them is alcohol abuse in the workplace. The aim of the research in article [12] was to identify the main problems related to alcohol consumption at work among construction industry employees, with particular emphasis on workplaces on construction scaffolding. This study confirmed that excessive alcohol consumption is the cause of many serious and fatal accidents. Of the 219 reported accidents related to work on building scaffolding, 17.4% indicated alcohol consumed at the workplace as the cause of the accident.

4. Selected Decision-Making Problems in the Construction Process

Due to the large variety of used building materials, the possible construction solutions, and the techniques used for the construction of buildings, the problem of choosing the best solution from among many that are possible often arises. In solving such problems, multi-criteria decision analysis methods are helpful.

Article [13] proposes a methodology for selecting the best solution for the construction of retaining walls located in various environments. In the developed methodology, the authors identified various types of retaining walls and defined the selection criteria that take into account: external and construction requirements, terrain characteristics, and economic criteria. The best solution is determined by the successive application of various multi-criteria methods of decision making.

In turn, article [14] includes a concept for supporting decisions when selecting the best contractor for the project. In the developed methodology, a combination of Analytical Hierarchy Process (AHP) and PROMETHEE methods was used. The proposed management procedure enables the demands of opposing stakeholders to be taken into account; it increases the transparency of the decision-making process and its coherence; it also increases the legitimacy of the final result. It is a new scientific approach with a great potential for being applied to similar decision-making problems.

During exploitation, construction objects deteriorate and require renovation. Renovation may be needed due to the poor technical condition of a building's elements, the ending of a material's durability, a building's location, or the protection of cultural heritage. Neglecting renovation is one of the main reasons for the decline in the technical value of buildings. Article [15] proposes a new and original methodology for determining needs in the field of the rehabilitation of buildings constructed using traditional technology. The implementation of the Analytical Hierarchy Process (AHP) method was used to set renovation priorities. The developed multi-criteria methodology for supporting decisions in the field of building renovation may be a tool for determining the correct sequence of renovation works while taking into account the technical condition of facilities, the preparation of work schedules, and the planning of renovation investment costs.

Conflicts of Interest: The authors declare no conflict of interest.

References

1. Niewiadomski, P.; Stefaniuk, D. Creep Assessment of the Cement Matrix of Self-Compacting Concrete Modified with the Addition of Nanoparticles Using the Indentation Method. *Appl. Sci.* **2020**, *10*, 2442. [CrossRef]
2. Góra, J.; Szafraniec, M. Influence of Maximum Aggregate Grain Size on the Strength Properties and Modulus of Elasticity of Concrete. *Appl. Sci.* **2020**, *10*, 3918. [CrossRef]
3. Wioletta Raczkiewicz, W.; Wójcicki, A. Temperature Impact on the Assessment of Reinforcement Corrosion Risk in Concrete by Galvanostatic Pulse Method. *Appl. Sci.* **2020**, *10*, 1089. [CrossRef]
4. Reza, K.; Hosein, N.; Houman, E.K.; Jakubczyk-Gałczyńska, A.; Jankowski, R. A Proposed Soft Computing Model for Ultimate Strength Estimation of FRP-Confined Concrete Cylinders. *Appl. Sci.* **2020**, *10*, 1769. [CrossRef]
5. Sakami, S.; Boughattem, L.; Boumhaout, M.; Benhamou, B. Development of Alfa Fiber-Based Mortar with Improved Thermo-Mechanical Properties. *Appl. Sci.* **2020**, *10*, 8021. [CrossRef]
6. Kolesnikov, G.; Gavrilov, T. Modeling the Drying of Capillary-Porous Materials in a Thin Layer: Application to the Estimation of Moisture Content in Thin-Walled Building Blocks. *Appl. Sci.* **2020**, *10*, 6953. [CrossRef]
7. Biruk, S.; Jaśkowski, P. Selection of the Optimal Actions for Crashing Processes Duration to Increase the Robustness of Construction Schedules. *Appl. Sci.* **2020**, *10*, 8028. [CrossRef]
8. Konior, J.; Szóstak, M. The S-Curve as a Tool for Planning and Controlling of Construction Process—Case Study. *Appl. Sci.* **2020**, *10*, 2071. [CrossRef]
9. Leśniak, A.; Górka, M. Structural Analysis of Factors Influencing the Costs of Facade System Implementation. *Appl. Sci.* **2020**, *10*, 6021. [CrossRef]
10. Plebankiewicz, E.; Malara, J. Analysis of Defects in Residential Buildings Reported during the Warranty Period. *Appl. Sci.* **2020**, *10*, 6123. [CrossRef]
11. Dmochowski, G.; Szolomicki, J. Technical and Structural Problems Related to the Interaction between a Deep Excavation and Adjacent Existing Buildings. *Appl. Sci.* **2021**, *11*, 481. [CrossRef]
12. Sawicki, M.; Szóstak, M. Impact of Alcohol on Occupational Health and Safety in the Construction Industry at Workplaces with Scaffoldings. *Appl. Sci.* **2020**, *10*, 6690. [CrossRef]
13. Muñoz-Medina, B.; Ordóñez, J.; Romana, M.G.; Lara-Galera, A. Typology Selection of Retaining Walls Based on Multicriteria Decision-Making Methods. *Appl. Sci.* **2021**, *11*, 1457. [CrossRef]
14. Marović, I.; Perić, M.; Hanak, T. A Multi-Criteria Decision Support Concept for Selecting the Optimal Contractor. *Appl. Sci.* **2021**, *11*, 1660. [CrossRef]
15. Nowogońska, B. A Methodology for Determining the Rehabilitation Needs of Buildings. *Appl. Sci.* **2020**, *10*, 3873. [CrossRef]

applied sciences

MDPI

Article

A Multi-Criteria Decision Support Concept for Selecting the Optimal Contractor

Ivan Marović [1,*], Monika Perić [1] and Tomaš Hanak [2]

[1] Faculty of Civil Engineering, University of Rijeka, HR-51000 Rijeka, Croatia; mperic@student.uniri.hr
[2] Faculty of Civil Engineering, Brno University of Technology, 602 00 Brno, Czech Republic; hanak.t@fce.vutbr.cz
* Correspondence: ivan.marovic@uniri.hr; Tel.: +385-51-265-921

Abstract: A way to minimize uncertainty and achieve the best possible project performance in construction project management can be achieved during the procurement process, which involves selecting an optimal contractor according to "the most economically advantageous tender." As resources are limited, decision-makers are often pulled apart by conflicting demands coming from various stakeholders. The challenge of addressing them at the same time can be modelled as a multi-criteria decision-making problem. The aim of this paper is to show that the analytic hierarchy process (AHP) together with PROMETHEE could cope with such a problem. As a result of their synergy, a decision support concept for selecting the optimal contractor (DSC-CONT) is proposed that: (a) allows the incorporation of opposing stakeholders' demands; (b) increases the transparency of decision-making and the consistency of the decision-making process; (c) enhances the legitimacy of the final outcome; and (d) is a scientific approach with great potential for application to similar decision-making problems where sustainable decisions are needed.

Keywords: contractor selection; multi-criteria decision making; decision support concept; AHP; PROMETHEE; construction procurement

Citation: Marović, I.; Perić, M.; Hanak, T. A Multi-Criteria Decision Support Concept for Selecting the Optimal Contractor. *Appl. Sci.* **2021**, *11*, 1660. https://doi.org/10.3390/app11041660

Academic Editors: Božena Hoła and Anna Hoła

Received: 30 December 2020
Accepted: 9 February 2021
Published: 12 February 2021

1. Introduction

Selecting the optimal contractor for construction projects can be seen as the most important strategic decision in such an investment, one which can have long lasting effects that may emerge not only during the particular project, but assuredly during its exploitation phase. At the same time, it is one of the most important decisions made by the clients. Often, decision-making and decision support in civil engineering is solely based on cost-benefit analysis (CBA). However, this has been found to be highly inadequate, both in terms of incorporation and assessment of multiple-criteria like environmental and wider economic issues which are usually essentially difficult to quantify, and because traditional CBA relies heavily on estimating both demand forecasts and construction costs [1–3]. Over the years, various researchers dealt with such aspects of project performance, claiming that demand forecasts and construction cost estimations in particular are subject to a large degree of uncertainty—commonly referred to as optimism bias [4–12].

In order to minimize uncertainty and achieve the best possible project performance, two EU Directives were implemented in 2004 that allowed a codification of rules and procedures across EU countries regarding public procurement—Directives 2004/18 /EC and 2004/17/EC. These directives guided contracting authorities, i.e., clients, to approach their projects in a more strategic and forward-looking way in order to achieve successful and thus sustainable projects. Accordingly, public procurement should be based on disinterested criteria [13] that ensure compliance with transparency, nondiscrimination, equal treatment, and with guarantees that tenders are evaluated in circumstances of effective competition. Such can be achieved by two approaches: "the lowest price" and "the most economically advantageous tender." Both approaches are present in EU countries, as each country often builds in some specificities, but almost as a rule it comes down to a single criterion, i.e., price. Dealing with a number of criteria directly implies the need to use multi-criteria methods that are usually perceived as "difficult to understand." In the latter approach, the criteria related to the particular public procurement (article 53 of Directive 2004/18/EC and 55 of Directive 2004/17/EC) are in the hands of the clients, the contracting authorities, and therefore vary from one tender to another. As it is a multi-criteria problem, the use of any multi-criteria decision-making method seems to be the right choice if the decision-maker is aiming toward a consistent decision-making process from beginning to end.

In general, the field of strategic management is not defined by a particular theoretical paradigm, but rather by its focus on a particular dependent variable—overall organizational performance—and the role of managers in shaping that performance [14], but also by extending, clarifying and applying such theories in new and interesting ways [15]. The strategic management process advocated by [16] has been defined as comprising a sequential set of analyses and choices that can increase the likelihood that a company will choose a strategy that generates competitive advantage. This can also be applied to projects, programs, and portfolios.

Similarly, such strategic thinking can be applied to the problem of selecting the optimal contractor for a particular construction investment by applying multi-criteria decision analysis (MCDA) approach and a logic of decision support systems (DSS). Salling and Pryn [1] proposed a decision support model named SUSTAIN-DSS to bring informed decision support, both in terms of single aggregated estimates, i.e., deterministic calculation, and also in terms of interval results by certainty graphs, i.e., stochastic calculation. Such interaction enabled the analysts to investigate not only the feasibility of risk when assessing investment projects [5] but also to highlight the importance of expanding the decision-making process beyond the consideration of solely economic factors and point estimates. Various researchers [17–22] proposed different MCDA approaches based on value measurement using qualitative inputs from a ratifying stakeholder group via multiplicative analytic hierarchy process (AHP), which were found to be well suited for group decision-making.

An extensive literature review from 2000 to 2018 [23] led to the identification and classification of commonly used criteria in construction procurement, commonly used decision-making techniques, and the origins of researchers working on this topic. In the last decades there have been a number of papers dealing with outranking methods in construction project management, focusing on AHP and/or PROMETHEE methods [24]. While some authors focused on the AHP method [25–30], the analytic network process [31] or PROMETHEE methods [20,32–37] the important driver was given by [38] to use these methods in synergy to achieve the most in a multi-stakeholder environment [19,39–45]. To tackle the problem of selecting an appropriate contractor, various authors approach the problem from the stakeholder point-of-view using the AHP or group AHP [21,29–31,46,47], while others consider the problem as an overall approach of managing stakeholders such as the multi-actor multi-criteria analysis methodology, i.e., MAMCA [19,39], or decision support concept, i.e., DSC [32,34,35,42].

Regardless of the approach used, in order to determine "the most economically advantageous tender" it is important to address not only technical aspects, but also economic, social, environmental, and other aspects of the tenderers as well as the long-term impact of the project outcomes as a whole. Therefore, such requirements can be achieved be establishing adequate selection criteria during the procurement process. This has been done by all the previously mentioned researchers, but some focused on defining the main criteria in a more detail way [3,13,48–54] by using them to select tenderers in competitive tendering systems.

In this context, the main objective is to develop a decision support concept for the selecting the optimal contractor based on the synergy effect of the AHP (for the development of the hierarchical goal structure) and PROMETHEE methods (for the pairwise comparison of alternatives, i.e., tenderers/contractors). An additional aim is to define and implement a multi-stakeholder management procedure during the construction procurement process that: (a) allows the incorporation of opposing stakeholders' demands; (b) increases the transparency of decision-making and the consistency of the decision-making process; (c) enhances the legitimacy of the final outcome; and (d) is a scientific approach with great potential to be applied in similar decision-making problems where sustainable decisions are needed.

2. Methods and Methodology

To ensure that the construction project can be successfully completed regarding the projects' scope, time, costs and quality, the client must select the most appropriate contractor, regardless of the type of investment, private or public. This involves a procurement system that comprises several process elements (project packaging, invitation to compete, prequalification, short-listing and bid evaluation).

The existing literature on contractor selection mainly deals with how to identify and evaluate the criteria, thus providing the general lists of criteria for managing purposes in civil engineering. A more promising approach that classifies the criteria for contractor selection has been provided by Hatush and Skitmore [48,49], and Cheng and Li [31]. Taking their approach into account, i.e., focusing exclusively on the elements of prequalification and bid evaluation in construction procurement, served as the basis for the proposed decision support concept.

2.1. Data and Methods

In order to address how the existing body of knowledge in civil engineering has developed in the direction of construction procurement, especially the contractor selection problem, a systematic literature review was conducted in this study as well as direct correspondence and collaboration with experts.

A systematic literature review was conducted for the purpose of multi-stakeholder analysis and establishing the hierarchical goal structure. The review was conducted in the Scopus and Web of Science databases using selected keywords (group decision-making, multi-criteria, contractor selection, decision support, construction procurement, AHP, PROMETHEE), and their syntax derivatives. To ensure the high quality and novelty of the analyzed knowledge, only papers published in scientific journals between January 2000 and December 2020 were considered. This resulted in a list of seven criteria that are most commonly used to select the optimal contractor.

This list of criteria was used in collaboration with two different groups of experts (contractors and clients). The first group of experts, i.e., the contractor group, consisted of eight private contractors selected from the local area. All examinees from this group are experts in the field of construction procurement with 15 (2 examinee), 25 (5 examinees) and 30 (4 examinees) years of experience and work at strategic management levels at their companies. Some contractors were represented by more than one representative, but their opinion was used in the further analysis as a single one, i.e., company point of view. The second expert group, i.e., the client group, consisted of 13 public contractors selected from the local area representing local government (5), government agencies (3), and universities (5). All examinees from this group are experts in the field of construction management and/or construction procurement with 15 (4 examinee), 20 (6 examinees), 25 (5 examinees) and 30 (3 examinees) years of experience and work at tactical and/or strategic management levels. As some clients were represented by more than one representative, their opinion was used in further analysis as a single one, i.e., client point of view.

By means of structured interviews as well as workshops, both groups participated in collective decision-making by expressing their view on criteria using the AHP and Saaty scale. This served not only as a participatory process in decision-making where stakeholders adopt the decisions through a majority vote [46], but also in seeking the agreement of those who participate by generating consensus among them. This resulted in two points-of-view, that of the clients and that of the contractors, which will be discussed further in Section 3. Since the identified criteria are both quantitative and qualitative, another outranking method, PROMETHEE, was used for ranking of the tenderers as an appropriate MCDA method for solving such problems. For this purpose, experts from the client group were asked to evaluate each tenderer in relation to each criterion, resulting in a decision matrix that was used for prioritization.

The proposed decision support concept was tested on a case study, a small multistory residential building, while the multi-stakeholder analysis and multi-criteria decision analysis were tested by involving experts from public and private procurement as mentioned.

2.2. Concept Development

The proposed decision support concept for selecting the optimal contractor (DSC-CONT) consists of several processes, as shown in Figure 1. The focus of the proposed concept is a two-stage procurement procedure: (1) prequalification, and (2) evaluation of tenderers. To achieve the best possible outcome, the DSC-CONT uses the synergy of the AHP and PROMETHEE methods. This approach of using the synergy of the AHP and PROMETHEE has been previously tested in various multi-criteria problems [20,32,35–37] and showed promising results. This is due to the strength of AHP in creating a hierarchical goal structure and the strength of PROMETHEE in ranking alternatives according to criteria that are evaluated both quantitatively and qualitatively. Creating such operational synergies by strengthening PROMETHEE with AHP gives the robustness and consistency in the decision-making process of the DSC-CONT. This approach is preferred by the authors, based on their own experience with similar methodological approaches, but also because of the research of other authors [19,33,34,39,41–44,55–57].

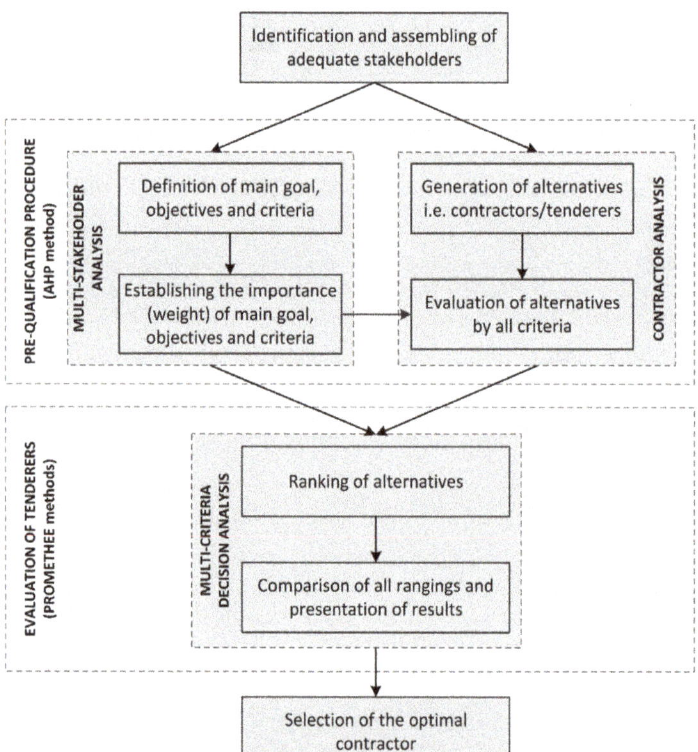

Figure 1. Decision support concept for selecting optimal contractor.

The novelty of the proposed concept is in its robustness and resilience to changes in the decision-making process, especially in allowing stakeholders to express their attitudes and their opposing demands. The methods used provide stakeholders with the opportunity to express their attitudes in a clear way. At the same time, the transparency of decision-making is increased and the legitimacy of the final outcome is strengthened. The advantage of such an approach is that even if there is a change in the structure of the decision-makers, the decision-making procedure itself remains intact and consistent. Moreover, the proposed concept takes into account EU directives and can be easily implemented in all public construction tenders regulated by Directives 2004/18 /EC and 2004/17/EC.

The DSC-CONT consists of two processes. During the prequalification process, it is important to compare key contractor-organizational criteria among a group of contractors desirous to tender. Such criteria can be identified in various ways. In general, this concept provides a hierarchical goal structure procedure (Figure 2) and brings stakeholders into the middle of the analysis. This is done by applying the AHP logic and giving stakeholders the opportunity to reach consensus in order to come up with a sustainable solution. The AHP [17,58] is used to determine the importance of the main goal, objectives and criteria of each stakeholder group (client and contractor). Depending on whether the aggregation is performed at the comparison level or at the priorities level, the procedure differs but the result remains the same, i.e., the hierarchical goal structure is formed with all weights. This is done by the multi-stakeholder analysis, while the contractor analysis offers the insight into the alternatives, i.e., contractors/tenderers, which leads to their evaluation according to previously defined criteria.

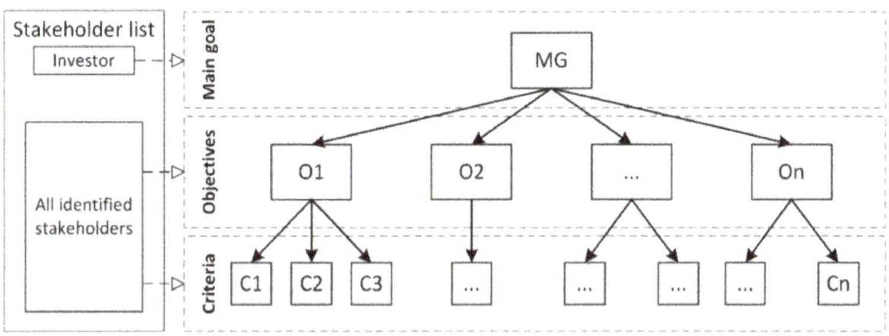

Figure 2. Hierarchical goal structure procedure [37].

The following process is the evaluation of tenderers and here essentially the multi-criteria decision analysis is carried out. Since the previously defined criteria can be both qualitative and/or quantitative, the DSC-CONT uses the strengths of the PROMETHEE methods for ranking the alternatives. Here, the PROMETHEE II method [59–62] is used to obtain a complete ranking, but before a final rank-list is produced, it is important to check the results using VisualPROMETHEE [63] features, such as PROMETHEE Diamond and/or PROMETHEE Network, as well. The rank-list provides the decision-maker with the basis for making a final decision, especially if it is presented graphically.

While the evaluation of tenderers process considers specific criteria that can measure the suitability of the tenderers, i.e., contractors, it is not equivalent to the contractor selection process, although in practice it is considered to be one. Since the evaluation of tenderers is the process of investigating or measuring specific project attributes, the contractor selection is referred to as the process of aggregating the results of the evaluation to identify the optimal choice. Cheng and Li [31] also highlighted this: "In practice, these two processes are always grouped together to represent a single procedure to prioritize the contractors according to the project specific criteria". Overall, the DSC-CONT provides the decision-maker with a tool to identify, evaluate, and analyze, but the final decision is always in their hands.

2.2.1. Building the Hierarchy

The stakeholder management is often seen as the most important part of construction project management [37], directly affecting the projects' scope, time, cost and quality. Therefore, to manage them proactively by capturing their attitudes, the hierarchical goal structure (HGS) procedure (Figure 2) is applied. This particular procedure has been used in some previous research [20,24,36,37] and showed promising results in multi-stakeholder analysis. The main advantages are the clear goal hierarchy by allowing stakeholders, i.e., experts, to participate in the creation of the hierarchical goal structure, but also to express their attitude towards each criterion. Assessing weights, i.e., stating attitudes, often seems to be the weakest element due to the subjective approach, but in the case where the search for consensus on weighting of each criterion is a necessity, this leads to a consensus weighting of all involved stakeholders and can therefore be considered objective. Nevertheless, the responsibility is in the hands of the decision-maker and his ability to involve all relevant stakeholders in the HGS procedure.

The proposed HGS procedure ensures insight into the definition of objectives (O) and criteria (C) of the defined main goal (MG). Since stakeholder relationships are not static, but on the contrary dynamic and in constant change [42], their attitudes and actions may change at different project stages, and endanger the overall performance of the project. Since the hierarchical goal structure procedure is an iterative process that ends when all stakeholders agree, the decision-maker can be sure that if the procedure is followed, all stakeholders' attitudes are embedded in the criteria, the objectives and the main goal.

The result of this procedure is a list of criteria, as shown in Table 1, and gives all stakeholders involved a clear insight into the HGS and how each element is described, evaluated and preferred. One can be assured that by completing and fulfilling each criteria, the main goal will be achieved as an outcome of the process. In addition, this becomes a transparent tool for the weighting phase.

Table 1. Criteria with short description, evaluation technique, and preference.

Criteria Label	Criteria Name	Short Description of Criteria and Evaluation Technique	Preference (Min/Max)
C1	Tender price	Tender price that includes construction/reconstruction costs according to bill of quantities and technical documentation; Expressed in 1.000 €	Min
C2	Expected duration	Expected duration of construction/reconstruction according to bill of quantities and dynamic plan; Expressed in weeks	Min
C3	Quality	Expert assessment taking into account quality of the work to be done; Qualitative scale 1 (min) to 5 (max)	Max
C4	Past relationship	Expert assessment taking into account past relationship of the tenderer; Qualitative scale 1 (min) to 5 (max)	Max
C5	Resources	Expert assessment taking into account tenderers' capabilities of possible allocation of the resources for finishing construction/reconstruction in given project constraints; Qualitative scale 1 (min) to 5 (max)	Max
C6	WLCC	Expert assessment of project's impact on whole life-cycle costs (WLCC); Qualitative scale 1 (min) to 5 (max)	Min
C7	Past experience	Expert assessment taking into account past experience of the tenderer; Qualitative scale 1 (min) to 5 (max)	Max

As mentioned earlier, this list of criteria (Table 1) resulted from the systematic literature review and the stakeholders were asked to state their attitudes only about them. It is important to emphasize that due to the differences in construction projects and tenders, the proposed HGS procedure offers the possibility to update this list or to create a completely new, i.e., customized, list of criteria that provides the best results in terms of the projects' scope, time, cost and quality.

2.2.2. Weighting Phase

Once the HGS is made, it is necessary to determine their importance, i.e., their weights. In a multi-stakeholder environment, this can be achieved in various ways. In this particular case, each stakeholder group (contractors and clients) has been given the opportunity to express their point-of-view. Typically, stakeholders think that their own expectations have not been taken properly into account. Therefore, it is of utmost importance that this procedure is transparent and all their attitudes and actions are considered as a part of the collaborative governance [46,64,65].

The AHP method and Saaty scale (1–9) were used for weighting. Since there may be multi-stakeholders in each group, we proposed the weight aggregation at the comparison level of each group. The multiplicative AHP is useful for stakeholders and decision-makers to align common viewpoints and ultimately reach an agreement, i.e., consensus. Each group can be further analyzed as a separate scenario and its consensus as a standalone scenario, if needed.

2.2.3. Ranking Procedure

While the AHP was used for the definition of HGS and weighting, the PROMETHEE methods are recommended as appropriate ones for the MCDA of the proposed decision concept. It is supported by the fact that there are different types of criteria which can be both qualitative and quantitative. Such cases are very common when dealing with criteria that involve various technical, economic, social, and environmental aspects. Since the general objective of this process is to rank and compare all alternatives, it is of utmost importance for decision analysts to prepare the results as graphically as possible. In this case, the use of PROMETHEE II results should be supported by graphical representation of PROMETHEE Diamond and/or PROMETHEE Network. The above is explained in more detail in the following section.

3. Results and Discussion

Once the HGS is created, it enables collaboration with the identified stakeholders. In this case, two stakeholder groups have been identified; contractors and clients. Stakeholders from both groups were interviewed about the HGS, especially ranking criteria. For the purpose of this study, the proposed concept is tested on a case study of a small multistory residential building. The central issue is to show the possibilities offered by the DSC-CONT, rather than the selection of the contractor in an actual tender. Therefore, in order to present the procedure, the criteria have been defined as previously described. At the same time, the procedure of creating HGS is also presented. This will allow decision-makers the opportunity to create HGS according to the specifics of their tender.

To provide insight into DSC-CONT and achieve the defined goals of the study, this section begins with the prequalification process and multi-stakeholder and contractor analysis. The interviews were conducted in one-on-one sessions where each stakeholder had the opportunity to reflect on given criteria and assign the weights. Each stakeholder made the pairwise comparisons for the defined criteria. Different scales were proposed to them to transform their judgments into numbers of the pairwise comparison. The one that was used in the end was Saaty's linear scale, where the values of comparison range from 1 (indifference) to 9 (extreme preference). This stage corresponds to the collaborative part of the governance process, where all the preferences, likes, and desires, i.e., attitudes of the stakeholders are included in matrices of pairwise comparison. By collecting their judgment in square matrices, the relative dominance of one criterion over the other is generated. Each stakeholder participated in the elaboration of the matrices among the experts who designed HGS and the final result was presented to them at the end.

The first group, i.e., contractors, consisted of eight selected contractors from the local area, as we saw in Section 2.1. They were all technical managers and/or general managers in small and medium-sized enterprises (SMEs). This group was asked to weight the criteria as they would like them to be evaluated in future tenders. Their respective weightings are shown in Figure 3.

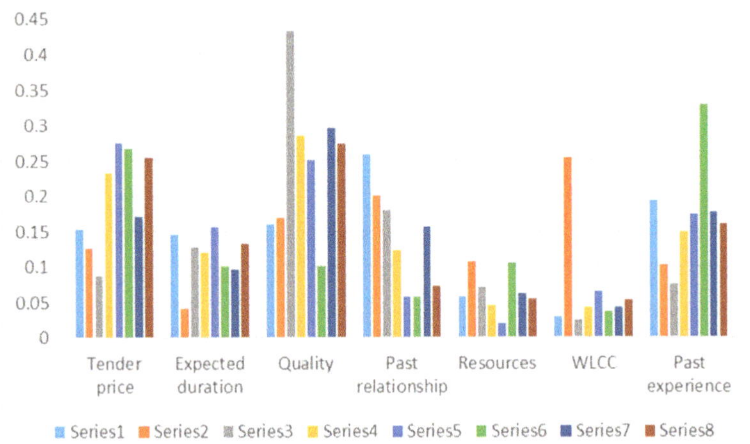

Figure 3. Weights of each criterion—contractors' point-of-view.

As mentioned earlier, some contractors saw certain criteria differently. From their point of view, the three most important criteria were defined as quality, tender price, and past experience. At the same time, three criteria have relative peaks in contrast to certain attitudes. Those criteria are quality (Figure 3, Series 3), whole life-cycle costs, i.e., WLCC (Figure 3, Series 7), and past experience (Figure 3, Series 8). It is interesting to see that even with a small number of experts involved, their attitudes differ significantly. In this case, the reasons can be found in their specializations. Consequently, in Series 3 the experts come from a company specialized in prefabricated buildings, in Series 7 the experts come from a company specialized in Design-Build projects, and in Series 8 the experts come from a company with a 55 year tradition in civil engineering. In summary, even with a small poll of stakeholders, the DSC-CONT provides the opportunity to incorporate opposing stakeholders' demands while increasing the transparency of the decision-making process.

When applying the AHP method, the consistency ratio (CR) must be considered. For all the contractors' matrices, the inconsistency found was less than 0.1, which means that the weights were calculated correctly. In order to evaluate them as a single group, i.e., as a scenario, the overall matrix was created with aggregated values (Table A1). The aggregation of each pairwise was done using the median, and the final weights are presented in Table 2. The CR is 0.08. These weights were used for the following evaluation of tenderers and presents Scenario 1—contractor group. To conclude, this approach additionally gives transparency during aggregation as all stakeholders' demands are included in the decision-making process.

Table 2. Contractors' aggregated weights.

	Tender Price	Expected Duration	Quality	Past Relationship	Resources	WLCC	Past Experience
w_i	21.9	11.6	25.7	13.1	4.2	4.9	18.6

The same approach was carried out for the second group, i.e., clients, which consisted of 13 selected clients from the local area, as we saw in Section 2.1. They were all public sector clients and mandatory users of public procurement. Five examinees represent the university experts' point of view, five represent local government on city municipality regions, and three at the regional government agencies. This group was asked to weight criteria in order to select the best contractor in the future tenders. Their respective weightings are shown in Figure 4.

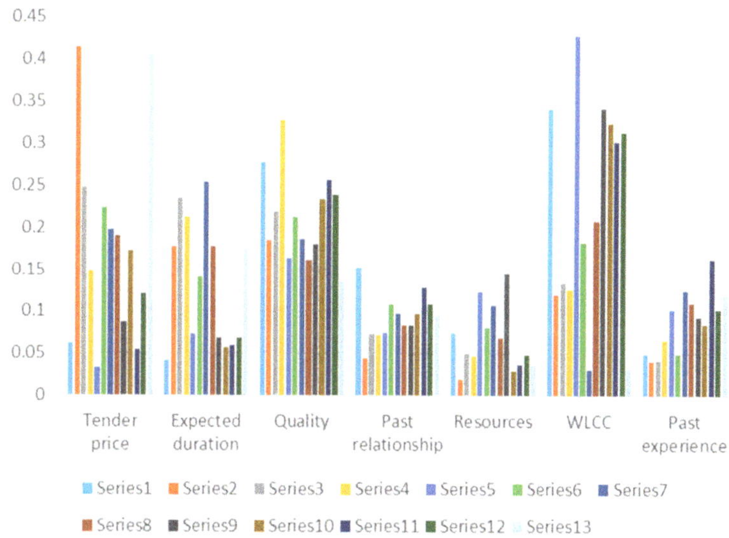

Figure 4. Weights of each criterion—clients' point-of-view.

As mentioned earlier, some clients see certain criteria differently. Their view resulted in defining the three most important criteria, namely WLCC, quality, and tender price. At the same time, two criteria have relative peaks in contrast to the given attitudes. Those criteria are tender price (Figure 4, Series 2 and 13), and WLCC (Figure 4, Series 1 and 5). It is interesting to see that even with a small number of experts involved, their attitudes differ significantly. In this case, the reasons can be found in their prior experience with construction projects. Consequently, in Series 2 and 13 the experts' prior experience indicate that they are more oriented towards traditional budgeting and more likely to see WLCC and nontraditional budgeting approaches such as public−private partnership (e.g., Series 1 and 5). In summary, even with a small poll of stakeholders, the DSC-CONT provides the opportunity to incorporate opposing stakeholders' demands at the same time increasing the transparency of the decision-making process.

For all their matrices, the determined inconsistency was less than 0.1, which means that the weights were calculated correctly. In order to evaluate them as a single group, i.e., scenario, the overall matrix was created with aggregated values (Table A2). The aggregation of each pairwise was done by the median, and the final weights are presented in Table 3. The CR is 0.05. These weights were used for the following evaluation of tenderers and presents Scenario 2—client group. To conclude, this approach additionally gives transparency during aggregation as all stakeholders' demands are included in the decision-making process.

Table 3. Clients' aggregated weights.

	Tender Price	Expected Duration	Quality	Past Relationship	Resources	WLCC	Past Experience
w_i	17.8	10.5	23.0	9.8	6.5	24.9	7.5

With the weighted HGS in place, the multi-stakeholder and contractor analysis ended. This allowed the use of DSC-CONT to perform a multi-criteria decision analysis using PROMETHEE methods. Since each group was analyzed as a separate scenario, it was important to create separate decision matrices. Therefore, Figures A1 and A2 present a decision matrix for each scenario. The main difference between these matrices lay in the preference section. As already described in Table 1, each criterion was unique. Some of them were quantitative (tender price and expected duration) and the others were qualitative (quality, past relationship, resources, WLCC, and past experience). In the case where a global consensus was reached, an additional aggregation of both contractors and clients' weights had to be performed. For this particular case, further results and discussions from both groups are presented.

To begin with the evaluation of tenderers process, the VisualPROMETHEE software was used. When using PROMETHEE methods, it is important to assign a preference function to each criterion. The preference functions can be randomly assigned as one of six predefined ones, but this is not recommended. The choice of a good preference function depends on the scale of the underlying criterion. For the purpose of evaluating tenderers, to all quantitative criteria the Linear preference function was assigned, while the Usual, Level and U-shape preference functions were assigned to the qualitative criteria.

As mentioned earlier, when using PROMETHEE methods such as PROMETHEE II it is important to assign a weight to each criterion. Since PROMETHEE methods lack consistent and transparent structuring of hierarchical goals, this is where the strength of the AHP comes into play. By using the AHP in the multi-stakeholder analysis, we now had a specific stakeholder weighting that could be implemented in PROMETHEE. The key point was that these weights represented the actual attitudes of all involved stakeholders and represented their consensus. This is particularly important when there are a number of stakeholders who see the problem differently. It must be stressed that this enhances legitimacy of the final outcome of decision-making process.

Taking all these into account, the PROMETHEE II was used and resulted in a complete ranking of all alternatives, i.e., tenderers, (Figures 5a and 6a) in terms of their group opinions, expressed by the criteria weights and by selecting an appropriate preference function for each criterion. The Phi net flow of each alternative was also visible. The higher the Phi net flow of a given alternative, the better it was, the same goes for the lower Phi net flow. From Figures 5a and 6a, it was evident that out of the five alternatives (Contractor A, B, C, D, and E), their rank remained almost the same, with the best alternative being Contractor B and the worst being Contractor C. These alternatives were used to simulate possibilities in the decision-making process and were not part of the any real tender.

The overall spread between the best and worst tender had shrunk slightly, while the close alternatives (Contractor A and D) had swapped rank positions. This sort of thing sometimes happens when the alternatives are similarly valued according to criteria (see decision matrices in Appendix A). This is very often the case in construction procurement, as the contractors' bids are very close to each other. With the proposed DSC-CONT, this brings consistency, transparency, and clarity to the decision-making process and can identify those very differences and help the decision-makers with their decision. At the same time, it is known that the final decision is based on the opinions of all parties involved and thus can be considered as the best or optimal decision.

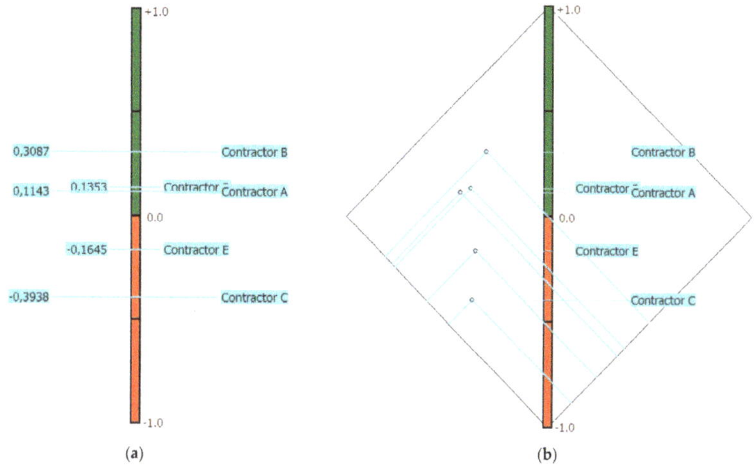

Figure 5. The ranking of alternatives with the contractors' weighting: (**a**) PROMETHEE II; (**b**) PROMETHEE Diamond.

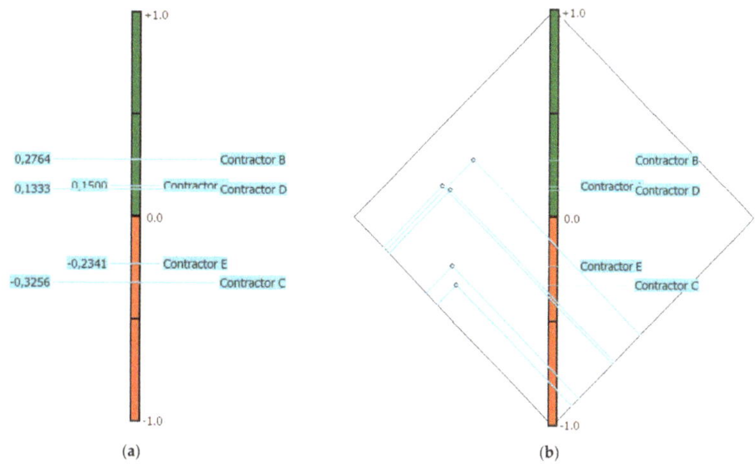

Figure 6. The ranking of alternatives with the clients' weighting: (**a**) PROMETHEE II; (**b**) PROMETHEE Diamond.

As previously mentioned, these tools give a graphical representation of the complete ranking and should additionally be checked with PROMETHEE Diamond and/or PROMETHEE Network. Figures 5b and 6b give an insight into PROMETHEE Diamond. In PROMETHEE Diamond each alternative is represented as a point in the Phi plane angled at 45° degrees so that the vertical dimension (green-red axis) corresponds with the Phi net flow axis from PROMETHEE II. The point of each alternative in the Phi plane is presented with Phi+ and Phi-, i.e., the results of the PROMETHEE I partial ranking.

Since the point of each alternative is a coordinate (Phi+, Phi-), it outlines a certain cone. When one alternative cone overlaps another, it means that the alternative is preferred over the other, while intersecting cones correspond to incomparable alternatives. When such a thing occurs, it does not mean that two alternatives cannot be compared, but that the comparison is difficult. In such case it is appropriate to examine the PROMETHEE

Network as a representation of the partial ranking resulting from the PROMETHEE I as it allows incomparability between the alternatives.

In the example in Figure 5b, it is evident that the cone of alternative Contractor B overlaps all the other alternatives, whereas in Figure 6b this is not the case. In such cases, the difficulty of comparing alternatives is emphasized and this helps the decision-maker to focus on these alternatives in detail. Therefore, a PROMETHEE Network of Scenario 2 is presented in Figure 7.

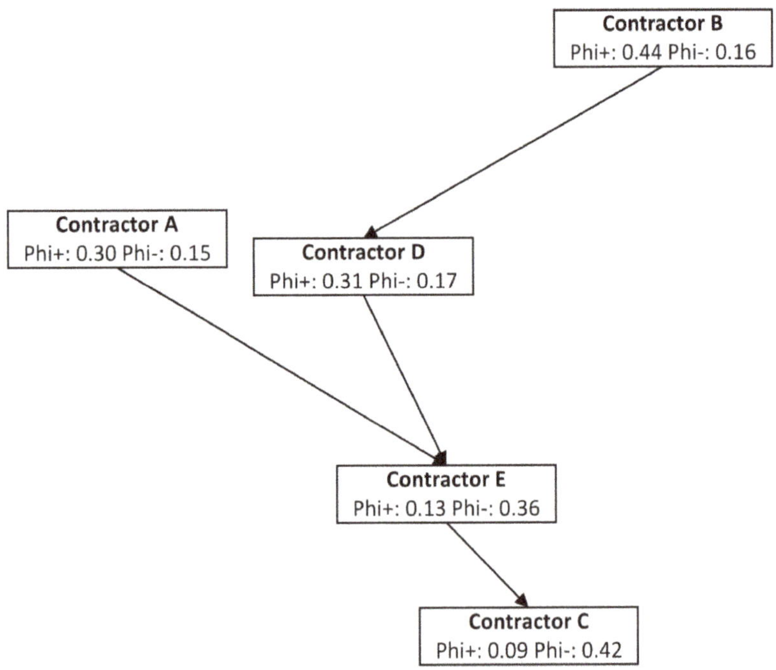

Figure 7. The ranking of alternatives with the clients' weighting—PROMETHEE Network.

From this additional insight (Figure 7), one can see the relative position of each alternative in PROMETHEE II and PROMETHEE Diamond, as well as the preferences represented by the arrows. This insight can further help the decision-maker not only to make the decision based on a complete ranking of alternatives, but also consider in detail whether certain alternatives are incomparable. From Figures 6 and 7, it can be concluded that the Contractor B alternative is the optimal one.

As we saw in this chapter, the synergy of the AHP method and PROMETHEE methods cope efficiently with the problem of selecting the optimal contractor if they are used adequately. Even the limitations that some of the above methods have when they are used solely in this approach cease to exist. The proposed decision support concept for selecting the optimal contractor showed its robustness, resilience, and consistency in the decision-making process even when changes occur.

4. Conclusions

The presented decision support concept for selecting the optimal contractor (DSC-CONT) shows a scientific approach for coping with the multi-stakeholder and multi-criteria decision-making environment in construction project management during the procurement process, focusing on (1) prequalification, and (2) evaluation of tenderers. In order to achieve the optimal solution, the concept is based on the synergic effect of the AHP (for the development of the hierarchical goal structure) and PROMETHEE (for the pairwise comparison of alternatives, i.e., tenderers/contractors) methods, each applied at different stages of the procurement procedure.

The advantage of the presented DSC-CONT is that it is easy to implement in any public construction tender regulated by Directives 2004/18 /EC and 2004/17/EC. The concept is robust and resilient to changes in stakeholders and allows for their opposing demands, at the same time it increases the transparency of decision-making and enhances the legitimacy of the final outcome. The advantage of such an approach is that even if there is a change in the structure of decision-makers, the decision-making procedure itself remains intact and consistent.

The limitations of this study are the given criteria. At the moment, they serve to validate the proposed decision support concept, especially the decision-making framework. Therefore, future directions are in expanding the dataset of stakeholders' attitudes towards specific types of building projects and providing lists of statistically significant criteria for particular tenders in civil engineering. This will potentially help the decision-makers to further speed up the process of defining criteria and focus energy on the criteria weighting and evaluating tenderers to select the optimal one.

Author Contributions: Conceptualization, I.M. and T.H.; methodology, I.M. and M.P.; software, I.M. and M.P.; validation, I.M. and M.P.; formal analysis, I.M., M.P. and T.H.; investigation, I.M. and M.P.; resources, I.M. and M.P.; data curation, I.M. and M.P.; writing—original draft preparation, I.M., M.P. and T.H.; writing—review and editing, I.M., M.P. and T.H.; visualization, I.M. and M.P.; supervision, I.M. and T.H.; project administration, I.M.; funding acquisition, I.M. All authors have read and agreed to the published version of the manuscript.

Funding: The APC was funded by the authors and the Faculty of Civil Engineering in Rijeka.

Institutional Review Board Statement: Not applicable.

Informed Consent Statement: Informed consent was obtained from all subjects involved in the study.

Data Availability Statement: The data presented in this study are available on request from the corresponding author.

Acknowledgments: The authors would like to thank all involved experts from private and public sectors for their engagement during conducted interviews and workshops. This research has been fully supported by the University of Rijeka under the project number uniri-pr-tehnic-19-18.

Conflicts of Interest: The authors declare no conflict of interest.

Appendix A

The AHP decision tables as an overview of contractors' and clients' points-of-view are given in the following tables.

Table A1. Overall matrix for Scenario 1—contractor group.

	Tender Price	Expected Duration	Quality	Past Relationship	Resources	WLCC	Past Experience
Tender price	1	4	2	1	3	5	1/3
Expected duration	1/4	1	1	1	3	3	1/2
Quality	1/2	1	1	3	7	3	4
Past relationship	1	1	1/3	1	4	3	1
Resources	1/3	1/3	1/7	1/4	1	1	1/3
WLCC	1/5	1/3	1/3	1/3	1	1	1/2
Past experience	3	2	1/4	1	3	2	1

Table A2. Overall matrix for Scenario 2—client group.

	Tender Price	Expected Duration	Quality	Past Relationship	Resources	WLCC	Past Experience
Tender price	1	3	1	3	1	1	1
Expected duration	1/3	1	1/3	3	1	1/3	2
Quality	1	3	1	3	5	1	3
Past relationship	1/3	1/3	1/3	1	3	1/3	3
Resources	1	1	1/5	1/3	1	1/5	1/2
WLCC	1	3	1	3	5	1	5
Past experience	1	1/2	1/3	1/3	2	1/5	1

Scenario1	Tender ...	Expecte...	Quality	Past rela...	Resources	WLCC	Past exp...
Unit	unit	unit	5-point	5-point	5-point	impact	5-point
Cluster/Group	◆	◆	◆	◆	◆	◆	◆
Preferences							
Min/Max	min	min	max	max	max	min	max
Weight	21,90	11,60	25,70	13,10	4,20	4,90	18,60
Preference Fn.	Linear	Linear	Usual	Level	Usual	U-shape	Usual
Thresholds	absolute	absolute	absolute	absolute	absolute	absolute	absolute
- Q: Indifference	€ 24,82	4,50	n/a	0,00	n/a	1,00	n/a
- P: Preference	€ 62,82	9,00	n/a	1,00	n/a	n/a	n/a
- S: Gaussian	n/a	n/a	n/a	n/a	n/a	n/a	n/a
Statistics							
Minimum	€ 680,00	65,00	2,00	3,00	3,00	2,00	2,00
Maximum	€ 750,00	80,00	4,00	4,00	4,00	4,00	4,00
Average	€ 716,00	74,00	2,80	3,40	3,40	3,00	3,20
Standard Dev.	€ 28,71	5,83	0,75	0,49	0,49	0,63	0,75
Evaluations							
Contractor A	€ 700,00	70,00	average	average	good	moderate	average
Contractor B	€ 750,00	65,00	average	good	good	low	good
Contractor C	€ 680,00	75,00	bad	average	average	high	bad
Contractor D	€ 700,00	80,00	good	average	average	moderate	average
Contractor E	€ 750,00	80,00	bad	good	average	moderate	good

Figure A1. Contractors' decision matrix—input for conducting PROMETHEE II.

Scenario2	Tender ...	Expecte...	Quality	Past rela...	Resources	WLCC	Past exp...
Unit	unit	unit	5-point	5-point	5-point	impact	5-point
Cluster/Group	◆	◆	◆	◆	◆	◆	◆
Preferences							
Min/Max	min	min	max	max	max	min	max
Weight	17,80	10,50	23,00	9,80	6,50	24,90	7,50
Preference Fn.	Linear	Linear	Usual	Level	Usual	U-shape	Usual
Thresholds	absolute	absolute	absolute	absolute	absolute	absolute	absolute
- Q: Indifference	€ 24,82	4,50	n/a	0,00	n/a	1,00	n/a
- P: Preference	€ 62,82	9,00	n/a	1,00	n/a	n/a	n/a
- S: Gaussian	n/a	n/a	n/a	n/a	n/a	n/a	n/a
Statistics							
Minimum	€ 680,00	65,00	2,00	3,00	3,00	2,00	2,00
Maximum	€ 750,00	80,00	4,00	4,00	4,00	4,00	4,00
Average	€ 716,00	74,00	2,80	3,40	3,40	3,00	3,20
Standard Dev.	€ 28,71	5,83	0,75	0,49	0,49	0,63	0,75
Evaluations							
Contractor A	€ 700,00	70,00	average	average	good	moderate	average
Contractor B	€ 750,00	65,00	average	good	good	low	good
Contractor C	€ 680,00	75,00	bad	average	average	high	bad
Contractor D	€ 700,00	80,00	good	average	average	moderate	average
Contractor E	€ 750,00	80,00	bad	good	average	moderate	good

Figure A2. Clients' decision matrix—input for conducting PROMETHEE II.

References

1. Salling, K.B.; Pryn, M.R. Sustainable transport project evaluation and decision support: Indicators and planning criteria for sustainable development. *Int. J. Sustain. Dev. World Ecol.* **2015**, *22*, 346–357. [CrossRef]
2. Burcar Dunović, I.; Radujković, M.; Vukomanović, M. Internal and external risk based assessment and evaluation for the large infrastructure projects. *J. Civ. Eng. Manag.* **2016**, *22*, 673–682. [CrossRef]
3. Ng, S.T.; Skitmore, R.M. Contractor selection criteria: A cost-benefit analysis. *IEEE Trans. Eng. Manag.* **2001**, *48*, 96–106. [CrossRef]
4. Flyvbjerg, B.; Skamris, M.K.; Buhl, S.L. How common and how large are cost overruns in transport infrastructure projects? *Transp. Rev.* **2003**, *23*, 71–88. [CrossRef]
5. Salling, K.B.; Banister, D. Assessment of large infrastructure projects: The CBA-DK model. *Transp. Res. Part A Policy Pract.* **2009**, *43*, 800–813. [CrossRef]
6. Žujo, V.; Car-Pušić, D.; Brkan-Vejzović, A. Contracted price overrun as contracted construction time overrun function. *Teh. Vjesn.* **2010**, *17*, 23–29.
7. Chen, Q.; Jin, Z.; Xia, B.; Wu, P.; Skitmore, M. Time and cost performance of design-build projects. *J. Constr. Eng. Manag.* **2016**, *142*, 04015074. [CrossRef]
8. Cajzek, R.; Klanšek, U. Cost optimization of project schedules under constrained resources and alternative production processes by mixed-integer nonlinear programming. *Eng. Constr. Archit. Manag.* **2019**, *26*, 2474–2508. [CrossRef]
9. Namazian, A.; Yakhchali, S.H. Modeling and solving project portfolio and contractor selection problem based on project scheduling under uncertainty. *Procedia Soc. Behav. Sci.* **2016**, *226*, 35–42. [CrossRef]
10. Hanak, T.; Korytarova, J. Subsidy risk related to construction projects: Seeking causes. *Open Eng.* **2018**, *8*, 484–489. [CrossRef]
11. Tijanić, K.; Car-Pušić, D.; Čulo, K. Impact of funding on cost-time aspects of public and social buildings. *Građevinar* **2019**, *71*, 21–32. [CrossRef]
12. Pesamaa, O.; Larsson, J.; Eriksson, P.E. Role of performance feedback on process performance in construction projects: Client and contractor perspectives. *J. Manag. Eng.* **2018**, *34*, 04018023. [CrossRef]
13. Huang, W.H.; Tserng, H.P.; Liao, H.H.; Yin, S.Y.L.; Chen, P.C.; Lei, M.C. Contractor financial prequalification using simulation method based on cas flow model. *Autom. Constr.* **2013**, *35*, 254–262. [CrossRef]
14. Rumelt, R.P.; Schendel, D.; Teece, D.J. Strategic management and economics. *Strateg. Manag. J.* **1991**, *12*, 5–29. [CrossRef]
15. Makadok, R.; Burton, R.; Barney, J. A practical guide for making theory contributions in strategic management. *Strateg. Manag. J.* **2018**, *39*, 1530–1545. [CrossRef]
16. Barney, J.B.; Hesterly, W.S. *Strategic Management and Competitive Advantage*, 5th ed.; Pearson: Upper Saddle River, NJ, USA, 2014.
17. Saaty, T.L. *Decision Making for Leaders: The Analytical Hierarchy Process for Decision in a Complex World*; The Analytical Hierarchy Process Series; University of Pittsburg: Pittsburg, PA, USA, 2001.
18. Phillips, L.D.; Bana e Costa, C. *Transparent Prioritization, Budgeting and Resource Allocation with Multi-Criteria Decision Analysis and Decision Conferencing*; Working Paper (LSEOR 05.75); London School of Economics and Political Science: London, UK, 2005.
19. Macharis, C.; Turcksin, L.; Lebeau, K. Multi actor multi criteria analysis (MAMCA) as a tool to support sustainable decisions: State of use. *Decis. Support Syst.* **2012**, *54*, 610–620. [CrossRef]
20. Marović, I.; Završki, I.; Jajac, N. Ranking zones model—A multicriterial approach to the spatial management of urban areas. *Croat. Oper. Res. Rev.* **2015**, *6*, 91–103. [CrossRef]
21. Ishizaka, A.; Labib, A. Selection of new production facilities with the group analytic hierarchy process ordering method. *Expert Syst. Appl.* **2011**, *38*, 7317–7325. [CrossRef]
22. Erdogan, S.A.; Šaparauskas, J.; Turskis, Z. A multi-criteria decision-making model to choose the best option for sustainable construction management. *Sustainability* **2019**, *11*, 2239. [CrossRef]
23. Khoso, A.R.; Md Yusof, A. Extended review on contractor selection in construction projects. *Can. J. Civ. Eng.* **2019**, *47*. [CrossRef]
24. Marović, I.; Tijanić, K.; Šopić, M.; Car-Pušić, D. Group decision-making in civil engineering based on AHP and PROMETHEE methods. *Sci. Rev. Eng. Environ. Sci.* **2020**, *29*, 474–484.
25. Danesh, D.; Ryan, M.J.; Abbasi, A. Using analytic hierarchy process as a decision-making tool in project portfolio management. World Academy of Science, Engineering and Technology. *Int. J. Econ. Manag. Eng.* **2015**, *9*, 3770–3780.
26. Danesh, D.; Ryan, M.J.; Abbasi, A. A systematic comparison of multi-criteria decision making methods for the improvement of project portfolio management in complex organisations. *Int. J. Manag. Decis. Mak.* **2017**, *16*, 280–320.
27. Darko, A.; Chan, A.P.C.; Ameyaw, E.E.; Owusu, E.K.; Parn, E.; Edwards, D.J. Review of application of analytic hierarchy process (AHP) in construction. *Int. J. Constr. Manag.* **2019**, *19*, 436–452. [CrossRef]
28. Fong, P.S.-W.; Choi, S. K-Y. Final contractor selection using the analytical hierarchy process. *Constr. Manag. Econ.* **2000**, *18*, 547–557. [CrossRef]
29. Jaskowski, P.; Biruk, S.; Bucon, R. Assessing contractor selection criteria weights with fuzzy AHP method application in group decision environment. *Autom. Constr.* **2009**, *19*, 120–126. [CrossRef]
30. Oyatoye, E.O.; Odulana, A.A. A prototype AHP system for contractor selection decision. In *Applications and Theory of Analytic Hierarchy Process: Decision Making for Strategic Decisions*, 1st ed.; de Felice, F., Petrillo, A., Saaty, T.L., Eds.; IntechOpen: London, UK, 2016; pp. 297–311.
31. Cheng, E.W.L.; Li, H. Contractor selection using the analytic network process. *Constr. Manag. Econ.* **2004**, *22*, 1021–1032. [CrossRef]

32. Jajac, N.; Knezić, S. Marović, I. Decision support system to urban infrastructure maintenance management. *Organ. Technol. Manag. Constr.* **2009**, *1*, 72–79.
33. Jajac, N.; Bilić, I.; Mladineo, M. Application of multicriteria methods to planning of investment projects in the field of civil engineering. *Croat. Oper. Res. Rev.* **2012**, *3*, 113–124.
34. Jajac, N.; Bilić, I.; Ajduk, A. Decision Support Concept to Management of Construction Projects—Problem of Construction Site Selection. *Croat. Oper. Res. Rev.* **2013**, *4*, 235–247.
35. Jajac, N.; Marović, I.; Baučić, M. Decision support concept for managing the maintenance of city parking facilities. *Electron. J. Fac. Civ. Eng. Osijek E Gfos* **2014**, *9*, 60–69. [CrossRef]
36. Marović, I.; Car-Pušić, D.; Hrvatin, Z. Establishing a model to evaluate public administration projects. *Electron. J. Fac. Civ. Eng. Osijek E Gfos* **2014**, *5*, 56–66. [CrossRef]
37. Marović, I.; Hanak, T. Selection of adequate site location during early stages of construction project management: A multi-criteria decision analysis approach. *IOP Conf. Ser. Mater. Sci. Eng.* **2017**, *251*, 012044. [CrossRef]
38. Macharis, C.; Springael, J.; De Brucker, K.; Verbeke, A. PROMETHEE and AHP: The design of operational synergies in multicriteria analysis. Strengthening PROMETHEE with ideas of AHP. *Eur. J. Oper. Res.* **2004**, *153*, 307–317. [CrossRef]
39. Macharis, C.; de Witte, A.; Ampe, J. The multi-actor, multi-criteria analysis methodology (MAMCA) for the evaluation of transport projects: Theory and practice. *J. Adv. Transp.* **2010**, *43*, 183–202. [CrossRef]
40. Zavadskas, E.K.; Turskis, Z.; Tamošaitiene, J. Contractor selection of construction in competitive environment. *J. Bus. Econ. Manag.* **2008**, *9*, 181–187. [CrossRef]
41. Rogulj, K.; Jajac, N.; Simic, F. Decision support concept for construction design project—Selecting the type of glass facade. *Croat. Oper. Res. Rev.* **2017**, *8*, 333–350. [CrossRef]
42. Jajac, N.; Kilić, J.; Rogulj, K. An integral approach to sustainable decision-making within maritime spatial planning—A DSC for the planning of anchorages on the island of Šolta, Croatia. *Sustainability* **2018**, *11*, 104. [CrossRef]
43. Kilić Pamuković, J.; Rogulj, K.; Dumanić, D.; Jajac, N. A sustainable approach for the maintenance of asphalt pavement construction. *Sustainability* **2021**, *13*, 109. [CrossRef]
44. Jajac, N.; Marović, I.; Mladineo, M. Planning support concept to implementation of sustainable parking development projects in ancient Mediterranean cities. *Croat. Oper. Res. Rev.* **2014**, *5*, 345–359. [CrossRef]
45. Semaan, N.; Salem, M. A deterministic contractor selection decision support system for competitive bidding. *Eng. Constr. Archit. Manag.* **2017**, *24*, 61–77. [CrossRef]
46. Losada Maestre, R.; Sanchez Medero, R.; Berlanga de Jesus, A.; Molina Lopez, J.M. The application of analytic hierarchy process to implement collaborative governance process: The allocation of the urban structural funds in the city of Madrid. *J. Multi Criteria Decis. Anal.* **2020**. [CrossRef]
47. Lakhani, A.; Zeeman, H.; Wright, C.J.; Watling, D.P.; Smith, D.; Islam, R. Stakeholder priorities for inclusive accessible housing: A systematic review and multicriteria decision analysis. *J. Multi Criteria Decis. Anal.* **2020**, *27*, 5–19. [CrossRef]
48. Hatush, Z.; Skitmore, M.R. Criteria for contractor selection. *Constr. Manag. Econ.* **1997**, *15*, 19–38. [CrossRef]
49. Hatush, Z.; Skitmore, R.M. Evaluating contractor prequalification data: Selection criteria and project success factors. *Constr. Manag. Econ.* **1997**, *15*, 129–147. [CrossRef]
50. Russell, J.S.; Skibniewski, M.J. Decision criteria in contractor prequalification. *J. Manag. Eng.* **1988**, *4*, 148–164. [CrossRef]
51. Marzouk, M. A superiority and inferiority ranking model for contractor selection. *Constr. Innov.* **2008**, *8*, 250–268. [CrossRef]
52. Hanak, T.; Nekardova, I. Selecting and evaluating suppliers in the Czech construction sector. *Period. Polytech. Soc. Manag. Sci.* **2020**, *28*, 155–161. [CrossRef]
53. Hopfe, C.J.; Augenbroe, G.L.; Hensen, J.L. Multi-criteria decision making under uncertainty in building performance assessment. *Build. Environ.* **2013**, *69*, 81–90. [CrossRef]
54. Lesniak, A.; Plebankiewicz, E.; Zima, K. Design and build procurement system-contractor selection. *Arch. Civ. Eng.* **2012**, *18*, 1–14. [CrossRef]
55. Hazir, O.A. A review of analytical models, approaches and decision support tools in project monitoring and control. *Int. J. Proj. Manag.* **2015**, *33*, 808–815. [CrossRef]
56. Kim, J.I.; Kim, J.; Fischer, M.; Orr, R. BIM-based decision-support method for master planning of sustainable large-scale developments. *Autom. Constr.* **2015**, *58*, 95–108. [CrossRef]
57. Taticchi, P.; Garengo, P.; Nudurupati, S.S.; Tonelli, F.; Pasqualino, R. A review of decision-support tools and performance measurement and sustainable supply chain management. *Int. J. Prod. Res.* **2014**, *53*, 6473–6494. [CrossRef]
58. Saaty, T.L. *The Analytic Hierarchy Process*; McGraw-Hill International: New York, NY, USA, 1980.
59. Brans, J.-P.; De Smet, Y. *PROMETHEE Methods: Multiple Criteria Decision Analysis*; Springer: New York, NY, USA, 2016.
60. Brans, J.P.; Mareschal, B.; Vincke, P.H. *PROMETHEE—A New Family of Outranking Methods in Multicriteria Analysis*; Brans, J.P., Ed.; Operational Research IFORS 84: Amsterdam, The Netherlands, 1984; pp. 477–490.
61. Brans, J.P.; Vincke, P.H. A preference ranking organization method, the PROMETHEE method for MCDM. *Manag. Sci.* **1985**, *31*, 647–656. [CrossRef]
62. Behzadian, M.; Kazemzadeh, R.B.; Albadvi, A.; Aghdasi, M. PROMETHEE: A comprehensive literature review on methodologies and applications. *Eur. J. Oper. Res.* **2010**, *200*, 198–215. [CrossRef]
63. Visual PROMETHEE. Available online: http://www.promethee-gaia.net/software.html (accessed on 15 September 2020).

64. Doberstein, C. Designing collaborative governance decision-making in search of a "collaborative advantage". *Public Manag. Rev.* **2015**, *18*, 819–841. [CrossRef]
65. Johnston, E.W.; Hicks, D.; Nan, N.; Auer, J.C. Managing the inclusion process in collaborative governance. *J. Public Adm. Res. Theory* **2011**, *51*, 699–721. [CrossRef]

applied sciences

MDPI

Article

Typology Selection of Retaining Walls Based on Multicriteria Decision-Making Methods

Belén Muñoz-Medina [1,*], Javier Ordóñez [2], Manuel G. Romana [1] and Antonio Lara-Galera [1]

[1] School of Civil Engineering, Universidad Politécnica de Madrid, 28040 Madrid, Spain; manuel.romana@upm.es (M.G.R.); antoniolorenzo.lara@upm.es (A.L.-G.)

[2] School of Civil Engineering, University of Granada, 18071 Granada, Spain; javiord@ugr.es

* Correspondence: mariabelen.munoz@upm.es; Tel.: +34-91-0674146

Abstract: In civil engineering and construction, in the selection of the most adequate and sustainable alternative, all of the alternatives and selection criteria, such as the requirements of the construction process (which are often overlooked) and the preferences of designers, clients, or contractors, are not always taken into account. The purpose of this article is to suggest a methodology that may allow studying all of the possible alternatives to find the most ideal solution among all of the existing possibilities for the selection of retaining walls to be built in infrastructures in different environments. For this purpose, all typologies of retaining walls and selection criteria (external requirements, construction requirements, characteristics of the natural land and economic criteria) are first identified. Subsequently, a simple methodological method is proposed, allowing the relative importance of each criterion to be established and allowing us to select the most suitable solution for each situation by successively applying different multicriteria decision-making methods. Finally, the methodology developed is applied to two projects in different locations with different constraints. The results obtained provide a set of compromise solutions that remain as best-rank alternatives when the weights of the criteria change. Therefore, the methodology developed can be applied to the selection of typologies of other structures in future projects.

Keywords: multicriteria decision-making; retaining wall; selection criteria; construction requirements; AHP method; VIKOR method; TOPSIS method

Citation: Muñoz-Medina, B.; Ordóñez, J.; Romana, M.G.; Lara-Galera, A. Typology Selection of Retaining Walls Based on Multicriteria Decision-Making Methods. *Appl. Sci.* **2021**, *11*, 1457. https://doi.org/10.3390/app11041457

Academic Editors: Asterios Bakolas and Bożena Hoła

Received: 16 December 2020

Accepted: 2 February 2021

Published: 5 February 2021

1. Introduction

A retaining wall could be defined as any uninterrupted structure that, whether in a passive or in an active way, produces a stabilizing effect over a mass of land [1]. The earth retaining walls are those structures that retain a piece of land with a steeper angle than the angle of friction of the land [2]. There are different classifications of retaining walls according to different criteria: load support mechanism (externally or internally stabilized walls), construction concept (fill or cut), system rigidity (rigid or flexible), and service life (permanent or temporary) [3]. Thus, several different types of retaining walls exist with different performance characteristics as well as constructability characteristics, as well as different uses [2]. Retaining walls are expensive structures that are designed and constructed to support cut and fill slopes where space is not available for construction of flatter, more stable slopes [3]; therefore, the cost of construction, the environment and the space available will be criteria to take into account in its design and construction. Selecting a type of retaining wall is a complex process, considering the various geotechnical and non-geotechnical factors involved [4]. Moreover, during the selection process, it is necessary to consider all the criteria during the whole life cycle [5].

In decision-making, it is necessary to consider all the alternatives and criteria involved in the decision process [6]. In civil engineering and construction, the choice of the most adequate and sustainable alternative is not always made by studying all of the possible typologies nor the life cycle of the infrastructure: design, construction, maintenance, and dismantlement. Therefore, it is necessary to identify all the typologies, their characteristics, and the selection criteria without forgetting the requirements of the construction process. The methodology to be developed must establish a systematic process in the decision-making process. Finally, the solutions obtained must remain unchanged in the face of changes in the decision-maker's preferences or variation in the weight of criteria, i.e., provide strong solutions.

Multicriteria decision-making methods have become a tool to solve engineering problems because they allow complex problems to be modeled. These methods can be used to select the best alternatives when there are several conflicting criteria in a context of uncertainty [7]. In this paper, a methodology is developed for the selection of typology of retaining walls according to different criteria. In the developed methodology, the first step is taken has been to identify all of the different types of retaining walls and selection criteria (external requirements, construction requirements, characteristics of the natural land, and economic criteria) that may determine whether a solution is the best option or not, and to which extent. Subsequently, the most suitable solution is determined by successively applying different multicriteria decision-making methods.

2. Literature Review

2.1. Retaining Walls

Retaining walls are constructed to sustain the lateral pressure of the earth behind them. They are structures used to contain soil or other loose materials when their natural steeps are undesirable, in the case when building linear infrastructures such as railways or roads [5]. Retaining walls are often an overlooked critical asset of infrastructures because they are constantly around us. Each year, globally, millions of square meters of retaining walls are constructed for private and public projects. Retaining walls save space, reduce impacts, and allow owners to get the most out of a given property or right-of-way. Thus, retaining walls are an important part of development projects today [8].

There are many kinds of retaining walls, with different forms and structural characteristics according to the dimensions and location [9]. Over the last three to four decades, due to the development of materials and enhancement in technical understanding of geotechnical engineering, different types of soil retention systems have evolved [10]. There are several classifications of types of earth retaining structures.

In general terms, these structures may be classified into two groups, externally stabilizes walls and internally stabilized walls. The examples of first category are gravity walls, reinforced concrete cantilever and reinforced concrete counterfort walls. These walls are essentially characterized by the concept that the lateral earth pressures due to self-weight of the retained fill and accompanied surcharge loads are carried by the structural wall [10]. The construction sequence of these walls involves the casting of base and stem followed by backfilling with specified material. This requires a considerable amount of time as concrete has to be adequately cured and sufficient time spacing has to be allowed for concrete of previous lift to gain strength before the next lift is cast [10]. The internally stabilized walls include metal strip walls, geotextile reinforced walls and anchored earth walls. These walls comprise of horizontally laid reinforcements which carry most or all of the lateral earth pressure via soil-reinforcement interaction or via passive resistance from the anchor block [10]. This reduces the volume of concrete and steel reinforcement in the wall significantly, thus its construction is relatively fast speed. Retaining walls with relief shelves can be considered as a special type of retaining wall [11].

Another classification establishes three different groups: gravity walls, embedded walls and hybrid walls [2].

- Gravity walls: The purpose of a gravity wall is to avoid sliding and overturns throughout its own weight and by the friction of its base with the land. They are generally built over flat land before being refilled behind the wall. Given their characteristics, they are generally used in high lands, as per example, for retaining embankments in roads and railways. They can also be used in order to support excavations, which are under the natural level of the land. For these situations, the excavation is made in open air and the refilling is done behind the wall. These walls require an additional excavation and refill, as well as the occupation of the land during construction. One of the basic characteristics of the gravity walls is that the drainage is done through drain tubes located behind the wall with the purpose of reducing the water pressure.

- Embedded walls: An embedded wall is the one which base penetrates in the land to achieve side support of the passive resistance of the land on the lower front part of the wall. This resistance can be used as a sole resistance to provide stability to the wall or it can also be used with moorings, supports or other support sources of side support to increase the stability. In this case, they are considered anchored walls or underpinned walls. Unlike with gravity walls, building embedded walls requires specialized equipment and in-situ operations. These walls need to be built before starting the excavation. It is preferred to use embedded walls rather than gravity walls, when:
    - It is necessary to make deep excavations.
    - It is not possible to temporarily occupy adjoining lands in order to make temporary excavations in open air.
    - There are buildings or structures close to the excavation that need to be supported or protected.
    - If there is a high phreatic level and it needs an excessive pumping in order to eliminate the water for making a temporary excavation in open air.
- Hybrid walls: These are the ones that combine different aspects of both gravity walls and built-in walls or slurry walls for retaining the land. An example of a hybrid wall would be a concrete wall built forming an L or T, with pile foundations.

A subclassification of these types of retaining walls is included in Table 1.

Table 1. Types of retaining walls. Source: the authors' own research.

Type of Retaining Wall	Sub-Type of Retaining Wall	Main Characteristics
Gravity wall	Mass concrete wall	It uses its own weight for stabilizing, it does not use any traction
Gravity wall	Reinforced concrete wall	They can be constructed in an L or T shape with a stabilizing platform, with buttresses, made in situ or prefabricated
Gravity wall	Ashlar factory walls	They are stone walls, ashlar masonry, placed one together with another
Gravity wall	Masonry walls	Walls formed by coined rough stones
Gravity wall	Riprap walls	Walls formed by irregular stone blocks, which are generally obtained from a quarry, with different shapes and rough surfaces.
Gravity wall	Crib walls	Walls comprise a grillage of header and stretcher units placed on firm foundations, usually of mass or reinforced concrete
Gravity wall	Gabion walls	Walls formed by superimposed boxes of prismatic form of wired (galvanized) reinforcing mesh refilled with small rocks.
Gravity wall	Green walls	Solid masses of earth reinforced with geotextiles. They do not generally need any foundation. Complete landscape integration.
Gravity wall	Garden walls	Walls formed by a grid of concrete elements placed over a concrete base
Gravity wall	Wooden garden walls	Walls formed by wooden pieces put in place and which interior is refilled with earth
Gravity wall	Prefabricated concrete walls	Walls with the form of an L. The panel and footing have been conjointly prefabricated forming a sole element. They have a 4 meters height limit. Transport limitations
Gravity wall	Concrete prefabricated blocks walls	Green: Formed by prefabricated empty pieces to be filled with sand Blocks of solid mass, which have been fit together. There is a great typology of blocks
Gravity wall	Slat walls	Formed by prefabricated transversal plates, placed between vertical buttresses. They usually have a vegetal cover.

Table 1. *Cont.*

Type of Retaining Wall	Sub-Type of Retaining Wall	Main Characteristics
Gravity wall	Reinforced soil	It is a special gravity wall. The gravity structure is provided by reinforcing the ground itself rather than providing a separate structure
Embedded wall	Traditional embedded walls	Wall penetrates the ground at its base to gain lateral support from the passive resistance of the ground in front of the lower part of the wall. This resistance may be or act in conjunction with anchorages, props or other sources of lateral support for stability
Embedded wall	Sheet pile walls	Walls are constructed by driving, vibrating or pressing steel sheets into the ground without any material being removed
Embedded wall	Contiguous o secant bored pile walls	Pile wall consists of bored cast-in-place piles installed along the line of the wall
Embedded wall	Diaphragm walls	Diaphragm walls are interlocking reinforced-concrete panels. The panels are formed by excavation under fluid to support the surrounding ground and lowering a prefabricated reinforce-ment cage into the trench excavation

2.2. Multicriteria Decision-Making

V. Neumann-Morgenstern (1943) works represent the starting point of the scientific treatment of the individual decision-making process problems [12]. The decision-making process can be done by applying different methods and tools, as well as using different objectives [13]. The use of multicriteria decision-making (MCDM) methods constitutes an efficient tool for reducing subjectivity and systematizing the decision-making process [14]. They can be used in different stages of the process, either to decide on the importance of the criteria for each alternative, to select the most suitable alternative or to establish a ranking of alternatives. Thus, MCDM methods can be used to select the best alternatives when there are several conflicting criteria in a context of uncertainty [7]. MCDM methods have had a fast-growing in many disciplines [15].

In a decision-making problem there are always several elements: Decision criteria C = $\{C_1, C_2, \ldots C_n\}$, conditions which allow us to differentiate alternatives and to establish the preferences of the decision–maker; Weight or measurements of the importance of the criteria for the decision-maker, being each criteria vector associated to a weight vector $[w] = (w_1, \ldots w_n)$. Weights can be established by direct allocation methods or by other methods as Simos method [16], Delphi method [17] or by paired comparisons, analytic hierarchy process (AHP) [18] among others. Alternatives, different solutions to be adopted in a decision-making problem, which are assigned as, As = $\{A_1, A_2, \ldots A_m\}$ (I = 1, 2, … m) are the possible alternatives. And last, the assessment or decision matrix, by which, for all of the criteria taken into account and for each alternative of the choice ensemble, the decision-maker is able to give a numeric or symbolic a_{ij} value that expresses an assessment or opinion of the alternative A_i regarding criteria C_j [19].

The MCDM methods can be classified into different groups according to similar characteristics [13]. These groups are: (1) Scoring methods, their basis consists of assessing the alternatives using basic arithmetical operations, including the simple additive weighting (SAW) and the complex proportional assessment (COPRAS) methods obtain the sum of the weighted normalized values of all the criteria [20]; (2) distance-based methods, the basic principle of them is obtaining the distance among each alternative and a specific point (a hypothetical best alternative), including technique for order preference by similarity to an ideal solution (TOPSIS) and the viekriterijumsko kompromisno rangiranje (VIKOR) methods [21–23]. This group also includes others based on Euclidean distance measurement [24,25]. (3) Pairwise comparison methods are widely used for their ease of calculation when selecting different alternatives in case there are quantitative and qualitative criteria. They allow evaluating the different alternatives according to qualitative criteria when comparing them in pairs and they are sometimes used to weigh the selection criteria as it is the case in this paper. The best-known method of this group is the Analytic Hierarchy Process (AHP) method [18]. Other methods in this group are Analytic Network Process (ANP) [26,27] and Measuring Attractiveness by a Categorical Based Evaluation Technique (MACBETH) [28,29] methods. (4) Outranking methods, this term includes all those MCDM methods that revolve around the theoretical concept of relationships of achievement, proposed by a group of French researchers in the mid-1960s. The first representative of the methods of overcoming was the ELimination Et Choix Traduisant la REalité (ELECTRE) method [30,31]. Another widely used method included in the outranking methods is the Preference Ranking Organization METHod for Enrichment Evaluations (PROMETHEE) method, was introduced by Brans and Vincke (1985) and Brans et al. Since then, numerous applications with a special interest in location problems have appeared: hydropower plants, commercial facilities in a competitive environment, waste disposal sites, financial evaluation, etc. [32].

There already are some precedent uses of MCDM methods in order to choose the type of retaining wall, such as the case of the Transportation Department of South Carolina in USA [3]. In this case, the determination of the most acceptable type is made based on the Important Selection Factor, ISF, rating and the weighted rating each of the above selection factors is given for each retaining wall type. Where the ISF ratio varies between 1: least important factor and 3: most important factor, that is to say, it is a qualitative evaluation. Likewise, for the selection of the most suitable construction solution of retaining walls on construction lands, other precedent selection uses have been included in the bibliography – although not studied in this review-, such as decision trees using logistic regression analysis, for the selection of walls [4]. In this paper, also all decision criteria are identified including construction requirements or requirements of the construction process.

3. Methodology

Description of the problem:

- Question to be answered; which is the most suitable retaining wall to be chosen for certain infrastructures?
- We have different alternatives for the resolution (A_j), which are the different typologies of retaining wall constructions.
- The solution is conditioned by a series of criteria, project and construction requirements (C_i), although not all criteria have the same importance regarding the decision. Therefore, they must be analyzed for each case depending on weights (w_i).
- If we were to represent the problem based on a matrix, we would obtain a matrix by which a_{ji} would be the obtained assessments for each alternative (A_j), depending on the criteria (C_i).

Once the problem has been defined, Figure 1 includes an outline of the steps in the decision process [33]. This process is followed by the methodology developed and described in this section.

3.1. Selection Criteria

To make correct use of the MCDM method, apart from identifying all the alternatives (in this case the types of retaining walls), it is required to identify all characteristics and requirements of the project and the construction process that may influence the selection of the most suitable typology. The project and construction requirements can be divided into five main groups: external requirements, requirements of the construction process, characteristics of the natural land, environmental and economic criteria. These criteria have been listed in Table 2. Some criteria are quite easy to incorporate, analyze and assess, such as the cost criteria, or the performance, which have quantitative assessments, but others, such as the culture (based on the frequency with which certain types of retaining walls are chosen in certain geographical areas), or the influence of the drainage, is not as easy and clear to analyze or assess.

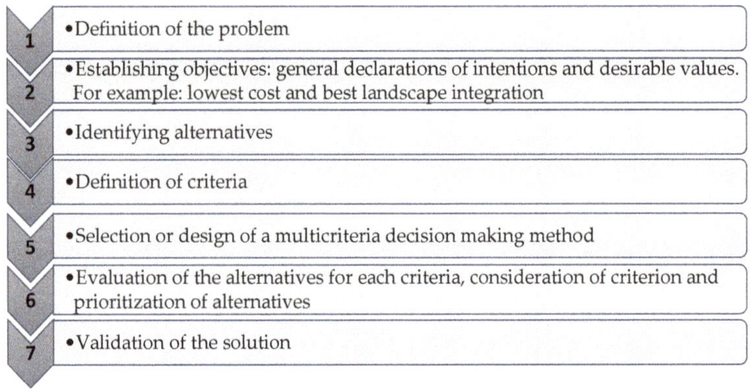

Figure 1. Steps or stages of the decision-making process.

Table 2. Selection criteria. Source: the authors' own research.

Group	Criteria	Type of Criterion	Relevance
External requirements	Retaining wall dimensions	Quantitative	High
External requirements	Space available	Quantitative	High
External requirements	Traffic restrictions	Qualitative	Medium
External requirements	Influence on public services	Qualitative	High
External requirements	Space available in the base of the wall	Quantitative	Medium
External requirements	Permissible settlement	Quantitative	High
External requirements	On-site overloading	Quantitative	High
External requirements	Seismicity	Quantitative	High
External requirements	Permeability	Quantitative	Medium
External requirements	Aesthetic appearance	Qualitative	Low
External requirements	Technical culture and customs	Qualitative	Medium
Environmental	Noise and vibration during construction	Quantitative	Medium
Environmental	Construction and demolition residues	Quantitative	High
Environmental	Landscape integration	Qualitative	High
Environmental	Energy use and greenhouse gas emissions (life cycle)	Quantitative	High
Construction requirements	Construction performance	Quantitative	High
Construction requirements	Elements inside the ground: round stones, gravel, debris	Qualitative	Low
Construction requirements	Surface disturbances	Qualitative	Low
Construction requirements	Machinery and specialized equipment	Qualitative	High
Construction requirements	Spaces for material storage	Quantitative	Medium
Construction requirements	Material availability	Quantitative	Medium
Characteristics of the soil natural	Density of the retained material	Quantitative	Medium
Characteristics of the soil natural	Interstitial water pressure	Quantitative	Medium
Characteristics of the soil natural	Soil resistance (internal angle of friction)	Quantitative	High
Economic criteria	Construction cost	Quantitative	High
Economic criteria	Preservation and maintenance necessities	Quantitative/Qualitative	High

3.2. MCDM Methods

A methodology for resolving the problem has been set out, combining two multiple-criteria decision-making methods, the analytic hierarchy process (AHP) and the VIKOR methods. The obtained selection will be compared with the one resulting when applying the TOPSIS method in substitution of the VIKOR method. Both methods are focused on trying to find the solution that is closest to the optimal solution but with a different process of evaluation [21].

By the analytic hierarchy process (AHP), the weight eigenvector is calculated for the criteria that determine which is the most ideal solution, by making a paired comparison of them for each project [34]. It must be taken into account that the weight eigenvector is not the same for each project, since certain criteria may have bigger importance in comparison to the others, depending on the characteristics of the project. We must remember that AHP measures the global inconsistency of the views by the consistency ratio (CR), calculated by dividing the consistency index (*CI*) and the random index (RI), and it should be of less than 10%. The consistency index measures the consistency of the comparison matrix [18].

$$CI = \frac{\lambda_{max} - n}{n - 1} \quad (1)$$

where λ_{max} is the biggest value of the transposed matrix of the paired comparison matrix, and n is the matrix range. The RI is an index which measures the consistency of a random matrix [18,35]. On the other hand, through paired comparisons we can establish the "behavior" of each alternative for each of the qualitative criteria that are part of the making-decision processes, to obtain a quantitative assessment for qualitative criteria.

Later, the VIKOR method will be applied for selecting the most suitable typology, based on a classification list of alternatives that shall provide us with one or more compromise solutions. "Viekriterijumsko kompromisno rangiranje" (VIKOR) is a Serbian term meaning multicriteria optimization and compromise solution [23]. It was developed to solve a decision problem with a limited number of alternatives (possible solutions) with conflicting criteria and with different units of measurement [22]. Therefore, the VIKOR method is suitable for solving decision-making problems with conflicted and noncommensurable criteria (which means, with different units) or when there are quantitative and qualitative criteria. The VIKOR method has been applied on many occasions for selection of alternatives in infrastructures, as reflected in the literature [36–40]. For normalization, converting the criteria into dimensional variables, a linear function is used in the VIKOR method that does not depend on a function of value of criteria as in the case of TOPSIS method [21]. The compromise solution shall be the one closest to the optimal solution [22]. To obtain a compromise solution (or solutions), we shall follow the next steps:

1. "Best" and "worst" value of each alternative is calculated for each criterion, as follows:

$$f_i^* = max_i f_{ji};\ \ f_i^- = min_i f_{ji};\ \text{If criterion i represents a benefit}.$$
$$f_i^* = min_i f_{ji};\ \ f_i^- = max_i f_{ji};\ \text{If criterion i represents a cost}.$$

   where i is the number of criteria and j is the number of alternatives, so f_{ji} is an evaluation of the alternative A_j with respect to criterion C_i.

2. Values S_j, R_j, and Q_j are calculated for each alternative as follows:

$$S_j = \sum_{i=1}^{n} w_i (f_i^* - f_{ji}) / (f_i^* - f_i^-) \quad (2)$$

$$R_j = \max_i |w_i (f_i^* - f_{ji}) / (f_i^* - f_i^-)| \quad (3)$$

   where w_i is the weight of criterion i relative to the rest, that is, it reflects the relative importance of each criterion. At this point, it is necessary to indicate that in the VIKOR method the normalized values are given by the expression: $(f_i^* - f_{ji})/(f_i^* - f_i^-)$. Thus, normalized values do not depend on the unity of the different criteria [21].

$$Q_j = \vartheta [(S_j - S^*)/(S^- - S^*)] + (1 - \vartheta)[(R_j - R^*)/(R^- - R^*)] \quad (4)$$

   where $S^* = \min_j S_j$, $S^- = \max_j S_j$, $R^* = \min_j R_j$, $R^- = \max_j R_j$, and ϑ represent weight of the strategy of "the majority of criteria" (or "the maximum group utility"). Consensus is for a value $\vartheta = 0.5$ [21]. Other authors have demonstrated how difficult it is to achieve consensus in situations of uncertainty and with large amounts of information and possible alternatives [41].

3. Different alternatives are sorted according to values *S*, *R* and *Q* and the results are three ranking lists.
4. Best-ranked alternative A(1) is identified as a compromise solution, i.e., the alternative with the lowest value of *Q*. This is true, if the following two conditions are met:

Requirement 1: Acceptable advantage.

$Q(A(2)) - Q(A(1)) \geq DQ$, where besides A(2) is the second alternative according to the value classification of *Q*, and DQ = 1/(J-1), where J represents the number of alternatives.

Requirement 2: Acceptable stability in the decision-making process.

Alternative A(1) will also be the best ranked according to the list of values of S and/or R. This alternative solution is stable within a decision-making process.

If one of the requirements is not met, the method offers a range of alternative solutions, which can consist of: Alternatives A(1) and A(2), if condition 2 is not met; Alternatives A(1), A(2), ... A(M), if condition 1 is not met; A(M) will be established taking into account the relation $Q(A(M)) - Q(A(1)) < DQ$. These alternatives are considered to be within the "closeness" [21,22].

The VIKOR method is an efficient tool to use as a multiple-criteria decision-making method when the decision-maker is not able to, or does not know how to, express their preferences at the beginning of the design process. The obtained compromise-solution may be approved by the decision-maker, given that it provides the biggest group usefulness to the majority, represented by the minimum *S*, and an individual minimum opposition represented by the minimum *R* [22].

Later, the TOPSIS method shall be applied in order to compare the results obtained by both methods. In order to do this we start with the decision matrix and we calculate the positive optimal solution and the negative optimal solution, and we will obtain the best ranked solution, being the one that is closer to the positive optimal solution, and further from the negative optimal solution [42,43]. To carry out the simulation and application of the VIKOR and TOPSIS methods, an algorithm has been developed in the MATLAB® software to automate the calculations.

4. Case Study

The described methodology is applied to two projects with different context and purpose. Project 1: Mountain road in a nature park in the province of Madrid (Spain) and Project 2: New highway under construction in Andalusia (Spain).

Validity will be verified by the behavior in the solution to be adopted when there are variations in the weight of criteria, the changes in the compromise solution will be determined using the VIKOR method. And lastly, the results will be compared to the obtained results using the TOPSIS method.

A small set of criteria and typologies have been taken into account to show the methodology simply. As alternatives are considered, four typologies included in Table 2: a reinforced concrete wall built in situ, garden wall, green wall reinforced with geotextiles, and riprap wall. These typologies have been selected given that they represent a bigger variety of retaining walls. In the decision-making process, we choose as determining criteria the following: construction cost in €/m, construction performance, m/day, landscape integration, technical culture and customs (construction frequency), and lastly, preservation and maintenance necessities. In a manner similar to that of the selection of alternatives, these five criteria were chosen because they are the most common and determining in retaining walls. All retaining walls have an average height of four meters. To apply the decision-making methodology, the following hypotheses are needed:

- All the retaining walls have a stable structure (the assessed solutions are stable).
- Logistic determinants have not been considered such as material availability or available space for gatherings.
- The drainage system works correctly in all cases

- Selection criteria are independent.

To evaluate the different alternatives concerning the selection criteria, firstly the cost and performance of construction has been determined based on the advice of suppliers, Spanish Ministry of Development [1,44], as well as the advice of designers and contractors who were consulted for this purpose. Next, for the quantitative evaluation of the different alternatives for the qualitative criteria: technical culture and customs and preservation and maintenance necessities, paired comparisons according to the Saaty scale of the AHP method [18,35] are used to determine such a quantitative assessment. In this way, the quantitative assessment of each alternative for each qualitative criterion varies between 0 and 1, with 0 being the lowest value and 1 the highest value for each criterion. The results obtained for each alternative respect to each selection criterion have been included in Table 3. If these results are analyzed, it can be seen that, for example, the alternative of reinforced concrete retaining wall is the one which obtains the lowest valuation concerning the criterion of landscape integration, is the one that is built most frequently and, on the contrary, is the alternative which requires the least conservation and maintenance actions throughout the life cycle.

Table 3. Data and values of alternative for each criterion. Source: the authors' own research.

Alternative	Construction Cost (€/m)	Construction Performance (m/Day)	Landscape Integration	Technical Culture and Customs	Preservation and Maintenance Neccesities
Reinforced concrete retaining wall	870	1.5	0.041	0.506	0.047
Garden retaining wall	640	2.48	0.099	0.257	0.316
Green retaining wall	430	2.75	0.608	0.065	0.564
Riprap retaining wall	608	2.8	0.251	0.172	0.073

Once the assessment of the alternatives has been determined, the values $f_i{}^*$ and $f_i{}^-$, being the best and worst values of each criteria function, are calculated according to the VIKOR method. The results are included in Table 4.

Table 4. Values $f_i{}^*$ and $f_i{}^-$, applicating step 1 of the viekriterijumsko kompromisno rangiranje (VIKOR) method Source: the authors' own research.

	Construction Cost (€/m) Minimum	Construction Performance (m/day) Maximum	Landscape Integration Maximum	Technical Culture and Customs Maximum	Preservation and Maintenance Necessities Minimum
$f_i{}^*$	430	2.8	0.608	0.506	0.047
$f_i{}^-$	870	1.5	0.041	0.065	0.564

It is important to highlight that the importance of each criterion depends on the location where the retaining wall is going to be constructed. Thus, for each project, and before using the VIKOR and TOPSIS methods, the weight vector is determined, by paired comparisons, and by applying the AHP. So, for project 1, given the environmental factors, when using the AHP method for obtaining the weight vector by paired comparisons, the weight vector is obtained, w = (0.03; 0.08; 0.54; 0.21; 0.13). It is important to remember that the consistency of the comparison matrix must be identified. After determining the consistency following equation (1), the consistency ratio = 0.094697, which is under 0.1, is obtained. Therefore, the assessments made can be considered as consistent.

Then, Equations (2)–(4) are applied to calculate the S_j, R_j and Q_j values. The alternatives classification list will be established according to the values of S, R and Q, to establish the solution or the ensemble of compromise solutions. To carry out the simulation and application of the VIKOR and TOPSIS methods, an algorithm has been developed in the

Matlab® software to automate the calculations. The results obtained by VIKOR method are shown in Table 5:

Table 5. Ranking of alternatives according to the VIKOR method. Project 1. Source: the authors' own research.

Alternative	Q	S	R
Green retaining wall	0.000	0.346	0.210
Riprap retaining wall	0.437	0.521	0.341
Garden retaining wall	0.917	0.711	0.487
Reinforced concrete retaining wall	0.925	0.657	0.542

Minimum value of Q can be observed for the green wall alternative (see Table 5). Both requirements of the VIKOR method are met: requirement 1, acceptable advantage, and requirement 2, stability of the decision-making process. Therefore, there is a compromise solution for the decision-making problem described here, being the alternative of green wall reinforced with geotextile the one most suitable to the determining criteria.

To verify the validity of the method, the VIKOR method is applied for another case varying the importance of the different criteria but keeping the criterion of landscape integration as the most important, with the weight between 0.48 and 0.54. Thus, it is confirmed as compromise solution the typology of the green wall, being the best-ranked option in the classification lists Q, S and R, although within an ensemble of compromise solutions in which a riprap wall is also included as the second option. When using TOPSIS, the green retaining wall is confirmed as the best-ranked solution, obtaining the following classification: Green retaining wall, $R_j = 0.9966$; Riprap retaining wall, $R_j = 0.5955$; Garden retaining wall, $R_j = 0.5227$; Reinforced concrete retaining wall, $R_j = 0.0034$.

For project 2, the process followed in the previous case must be repeated, calculating on first place the weight matrix for the selection criteria, by paired comparisons. The weight vector obtained is $w = (0.41; 0.03; 0.26; 0.11; 0.18)$. The consistency of the comparison matrix has been assessed, obtaining a Consistency Ratio of less than 0.1, being able now to conclude that the assessments made are consistent. In the same way, we complete the decision-making process by calculating the values S_j, R_j and Q_j and establishing the classification lists of alternatives, Table 6.

Table 6. Ranking of alternatives according to the VIKOR method. Project 2. Source: the authors' own research.

Alternative	Q	S	R
Green retaining wall	0.034	0.294	0.182
Riprap retaining wall	0.158	0.424	0.166
Garden retaining wall	0.514	0.598	0.237
Reinforced concrete retaining wall	1.000	0.707	0.409

As a result, a minimum of the Q, S and R values for the case of the green wall is obtained, but with similar values to those for riprap wall, from which we deduce that there is not a clear optimal solution, but an ensemble of compromise solutions that can be a solution to the problem in a more or less appropriate form. It is demonstrated this by applying requirement 1, acceptable advantage of the VIKOR method. $Q(A(1)) - Q(A(2))$, is lower than 0.333, therefore the requirement 1 of acceptable advantage is not met. Therefore, as a solution to the decision-making problem, we suggest an ensemble of compromise solutions formed by the alternatives green retaining wall and riprap retaining wall. It should be remembered that the VIKOR method proposes an ensemble of compromise solutions to those alternatives, A(1), A(2), ..., A(M), that make $Q(A(M)) - Q(A(1)) < DQ$. For project 2, when the most important criterion is the cost, the best-valued solution is the green wall, being a close second (admissible solution) the riprap wall, therefore being both valid with VIKOR. However, in the case the importance of the construction performance is increased (+3%), the optimal solution is the riprap retaining wall, and the garden retaining wall would be the second solution. With this analysis, we can see the sensibility of the method and its possible use when no characteristic is clearly predominant, like the landscape integration in the previous case. If the criteria weight is not defined, the strongest solution will be the riprap retaining wall, which is the best ranked or the second in all the cases. A strong solution is the one that is still admissible when the importance of the criteria changes.

When using TOPSIS, the green retaining wall is confirmed as the best-ranked solution and as the second option the riprap retaining wall, obtaining the following classification: Green retaining wall, $R_j = 0.9991$; Riprap retaining wall, $R_j = 0.5955$; Garden retaining wall, $R_j = 0.5227$; Reinforced concrete retaining wall, $R_j = 0.0005$. This way it is proven that the proposed method can be applied both in cases with different criteria importance and in cases in which the precedence is clear but the differences not so much.

5. Results

After applying the methodology (AHP + VIKOR methods) to two projects in different environments, in both cases, we obtained a compromise solution (or set of solutions) that remains stable in the case of changes in the weights of the selection criteria. In other words, the best-rank solution of type of retaining wall remains the best-ranked first or second whit different weighting criteria, which is why this solution is admissible when the importance of the criteria changes. In both projects, the ranking obtained by the VIKOR method is confirmed after the application of the TOPSIS method. Obtaining the green retaining as the first-rank solution and the rip rap retaining wall as the second-rank solution in the selection of retaining wall types for two different projects, one in a natural environment and other in design and construction of a new highway.

6. Study Implications and Contributions

The methodology developed first provides a systematic process for the identification of alternatives and selection criteria for the selection of types of retaining walls. It incorporates an exhaustive list of all the selection criteria that need to be considered and analyzed in the decision-making process, including the requirements of the construction process. This list of criteria, and their definition, may help other researchers and practitioners to include in the selection of best alternative in future projects for example for selection de typologies of bridges. Moreover, it is possible to systematically incorporate the importance of the different criteria according to the environment in which the structure is built since it is not the same as a structure being built in a natural environment as in a newly constructed highway with fewer environmental and space limitations.

7. Discussion

After applying the methodology to the case studies, it is appropriate to discuss whether some issues could improve the identification or weighting of criteria or the assessment of each alternative against each criterion. By that and concerning the selection of retaining wall typologies, the article provides a more objective process for the weighting of criteria than other previous investigations commented on in the review of the literature that is carried out by subjective assessments of the decision-maker and with qualitative scales. However, in any way, the weighting of criteria by paired comparisons (AHP method) has a subjective component, although this can be overridden with an appropriate number of decision-makers/experts or decision group. Therefore, for future research, it would be appropriate to modify the methodology for the selection of retaining walls by applying other methods of weighting selection criteria such as the entropy method, which does not involve the opinions of experts and increases objectivity, [45] and compare the results with the application of the AHP method with an appropriate number of decision-makers. On the other hand, the selection methodology has not taken into account the dependence of the criteria, so this issue should be analyzed to verify if the solutions obtained would vary and whether the methodology is appropriate [26,35].

In the paper, all the typologies of retaining walls have been identified, as well as all the selection criteria, however, and to simplify the application of the methodology, the selection method has been applied to a reduced group of alternatives and criteria. For this reduced set of alternatives and criteria, a ranking of stable compromise solutions when there are changes in the weight of the criteria has been obtained, but it would be convenient to extend the application to a greater number of alternatives and criteria and to analyze the results. On this option, the rank reversal process should be analyzed when adding new alternatives or criteria to the previously chosen set. In this phenomenon, the ordering of alternatives inverts when an alternative is added or eliminated from the list of alternatives [46–49].

The TOPSIS method is applied to confirm the best-rank solution previously classified by the VIKOR method; however, the TOPSIS method has several disadvantages. One of them is that it requires the normalization of the values of the decision matrix to avoid the effect of the dimensionality of the assessments of the alternatives with respect to each criterion [42,50]. Another disadvantage of the TOPSIS method is that, when using the Euclidean distance to determine the distance of each alternative to positive ideal solutions and negative ideal solutions, it does not consider the correlation between criteria [51].

8. Study Limitations and Future Research Directions

As discussed in previous sections, one of the limitations of the study is the determination of criteria weights, which has been done according to the AHP method assuming criteria independence. In future research, this issue should be addressed so that it is studied how the correlation of the criteria influences the weighting of the criteria and to consider other methods of decision-making that allow this correlation to be taken account. Furthermore, in future studies, the methodology can be improved to avoid the phenomenon of "rank reversal" when an alternative is added or removed.

9. Conclusions

The methodology developed first provides a systematic process for the identification of alternatives and selection criteria for the selection of types of retaining walls. Thus, the application of the methodology allows identifying all the determining criteria in the selection of retaining walls, both minor and major importance criteria, and identifying the construction requirements that are often overlooked in the phase of design and selection of alternatives. The relative importance of the criteria for different projects is determined through paired comparisons. As result, a ranking of constructive solutions of retaining walls that is still admissible when the importance of the criteria changes is obtained.

Case study validates the combined sequential use of two decision-making methods for selecting the best constructive solution for a wall in different and specific situations by proving that a solid and transparent ensemble of criteria can be taken into account, such as the environment, with the construction performance and the costs through an objective and transparent process, making clear which are the expressed preferences and their importance in the process. Moreover, verification of the validity of the methodology shows the stability of the solutions obtained, even if there are changes in the weighting of the criteria.

The suggested decision-making process is based on data that are easy to obtain and to allow the evaluation of alternatives according to qualitative criteria. As a result of this, the fact of obtaining a solution as a result of the application of a systematic process that is relatively objective will allow—in situations of disagreements among different groups of interests, stakeholders—to justify the adopted solution. As a result of the research, a methodology for the selection of types of retaining walls is provided, which can be useful for public administrators, designers, project managers, and constructors.

It should be noted that the methodology can be applied to the selection of other infrastructures in which the design and construction requirements may determine that particular alternatives are not suitable or are suitable to a lesser extent, for example, for the selection of bridge types.

Author Contributions: Conceptualization, B.M.-M., J.O., and M.G.R.; methodology, B.M.-M.; software, B.M.-M.; validation, B.M.-M., J.O., and M.G.R.; formal analysis, A.L.-G.; investigation, B.M.-M.; writing—original draft preparation, B.M.-M.; writing—review and editing, B.M.-M. and A.L.-G. All authors have read and agreed to the published version of the manuscript.

Funding: This research received no external funding.

Institutional Review Board Statement: Not applicable.

Informed Consent Statement: Not applicable.

Data Availability Statement: Data is contained within the article or supplementary material.

Conflicts of Interest: The authors declare no conflict of interest.

References

1. Spanish Ministry of Development. *Tipologia de Muros de Carretera*, 1st ed.; MINISTERIO DE FOMENTO: Madrid, Spain, 1999.
2. Anderson, S. Chapter 62 Types of retaining walls. In *ICE Manual*; ICE Publishing: London, UK, 2012; pp. 959–968.
3. South Carolina Department of Transportation (SCDOT). *SCDOT Geotechnical Design Manual*; South Carolina Department of Transportation (SCDOT): North Charleston, SC, USA, 2010.
4. Choi, M.; Lee, G. Decision tree for selecting retaining wall systems based on logistic regression analysis. *Autom. Constr.* **2010**, *19*, 917–928. [CrossRef]
5. Pons, J.J.; Penadés-Plà, V.; Yepes, V.; Martí, J.V. Life cycle assessment of earth-retaining walls: An environmental comparison. *J. Clean. Prod.* **2018**, *192*, 411–420. [CrossRef]
6. Liou, J.J.H.; Lu, M.T.; Hu, S.K.; Cheng, C.H.; Chuang, Y.C. A hybrid MCDM model for improving the electronic health record to better serve client needs. *Sustainability* **2017**, *9*, 1819. [CrossRef]
7. Chen, Y.; Kilgour, D.M.; Hipel, K.W. Screening in multiple criteria decision analysis. *Decis. Support Syst.* **2008**, *45*, 278–290. [CrossRef]
8. Anderson, S.A.; Alzamora, D.; Demarco, M.J. Asset management systems for retaining walls. In *Geo-Velopment: The Role of Geological and Geotechnical Engineering in New and Redevelopment Projects—Proceedings of the 2008 Biennial Geotechnical Seminar*; ASCE Library: Denver, CO, USA, 2008; pp. 162–177. [CrossRef]
9. Zastrow, P.; Molina-Moreno, F.; García-Segura, T.; Martí, J.V.; Yepes, V. Life cycle assessment of cost-optimized buttress earth-retaining walls: A parametric study. *J. Clean. Prod.* **2017**, *140*, 1037–1048. [CrossRef]
10. Khan, A.J.; Sikder, M. Design Basis and Economic Aspects of Different Types of Retaining Walls. *J. Civil Eng. (IEB)* **2004**, *32*, 17–34.
11. Shehata, H.F. Retaining walls with relief shelves. *Innov. Infrastruct. Solut.* **2016**, *1*, 1–13. [CrossRef]
12. Ríos, S. Algunos procesos y problemas en la Ciencia de la decisión. *Rev. Matemática Complut.* **1988**, *11*. Available online: http://www.mat.ucm.es/serv/revmat/vol11-1/vol11-1g.html (accessed on 11 July 2020).
13. Penadés-Plà, V.; García-Segura, T.; Martí, J.; Yepes, V. A Review of Multi-Criteria Decision-Making Methods Applied to the Sustainable Bridge Design. *Sustainability* **2016**, *8*, 1295. [CrossRef]
14. Cheney, J.N. *Utilizing the Decision Matrix to Introduce the Engineering Design Process; Center for Teaching Excellence, United States Military Academy, West Point*; United States Military Academy: Orange County, NY, USA, 2009.
15. Triantaphyllou, E. *Multi-Criteria Decision Making Methods: A Comparative Study*; Springer: Boston, MA, USA, 2000; Volume 44.
16. Siskos, E.; Tsotsolas, N. Elicitation of criteria importance weights through the Simos method: A robustness concern. *Eur. J. Oper. Res.* **2015**, *246*, 543–553. [CrossRef]
17. Sierra, L.A.; Yepes, V.; Pellicer, E. A review of multi-criteria assessment of the social sustainability of infrastructures. *J. Clean. Prod.* **2018**. [CrossRef]
18. Saaty, T.L. How to make a decision: The analytic hierarchy process. *Eur. J. Oper. Res.* **1990**, *48*, 9–26. [CrossRef]
19. Fülöp, J. *Introduction to Decision Making Methods*; BDEI-3 Workshop: Washington, DC, USA, 2005; pp. 1–15.
20. Podvezko, V. The comparative analysis of MCDA methods SAW and COPRAS. *Eng. Econ.* **2011**, *22*, 134–146. [CrossRef]
21. Opricovic, S.; Tzeng, G.H. Compromise solution by MCDM methods: A comparative analysis of VIKOR and TOPSIS. *Eur. J. Oper. Res.* **2004**, *156*, 445–455. [CrossRef]
22. Opricovic, S.; Tzeng, G.H. Extended VIKOR method in comparison with outranking methods. *Eur. J. Oper. Res.* **2007**, *178*, 514–529. [CrossRef]
23. Shumaiza; Akram, M.; Al-Kenani, A.N.; Alcantud, J.C.R. Group decision-making based on the VIKOR method with trapezoidal bipolar fuzzy information. *Symmetry* **2019**, *11*, 1313. [CrossRef]

24. Badi, I.; Ballem, M.; Shetwan, A. Site selection of desalination plant in Libya by using Combinative Distance-Based Assessment (CODAS) method. *Int. J. Qual. Res.* **2018**, *12*, 609–624.
25. Karaşan, A.; Boltürk, E.; Kahraman, C. A novel neutrosophic CODAS method: Selection among wind energy plant locations. *J. Intell. Fuzzy Syst.* **2019**, *36*, 1491–1504. [CrossRef]
26. Saaty, T.L. *Decision Making with Dependence and Feedback: The Analytic Network Process*; RWS Publications: Pittsburgh, PA, USA, 1996; p. 370, ISBN 0-9620317-9-8.
27. Saaty, T.L. Fundamentals of the analytic network process—Multiple networks with benefits, costs, opportunities and risks. *J. Syst. Sci. Syst. Eng.* **2004**, *13*, 348–379. [CrossRef]
28. Costa, C.A.B.E.; Vieira, A.C.L.; Nóbrega, M.; Quintino, A.; Oliveira, M.D.; Costa, J.B.E. Collaborative Value Modelling in corporate contexts with MACBETH. *Procedia Comput. Sci.* **2019**, *162*, 786–794. [CrossRef]
29. Giron, X.; Madrid, D. *Comparación de Dos Métodos de Toma de Decisiones Multicriterio: Macbeth y Promethée*; Universitat Rovira i Virgili: Tarragona, Spain, 2008.
30. Roy, B. Classement et choix en présence de points de vue multiples. *RAIRO Oper. Res. Rech. Opérationnelle* **1968**, *2*, 57–75. Available online: http://www.numdam.org/conditions (accessed on 16 July 2020). [CrossRef]
31. Roy, B. The outranking approach and the foundations of electre methods. *Theory Decis.* **1991**, *31*, 49–73. [CrossRef]
32. Brans, J.-P.; Smet, Y. *PROMETHEE Methods*; Chapter 1; Multiple Criteria Decision Analysis. International Series in Operations Research & Management Science; Springer: New York, NY, USA, 2016; Volume 233, ISBN 978-1-4939-3094-4. [CrossRef]
33. Baker, D.; Bridges, D.; Hunter, R.; Johnson, G.; Krupa, J.; Murphy, J.; Sorenson, K. *Guidebook to Decision-Making Methods*; U.S. Department of Energy: Washington, DC, USA, 2001.
34. Yepes, V.; García-Segura, T.; Moreno-Jiménez, J.M. A cognitive approach for the multi-objective optimization of RC structural problems. *Arch. Civ. Mech. Eng.* **2015**, *15*, 1024–1036. [CrossRef]
35. Saaty, R.W. *Decision Making in Complex Environments. The Analytic Network Process (ANP) for Decision Making with Dependence and Feedback*; RWS: Pittsburgh, PA, USA, 2016; ISBN 1-888603-00-3.
36. Belošević, I.; Kosijer, M.; Ivić, M.; Pavlović, N. Group decision making process for early stage evaluations of infrastructure projects using extended VIKOR method under fuzzy environment. *Eur. Transp. Res. Rev.* **2018**, *10*. [CrossRef]
37. Sennaroglu, B.; Varlik Celebi, G. A military airport location selection by AHP integrated PROMETHEE and VIKOR methods. *Transp. Res. Part D Transp. Environ.* **2018**, *59*, 160–173. [CrossRef]
38. Ramezanianpour, A.A.; Tabatabaei, S.A.; Pourlak, M.; Abareshi, M. AHP-VIKOR Bridge Structural System Selection in Urban Areas Tehran Interchanges Case Study. *Ambient Sci.* **2016**, *3*. [CrossRef]
39. Awasthi, A.; Omrani, H.; Gerber, P. Investigating ideal-solution based multicriteria decision making techniques for sustainability evaluation of urban mobility projects. *Transp. Res. Part A Policy Pract.* **2018**, *116*, 247–259. [CrossRef]
40. Mardani, A.; Jusoh, A.; Zavadskas, E.; Cavallaro, F.; Khalifah, Z. Sustainable and Renewable Energy: An Overview of the Application of Multiple Criteria Decision Making Techniques and Approaches. *Sustainability* **2015**, *7*, 13947–13984. [CrossRef]
41. Jelokhani-Niaraki, M.; Malczewski, J. Decision complexity and consensus in Web-based spatial decision making: A case study of site selection problem using GIS and multicriteria analysis. *Cities* **2015**, *45*, 60–70. [CrossRef]
42. Ceballos, B.; Lamata, M.T.; Pelta, D.; Sanchez, J.M. El método TOPSIS Relativo vs. Absoluto. In *Rect@ Rev. Electrónica Comun. y Trab. ASEPUMA, ISSN-e 1575-605X*; 2013; Volume 14, pp. 181–192. Available online: https://dialnet.unirioja.es/servlet/articulo?codigo=4743030&info=resumen&idioma=ENG (accessed on 15 December 2020).
43. San Cristóbal, J.R. Contractor Selection Using Multicriteria Decision-Making Methods. *J. Constr. Eng. Manag.* **2012**, *138*, 751–758. [CrossRef]
44. Spanish Ministry of Development. *Guía Para el Proyecto y la Ejecución de Muros de Escollera en Obras de Carretera*; Spanish Ministry of Development: Madrid, Spain, 2006.
45. Lee, H.C.; Chang, C. Comparative analysis of MCDM methods for ranking renewable energy sources in Taiwan. *Renew. Sustain. Energy Rev.* **2018**, *92*, 883–896. [CrossRef]
46. García-Cascales, M.S.; Lamata, M.T. On rank reversal and TOPSIS method. *Math. Comput. Model.* **2012**, *56*, 123–132. [CrossRef]
47. Mousavi-Nasab, S.H.; Sotoudeh-Anvari, A. A new multi-criteria decision making approach for sustainable material selection problem: A critical study on rank reversal problem. *J. Clean. Prod.* **2018**, *182*, 466–484. [CrossRef]
48. Ceballos, B.; Pelta, D.A.; Lamata, M.T. Rank Reversal and the VIKOR Method: An Empirical Evaluation. *Int. J. Inf. Technol. Decis. Mak.* **2018**, *17*, 513–525. [CrossRef]
49. Wang, Y.M.; Luo, Y. On rank reversal in decision analysis. *Math. Comput. Model.* **2009**, *49*, 1221–1229. [CrossRef]
50. Çelen, A. Comparative analysis of normalization procedures in TOPSIS method: With an application to Turkish deposit banking market. *Informatica* **2014**, *25*, 185–208. [CrossRef]
51. Villanueva Ponce, R.; García Alcaraz, J.L. Evaluación de Tecnología utilizando TOPSIS en Presencia de Multi-colinealidad en Atributos: ¿Por qué usar distancia de Mahalanobis? *Rev. Fac. Ing. Univ. Antioquia* **2013**, *67*, 31–42.

applied sciences

MDPI

Article

Technical and Structural Problems Related to the Interaction between a Deep Excavation and Adjacent Existing Buildings

Grzegorz Dmochowski and Jerzy Szolomicki *

Faculty of Civil Engineering, Wroclaw University of Science and Technology, 50-370 Wroclaw, Poland; grzegorz.dmochowski@pwr.edu.pl
* Correspondence: jerzy.szolomicki@pwr.edu.pl; Tel.: +48-505-995-008; Fax: +48-71-320-36-45

Abstract: Currently, new housing in city centers is more and more often developed on small plots of land, or existing buildings on such plots are rebuilt to such an extent that only their façade walls remain. In both cases, as a rule, a deep excavation is also made, either at the existing object or within its area. Serious damage often occurs because of the carried out work. It is not possible to accurately determine the response of a building to the deformation associated with the excavation due to the variability of many factors that influence it. As a result, the response of the building must be estimated on the basis of constant monitoring and approximate calculations. Depending on the size of the predicted ground displacements and the technical condition of buildings, it is often necessary to protect or strengthen their structural elements. In the paper, the authors analyzed various risk factors for the implementation of infill buildings and the revitalization of historic buildings using only their façade walls. In addition, examples of contemporary solutions for securing the walls of existing buildings, and the method of monitoring vertical deformations using the Hydrostatic Levelling Cell (HLC) system, are presented.

Keywords: deep excavation; adjacent buildings; HLC monitoring; temporary support structure

Citation: Dmochowski, G.; Szolomicki, J. Technical and Structural Problems Related to the Interaction between a Deep Excavation and Adjacent Existing Buildings. *Appl. Sci.* **2021**, *11*, 481. https://doi.org/10.3390/app11020481

Received: 29 November 2020
Accepted: 3 January 2021
Published: 6 January 2021

1. Introduction

It has recently become general practice to use small building plots located in dense urban housing. In such places, infill buildings are constructed, or historic buildings are revitalized using their façade walls for the construction of new objects [1,2]. In both cases, the construction projects involve deep excavation work. During the construction of infill buildings, these excavations can cause serious damage to existing buildings located in the surrounding neighborhood. However, when erecting a building that uses historic façade walls, the excavations have an impact on the structure of these walls and their foundations [3]. As a result of ground deformations additional forces are generated in the elements and their connections as well as additional deformations and displacements. It is the horizontal and vertical displacements of existing structures that are an important factor when assessing the degree of risk of structural collapse. The inclination of the building from the vertical axis is usually allowed up to a value of 3 per mille [4], and its settlement to a value ranging from 5 to 15 mm [5]. Inclinations of existing buildings from their vertical axis cause additional horizontal forces (which are the components of vertical loads) in the elements of the structures. The horizontal forces that act on the entire structure must be transferred by its load-bearing systems under the required conditions of stability and strength, or by external designed supports.

In general, an investment in a dense downtown housing development, in order to ensure the stability of existing buildings, should include [6]:

- the identification of soil and water conditions under the designed building structure and in the zone of its impact on surrounding buildings;
- the defining of the requirements for functional and spatial solutions of the underground part of the investment;

- the determination of the zones of impact of the designed object on neighboring buildings;
- the protecting and strengthening of the walls of existing buildings;
- the monitoring of the vertical deformations of existing buildings.

The article presents examples of the innovative support (protection) of the walls of existing buildings in dense urban housing, and also presents the system of monitoring vertical deformations with the use of the hydrostatic levelling cell (HLC) system.

2. Methods

In dense urban developments, a common problem is deep excavations which are adjacent to existing buildings. In this case, it is important to assess the impact of excavation construction on the surrounding soil and adjacent buildings, and also to estimate soil settlement for the proper design of the new structure. From a technical point of view, the condition of adjoining buildings is usually critical due to the weakness of the structural system or the shallow depth of the foundations. As a result of deep excavations, lateral and vertical soil displacements occur. Lateral loads cause bending moments and foundation deflections, which can lead to dangerous increases in stress or damage to the structure. Many failures have occurred due to the inadequately supported excavations, walls and foundations of existing buildings. The choice of the support system is affected by many factors. These include, for example, the type of soil, the type of foundation, the height of the building under construction, the level of foundation of neighboring buildings, as well as the budget of the investment. In order to limit soil displacement and deformation as a result of deep excavation, various support systems are used. The authors of this paper present recommendations for the implementation of a deep excavation in the form an innovative system for supporting the walls of existing buildings and for monitoring them, which is especially important from the point of view of the engineers responsible for the construction of the object. The article uses a method that includes the following elements:

- collecting information on the implementation of infill buildings and the revitalization of historic buildings with the use of façade walls in the vicinity of deep excavations;
- interpretation and synthesis of collected information on the basis of literature and the authors' own design experience related to the applied protection and monitoring systems.

3. Methodology for Determining the Impact of Deep Excavation on Existing Buildings, the Designing of Their Protection, and the Monitoring of Their Technical Condition

When designing buildings that require a deep excavation in the surrounding neighborhood of existing urban housing, it is necessary to estimate the range of the excavation impact zone, as well as to determine which buildings are located in this zone. For these buildings, expert opinions regarding the technical condition of their structures should be made. This will enable the possibility of transferring additional loads by the structure of these buildings (due to the anticipated uneven ground movement in the area of their foundation) to be determined. The expert opinions should also define the permissible displacements for these buildings. There are many methods in the literature of determining ground movements. Some of these methods are empirical and include all the interactions involved in the erection of a building [7,8], wheras others only describe a specific aspect in which individual interactions combine [9–11]. Different approach represents numerical methods which concern the displacement-induced damages to existing structures located in the vicinity of excavation [12–14]. Most recently design methodologies were proposed to estimate the damage to buildings adjacent to deep excavations using probabilistic or semi probabilistic analyses [15–18].

The measures of the displacements of buildings that are adjacent to the construction of deep excavations include tilts, deflections, and cracks of structural elements [19]. The value of the vertical displacements of the terrain surface and the shape of the settling basin depend primarily on the type of used excavation casing and the type of ground foundation. The range of impact of the excavations depends on the deformability of the soil, the depth of the excavations, the dimensions of the plan of the excavations, the possible presence of groundwater and the stiffness of the excavation casing. The scopes of the applied construction protections are different and depend on a specific situation [20–22]. The most common external protections are in the form of a diaphragm wall, i.e., a Berliner wall (lagging wall), a sheet pile wall, and a retaining column wall made by jet grouting and mixed technologies (Figure 1). The stability of deep excavation casing made with one of these technologies is additionally ensured by bracing struts, ground anchors (Figure 2), or the floor slabs of underground storeys [23]. If there are relatively small spans, the bracing struts can be made of wide-flange I-sections. Larger spans require bracing that uses steel tubes with a diameter of 400–800 mm [24]. In addition to the compressive forces and the bending moment derived from excavation support load, very large impact on the axial forces comes from the thermal effects. In order to determine the real effect of temperature on the bracing strut, the susceptibility of their props needs to be determined.

Figure 1. Various structural solutions for deep excavation support: (**a**) diaphragm wall, (**b**) Berliner wall (lagging wall), (**c**) sheet pile wall, (**d**) retaining column wall made by jet grouting (developed by authors).

Figure 2. Structural solutions to ensure the stability of a deep excavation: (**a**) bracing struts, (**b**) ground anchors (developed by authors).

In the case when the building structure is not able to transfer these additional loads, it is necessary to design the structure's reinforcement and ground foundation [25].

Various methods can be used to increase the bearing capacity of foundations. One commonly used method is the classical method, which involves the excavating of the foundations followed by the underpinning of them, or the extending of the foundations while assuming that there is no water at the depth of the intended works, with the height of the underpinning not exceeding 3 m.

The walls of existing buildings can also be reinforced using injection micro piles and statically pressured micro-piles, as well as drill and screw micro-piles, the advantage of which is their high rigidity. These elements are shown in Figure 3.

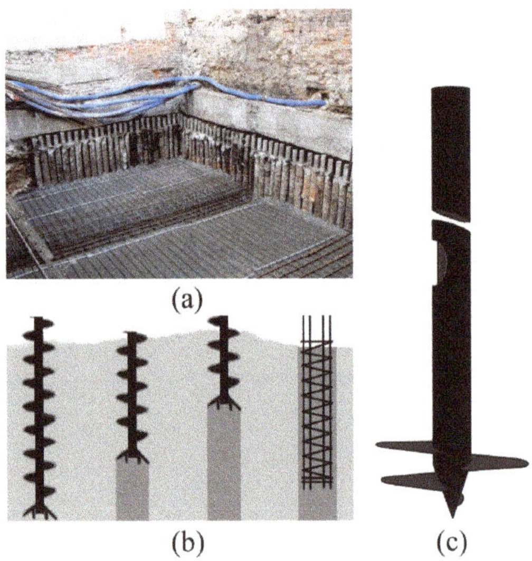

Figure 3. Additional structural elements that are used to transfer loads from the deeper layers of soil with a higher load-bearing capacity: (**a**) injection micro-piles, (**b**) drill micro-piles, (**c**) screw micro-piles (developed by authors).

Currently, the most common method of strengthening existing building, especially historic ones is underpinning by means of jet grouting injections (Figure 4).

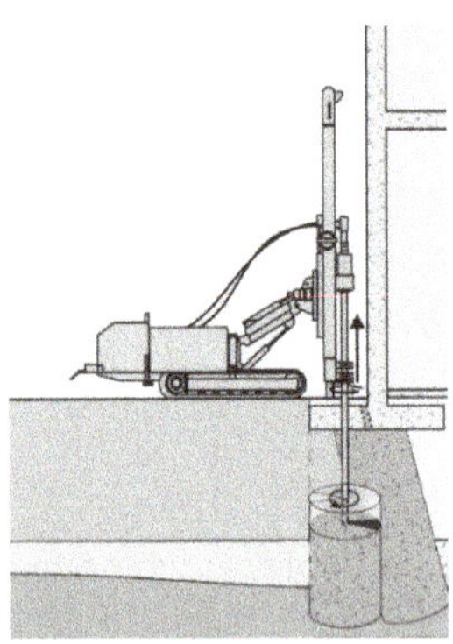

Figure 4. Increasing the bearing capacity of a foundation by means of jet grouting (developed by authors).

Estimating the impact of an excavation on neighboring buildings involves:

- the determination of the impact range of the excavation (impact zone),
- the determination of the vertical and horizontal displacements of the excavation casing and the adjacent area,
- the determination of the expected horizontal and vertical displacements of adjacent buildings;
- making sure that the expected displacements of existing buildings are lower than their maximum limit displacements.

The range of the excavation impact zone is determined with regards to the type of soil and the depth of the excavation. According to various studies presented in the literature, this range is a multiple of the excavation depth H and depends on the type of soil and the excavation casing, as shown in Table 1. The impact zone can also be determined with regards to the shape and width of the foundation of the erected building, the amount of pressure at its base, and the average value of the deformation modulus in the settling soil layer. The forecasted vertical displacements of the ground surface in the close vicinity of the excavations, which can be found in literature, are summarized in Table 2.

Table 1. The range of the excavation impact zone [23].

Authors [Reference]	Range of Interaction	Types of Soil and Other Conditions
Breymann and others [26]	1.5H ÷ 2H	non-cohesive soils, fine sands, medium and gravel
Simpson B. [27]	2H ÷ 2.5H	solid cohesive soil
O'Rourke T. [28]	2H ÷ 4H	London clay and boulder-clays
	2H	sands
Kotlicki W., Wysokiński L. [5]	2.5H	clays
	3H ÷ 4H	loam soil
TSN 50-302 standard	30 m	preliminary analysis
	5H	anchored structure which supports the excavation, but not more than 2 Lk (where: Lk is anchor length)
SP 22.13330 standard	4H	steel structure protecting the excavation, which operates as a cantilever, a structure with steel bracing struts, or in the case of executing an open excavation
	3H	structure made of diaphragm walls or walls made of piles that work as a cantilever or which are supported by steel bracing struts
	2H	structure made of diaphragm walls or walls made of piles, and the excavation is executed using the ceiling method

Table 2. Summary of the values of vertical displacements [23].

Authors [Reference]	Vertical Displacements	Types of Soil
Burland J.B. and others [29]	0.002H	non-cohesive compact soils
	0.005H	non-cohesive loose soils
	0.0015H ÷ 0.002H	cohesive, hard-plastic, and semi-compact soils
Simpson B. and others [27]	0.001H ÷ 0.002H	cohesive soils
Michalak H. [30]	0.5U0 ÷ 0.75U0 where U0: horizontal displacement of excavation casing	non-cohesive and cohesive soils
Long M. [31]	0.002H	non-cohesive and cohesive soils
Smoltczyk U. [32]	0.002H	non-cohesive and cohesive soils

Vertical displacements of the soil in the zone adjacent to an erected building are the result of the superposition of displacements from individual stages of works, including the execution of the casing, the deepening of the excavation and the supporting of its casing, the implementing of the underground part of the building, the erecting of the entire aboveground structure, and also the impact of the conditions of using the building. For existing buildings, two types of allowable limit displacements are defined-the first $[s_k]_u$ due to the serviceability limit state, and the second $[s_k]_n$ due to the ultimate limit state. For the first condition, cracks of a small width are allowed, which may then be painted or smoothed. In the second condition, the appearance of cracks and failures that require repair are allowed, but they must not threaten the safety of the building. Approximate values of the maximum limit displacements for buildings according to [5] are given in Table 3.

Table 3. Approximate values of the maximum limit displacements for buildings according to [5].

Type of Building Structure	Limit Displacement (SLS) $[s_k]_u$, mm	Limit Displacement (ULS) $[s_k]_n$, mm
Masonry buildings with timber floor slabs or Klein's floor slabs	5–7	15–18
Masonry buildings with rib and slab floors or RC floors, prefabricated building	7–9	20–25
Monolithic reinforced concrete buildings	9–11	23–35

4. Monitoring of the Technical Condition of Existing Buildings

Monitoring of the technical condition of existing buildings is the most important element when controlling construction works and ensuring the safety of these objects [33,34]. The scope of observation of the structures of existing buildings depends on their location and the distance from a new investment. It usually includes the geodetic measurements of horizontal and vertical displacements for walls in the direct vicinity of the executed excavation, building observation, and possibly the measuring of the width of existing cracks with feeler gauges. An exemplary instrument for measuring the width of cracks is shown below in Figure 5. Other less frequently used monitoring methods are strain gauge, which measures deformation using foil strain gauges, and piezometric which is installed to control groundwater levels.

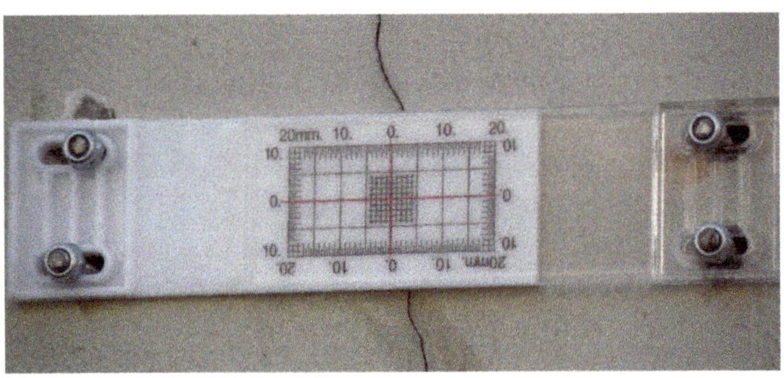

Figure 5. An exemplary instrument used to measure the width of cracks (photograph by authors).

The frequency of measurements should be adapted to the course of the construction works and the current technical condition of the building. Conducting traditional geodetic measurements allows the results of displacements to be obtained in quite significant time intervals, which is often insufficient in relation to the significant dynamics and the speed of the phenomena occurring in the excavation. Therefore, systems that allow for the continuous measurement of displacements of existing buildings and the obtaining of measurement results on-line are becoming more and more popular. One of them is the innovative hydrostatic leveling cells system, which measures the vertical displacements of structures in real time with high accuracy using a network of connected sensors [35].

The system is characterized by high accuracy measurement and was designed to measure the vertical movement of structural elements. Its measuring range applies to displacements of up to ±500 mm. The mounted sensors create a map of the 2D measurement network, which can then monitor the vertical displacements of the structure over the entire surface. Contour line plots can be generated over time to directly detect which area should be analyzed. All the sensors of the system are connected to each other with the same liquid pipe and then connected to the liquid expansion tank, which is located above the entire measuring system. This method of system construction allows the appropriate pressure to be forced inside. The vertical displacements of individual sensors mounted on the structure cause pressure changes inside the system, which are measured in the place of the installed sensors. When the sensor moves vertically, the corresponding liquid height changes accordingly, independent of all the other sensors. As liquid pressure can be measured very accurately at each sensor, vertical movement can be calculated from the density of the used liquid. Each sensor is also connected to each other sensor by an air pipe in order to normalize the reference pressure between the sensors. When the structure settles down, the measuring cells settle with the structure, in turn increasing the pressure in the system. Therefore, a settlement measurement can be performed. Lifting of the structure is similarly detected, i.e., when the liquid pressure in the system is reduced. The HLC system is fully integrated with the software system, e.g., the Quickview interface. Data is streamed online within minutes, so movements of the structure can be monitored in real time. Supervision over the measured structure is clearly achieved by means of a simple graphic visualization and automatically generated reports. If the measurement shows that the displacements of the structure have exceeded the previously set alarm thresholds, the system can send automatic notifications. An example of an HLC system application is shown in Figures 6–8. This system was installed along the facades of buildings and around the planned earthworks (excavations) related to the revitalization of a historic hotel building in the center of Wroclaw.

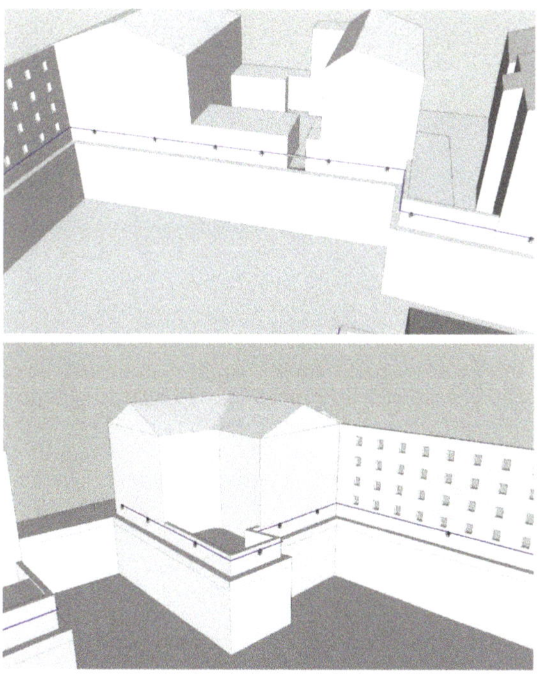

Figure 6. Automatic measurement of vertical deformations (settlements) using a hydrostatic leveling cell system (HLC) in installed along façades of buildings around the planned earthworks (developed by authors).

Figure 7. HLC system for baseline measurements with data transfer to the GETEC server (photograph by authors).

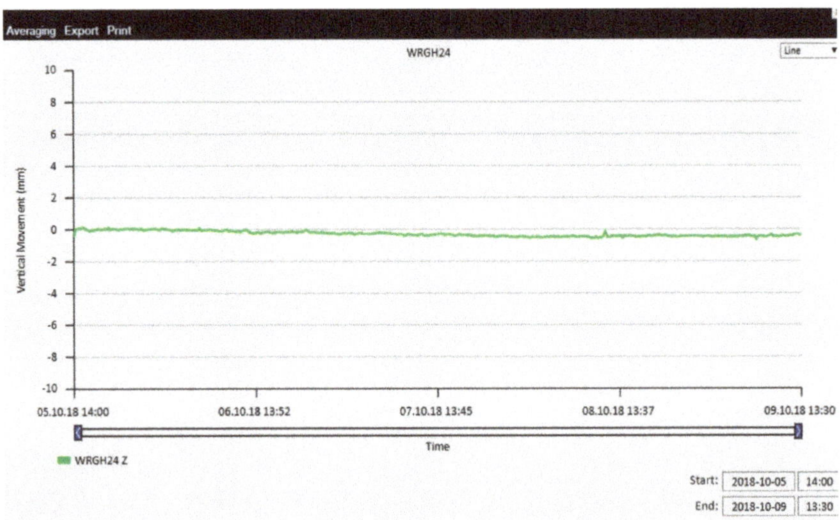

Figure 8. Graph of vertical displacements from monitoring and Quickview GETEC software (developed by the authors).

An additional system for monitoring a building's geometry is a system of wireless inclinometers installed on the façades of the building (Figures 9 and 10). This system consists of a control and monitoring module, and also sensor modules to which an inclinometer and a thermometer are connected. The inclinometer is a high-precision measuring device that is used in the construction of cranes, wind turbines and ships. Temperature measurement is used to assess its influence on the operation of the inclinometer and the measured structure. The sensor modules record data from the inclinometer and thermometer and save them in random access memory. On the command sent from the monitoring and control module, they perform the required calculations and send their results to the database.

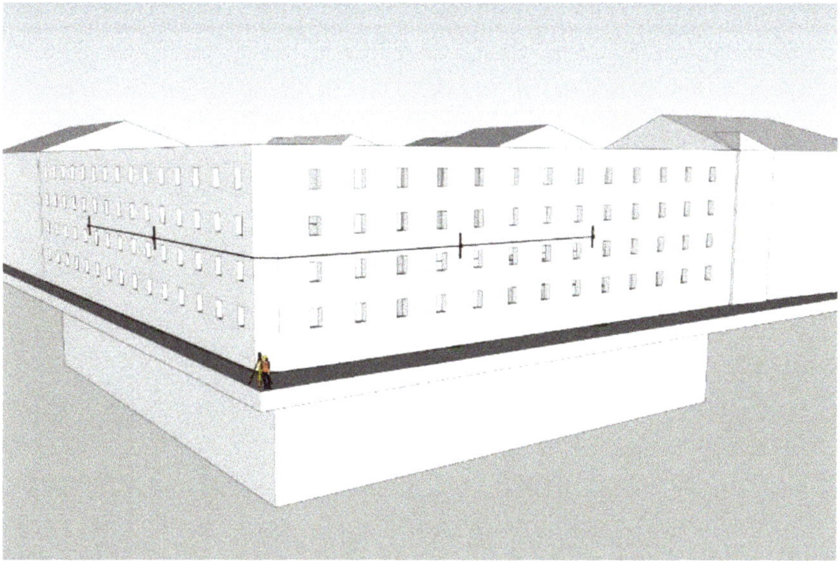

Figure 9. Automatic measurement of changes in façade inclinations using wireless, automatic inclinometers (developed by authors).

(a) (b)

Figure 10. The wireless inclinometer system: (**a**) installation on two façades of the building, (**b**) a control computer for transmitting measurement data located on the top storey at the corner point between two façades (photograph by authors).

5. Protection of the Walls of Existing Buildings

In many building revitalization projects, the existing brick or stone façade is kept and is temporarily supported by a steel structure while the rest of the building is demolished. In the internal space, a new supporting structure of the building is erected, and then adapted to the modern technical requirements and the expectations of future users. The façade walls are anchored to the new support structure and carefully restored. In this way, the appearance of the building does not change, but its usability is significantly improved.

However, before this happens, the stability of the walls, often of considerable height, must be supported during the process of construction. The support may be realized by means of horizontal steel ties made of bars and located on several levels, or by means of stiffening or bracing elements [36,37].

Examples of supporting an existing masonry façade wall with an external temporary steel framework are shown in Figures 11–14. Figure 11 shows the temporary supporting structure for the façade of a revitalized textile factory building that is having its function changed into a cultural, recreational, and commercial center. The supporting structure is anchored to concrete blocks.

Figure 11. Temporary protection for the façade wall in the form of a strut-and-tie steel structure [3].

Figure 12. Temporary protection for the façade wall in the form of a steel frame structure [1].

Figure 13. Temporary protection for a façade wall in the form of a steel frame structure [22].

Figure 14. Temporary protection for a façade wall in the form of a steel buttresses (photograph by authors).

The temporary protection for façade walls in the form of a steel spatial frame is shown in Figure 12. This design required the strengthening of the foundations using micro piles. In this case, the façade walls were adapted to the new architectural concept, which involved incorporating the historic building into a complex of two high-rise residential towers.

Another example of the application of the temporary protection of façade walls with the use of a steel space frame is shown in Figure 13.

Figure 14 shows temporary protection in the form of steel buttresses with horizontal sections attached to them, which are fixed to the wall with steel connectors Steel buttresses can be constrained to the ground with the use of piles or concrete blocks that operate as a ballast.

The main disadvantage of the above-mentioned typical solutions for supporting façade walls is the very high consumption of steel for the buttresses, as well as the occupation of the area around the walls. In dense downtown housings, especially for buildings located on busy streets, it is often impossible to support the walls on the side of the street or sidewalk. On the other hand, the arrangement of the steel supporting structure from the inside of the building makes it impossible to carry out excavation work or erect new foundations. An innovative method of supporting walls, which is used for the reconstruction of a historic hotel in the center of Wroclaw, is presented below. Figure 15 shows the Push-Pull Props support structure used for the remaining façade walls of the modernized building. This structure is extendable with an extension length of up to 14 m.

Figure 15. Temporary protection for a façade wall made of the PERI Push-Pull Props support structure (photograph by authors).

The PERI Push-Pull Props system was used for bracing the walls and was anchored in points in specially made micro piles. Due to the large height of the remaining walls, the support was made in 3 levels, as shown in Figure 16.

Figure 16. Supporting the walls of the existing building in 3 levels with the PERI Push-Pull Props system (photograph by authors).

The steel support structures were anchored in the previously made heads of reinforced concrete piles. This enabled the appropriate load-bearing mounting of these supports, and at the same time, collision-free work related to the excavation and construction of the foundation slab. After the slab floors in the new building are finished, the support will be dismantled. Then, after sealing the pile contact with the foundation slab, they will be cut out. Figure 17 shows the details of anchoring the PERI Push-Pull Props system to the pile head.

Figure 17. Anchoring of the PERI Push-Pull Props system to the pile head (photograph by authors).

During construction works in the vicinity of existing buildings, especially deep excavations for garages designed in new buildings, excessive movements of these objects, despite the applied protection systems described in Section 2, often occur. This results in significant damage and excessive deflection of the walls, which may lead to the collapse of the construction. In this case, it is necessary to quickly support the walls in order to stabilize them. In such cases, a common solution is to use temporary steel braces of the strut-cantilever or tie type. An example of this type of bracing is shown in Figure 18.

Figure 18. Temporary steel bracing of the strut-cantilever type (photograph by the authors).

A disadvantage of such solutions is the high consumption of steel, but also the time-consuming implementation, which can be essential when you need to quickly support a wall. An alternative solution that can be used in such cases is presented below. It also uses the PERI Push-Pull Props system [38], which is fastened on one side to the supported wall, and on the other to e.g., reinforced concrete road slabs laid at the bottom of the excavation. The proposed solution, which is currently being implemented at a historic tenement house in a city center location, is shown schematically in Figure 19. This solution enables the wall support to be made almost immediately.

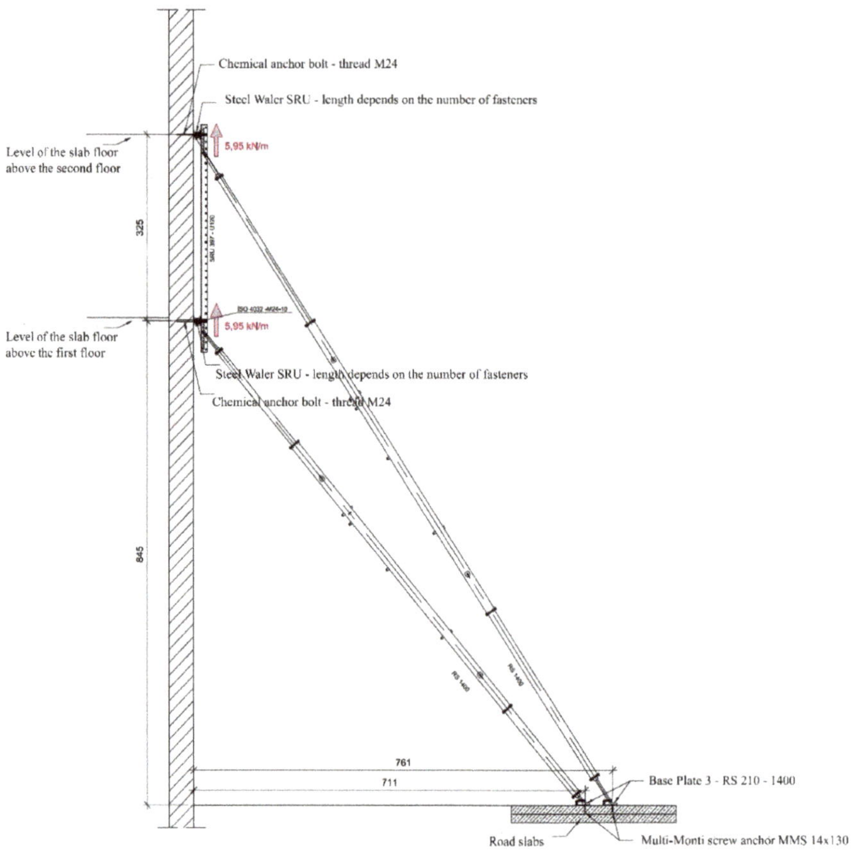

Figure 19. An exemplary way of supporting an existing wall with the use of the PERI Push-Pull Props system (picture courtesy of PERI).

6. Case Study

The numerical analysis of the supporting structure–using PERI Push-Pull Prop system–of the walls of the rebuilt, historical building, situated around a deep foundation excavation, was presented in the paper (Figures 20 and 21). The building undergoing modernization, erected at the end of the 19th century was rebuilt in the fifties of the 20th century. In plan, the building has dimensions 23.55 m × 16.10 m. The building has a basement, four overground storeys and an attic. On the south-west side, a corner building is adjacent to the gable wall of the analyzed building, while on the north-eastern side there is an empty area, where construction works of another residential development have started.

The structural system of the building is longitudinal, with floor slabs supported at its front, rear and internal longitudinal walls.

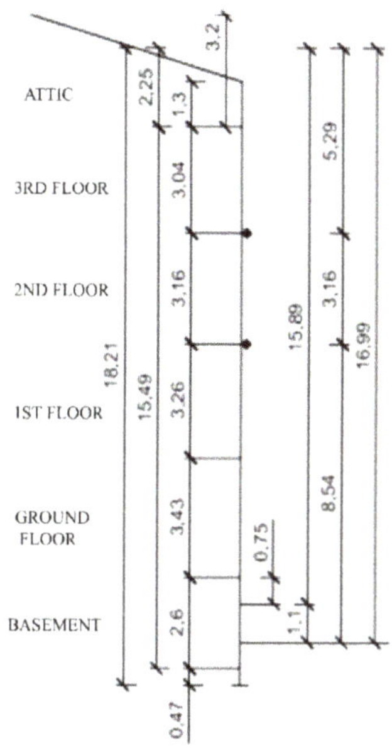

Figure 20. The geometry of the gable wall.

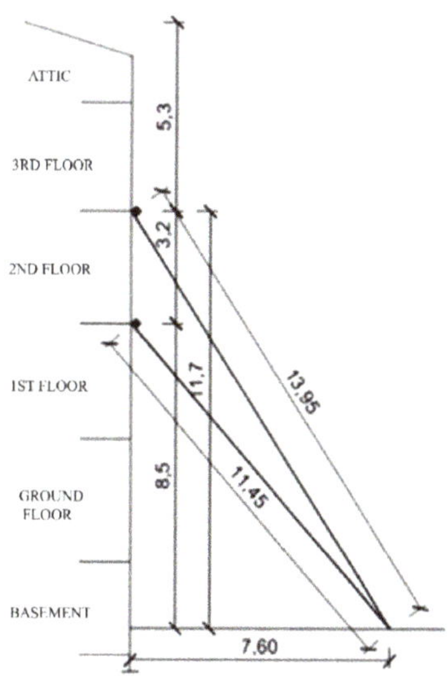

Figure 21. Scheme of the gable wall support (PERI Push-Pull Props system).

In the conducted calculations loads from the wind suction, acting on the entire surface of the gable wall, as well as the horizontal component of the wall self-weight, resulting from the assumed 50 mm displacement at its top level, were applied (Figure 22). As a result of the numerical analysis optimal wall support was obtained at the level of the first and second floors using PERI RS 1400 brace struts with maximum reach of 14.00 m. Reinforced road-slabs, stabilized by their self-weight, were used to support the bracing struts.

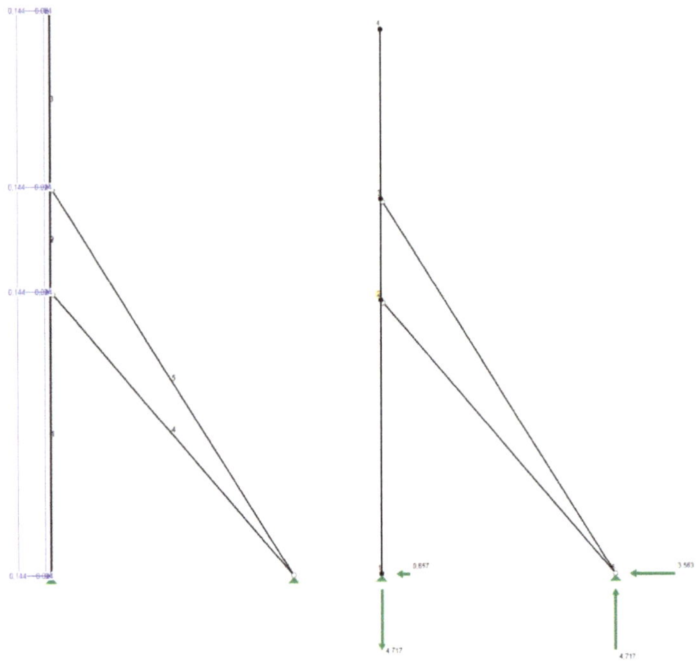

Figure 22. The numerical model of the supporting structure of the gable wall.

7. Discussion

The search for space in the conditions of dense urban development sets the trend for deep excavations near densely arranged objects. Additionally, the presence of underground utilities and the necessity to limit the lateral movements of the soil make it extremely difficult to properly protect such a trench. Therefore, it requires combination of geotechnical, hydrological, structural, civil engineering and waterproofing expertise. The protections are implemented with diaphragm walls, steel sheet pile walls, large-diameter struts, walls made of columns by jet grouting, etc. These solutions are designed so that the protections are stable and that they limit the deformation caused by excavations to levels acceptable for neighboring buildings. To achieve this, computer simulations using the finite element method (PLAXIS 2D, MIDAS/GTS) are performed to demonstrate the behavior of these protections in terms of stress distribution and soil deformation characteristics under adjacent buildings in sensitive areas [39–41]. The model used for soil simulation is the Mohr-Coulomb model for an elastic-perfectly plastic material.

Another important aspect of this type of construction is the optimal reinforcement and protection of the walls and foundations of existing buildings. This reinforcement can be executed, e.g., using injection, drilled and screwed micro-piles. Traditionally, temporary steel spatial frames or strut-and-beam structures are used to support the walls of existing buildings. These solutions are not optimal due to the high consumption of steel and the large space around the building, which makes construction work very difficult. The recommended solution by the authors of the article is the Push-Pull Props system (PERI, Wroclaw, Poland), which, in combination with its fastening to reinforced concrete piles, is characterized by low steel consumption, quick assembly and disassembly, and no restriction of construction works inside the building. Monitoring is a very important issue related to the technical problems of construction near deep excavations and the ensuring of safety. Currently, the most innovative monitoring system is the hydrostatic leveling cell (HLC) system connected with a recording computer and an LTE modem for data transmission, which is used to measure the movement of structural elements. On the basis of real, continuous measurements made on-line, with the help of this system it is possible to create a database and develop accurate theoretical models of the response of existing buildings to construction works carried out nearby. The use of these models in the future will optimize the methods of protecting of existing buildings, and by introducing complete automation of deformation measurement, it will increase the safety of works.

8. Conclusions

The paper focuses on the problem of soil-structure interaction mechanisms for buildings located in the vicinity of deep excavations, as well as finding appropriate methods of designing their protection and reinforcements and monitoring their vertical deformations. During the realization of a new investment in the area of a dense housing development, in the case of infill buildings or revitalizing historic buildings using only their façade walls, there are always complex technical problems related to the safety of both the new and old buildings. The main purpose of using temporary steel bracing is to enable safe construction works and to prevent the uncontrolled collapse of a building or its parts. The main problem is that old buildings in new investments are not designed to transfer the loads associated with the implementation of these investments. The assessment of the reaction of buildings to deformations caused by deep excavations is influenced by the accuracy of determining the deformations and stress changes caused by these excavations. The interaction of the deep excavation and the existing building is bilateral. The movement of the soil causes deformation of the building and the possibility of cracks and other types of damage. Moreover, existing buildings modify the displacement of the ground directly below them. In this situation, geodetic measurements are very important, and in particular the monitoring of vertical displacements of existing buildings, which enables the objective determination of changes related to the safety of both old and new structures. The correct assessment of the impact of new buildings on the neighboring existing objects enables the use of appropriate protections and new technologies that allow the damage and failures of adjacent buildings to be eliminated. Currently, the leading solutions used in the modern structures of temporary bracings are steel bar braces of the Push-Pull Props system. A particularly innovative solution for supporting the walls of existing buildings is use of this system and its anchoring on reinforced concrete or steel micro-piles. The Push-Pull Props are characterized by a high ratio of strength to self-weight, thanks to which relatively small dimensions of both the bar sections and the supporting structures are obtained. In addition, their versatility is determined by the possibility of quick assembly and disassembly, a high degree of prefabrication, a high accuracy of matching individual elements to the dimensions of the existing structure, and the possibility of full recycling, which is in line with modern environmental protection requirements [42].

Author Contributions: Both authors contributed the same to the analysis of the problem, discussion and writing the paper. Both authors have read and agreed to the published version of the manuscript.

Funding: This research received no external funding.

Institutional Review Board Statement: Not applicable.

Informed Consent Statement: Not applicable.

Data Availability Statement: Not applicable.

Conflicts of Interest: The authors declare no conflict of interest.

References

1. Gordon, A.; Sousa, M.; Isherwood, B. Heritage Facade Support and Excavation Shoring—A Micropile Case History. In Proceedings of the 13th International Workshop on Micropiles, Vancouver, BC, Canada, 29 March–1 April 2017; pp. 1–19.
2. Highfield, D. *The Construction of New Buildings behind Historic Façades*; CRC Press: Boca Raton, FL, USA, 1991.
3. Seruga, T.; Płachecki, M.; Seręga, S. Structural Problems of Adaptation of Historic Post-Industrial Buildings to New Function. *Tech. Trans.* **2009**, *R. 106, z.2_B*, 317–325. (In Polish)
4. Runkiewicz, L. Wzmacnianie i zabezpieczanie istniejących obiektów w sąsiedztwie realizowanych budynków plombowych. *Prz. Bud.* **2008**, *4*, 28–38. (In Polish)
5. Kotlicki, W.; Wysokiński, L. Protection of buildings adjacent to deep excavations. *ITB Recomendations* **2002**. (In Polish)
6. Runkiewicz, L.; Sieczkowski, J. Problemy techniczne budowy obiektów na terenie istniejącej gęstej zabudowy. *Prz. Bud.* **2015**, *9*, 18–23.
7. Peck, R.B. Deep excavations and tunneling in soft ground (State of the Art Report). In Proceedings of the 7th International Conference of Soil Mechanics, Mexico City, Mexico, 25–29 August 1969; Volume 7, pp. 225–290.
8. Goldberg, D.T.; Jaworsky, W.E.; Gordon, M.D. Lateral Support Systems and Underpinning. Volume I: Design and Construction. In *Federal Highway Administration*; Report No. FHWA-RD-75-128; Bureau of Transportation Statistics, U.S. Department of Transportation: Washington, DC, USA, 1976; Volume 1. Available online: https://rosap.ntl.bts.gov/view/dot/14528 (accessed on 15 September 2020).
9. Clough, G.W.O.; Rourke, T.D. Construction induced movements of in-situ walls. In *Design and Performance of Earth Retaining Structures*; ASCE Special Publication No. 25; 1990; pp. 439–470. Available online: faculty.uml.edu/spaikowsky/14.533AdvFoundEng/Documents/ConstInducedMoments_Clough&O\T1\textquoterightRourke_1990.pdf (accessed on 15 October 2020).
10. Moorman, C.; Moorman, H.R. A study of wall and ground movements due to deep excavations in soft soil based on worldwide experiences. In *Geotechnical Aspects of Underground Construction in Soft Ground, Kastner, Emeriault, Dias, Guilloux (eds)*; 2002; pp. 477–482. Available online: https://www.issmge.org/uploads/publications/6/10/2002_070.pdf (accessed on 15 October 2020).
11. Konstantakos, D.C.; Regalado, A.J.W.C.; Scharner, B. Control of Ground Movements for a Multi-Level-Anchored, Diaphragm Wall During Excavation. In Proceedings of the Fifth International Conference on Case Histories in Geotechnical Engineering, New York, NY, USA, 16 April 2004; Paper No. 5.58. pp. 1–7.
12. Boscardin, M.D.; Cording, E.J. Building Response to Excavation-Induced Settlement. *J. Geotech. Eng.* **1989**, *115*, 1–21. [CrossRef]
13. Bentler, D.J. Finite Element Analysis of Deep Excavations. Ph.D. Thesis, Faculty of Virgnia Polytechnic Institute and State University, 1998. Available online: https://vtechworks.lib.vt.edu/bitstream/handle/10919/30767/ETD.PDF?sequence=19&isAllowed=y (accessed on 15 October 2020).
14. Roboski, J.; Finno, R.J. Distributions of ground movements parallel to deep excavations in clay. *Can. Geotech. J.* **2006**, *43*, 43–58. [CrossRef]
15. Hsiao, E.C.L.; Schuster, M.J.; Juang, C.H.; Kung, G.T.C. Reliability analysis and updating of excavation-induced ground settlement for building serviceability assessment. *J. Geoenviron. Eng.* **2008**, *134*, 1448–1458. [CrossRef]
16. Schuster, M.J.; Kung, G.T.C.; Juang, C.H.; Hashash, Y.M.A. Simplified model for evaluating damage potential of building adjacent to a based excavation. *J. Geotech. Environ. Eng. ASCE* **2009**, *135*, 1823–1835. [CrossRef]
17. Boone, S.J. Ground-Movement-Related Building Damage. *J. Geotech. Eng.* **1996**, *122*, 886–896. [CrossRef]
18. Castaldo, P.; Calvello, M.; Palazzo, B. Structural safety of existing buildings near deep excavations. *Int. J. Struct. Eng.* **2014**, *5*, 163–187. [CrossRef]
19. Bryson, L.S.; Kotheimer, M. Cracking in Walls of a Building Adjacent to a Deep Excavation. *J. Perform. Constr. Facil.* **2011**, *25*, 491–503. [CrossRef]
20. Matysiak, A.; Grochowska, E. Zabezpieczenie stateczności ścian zabytkowych budynków mieszkalnych z XVIII wieku za pomocą przypór. *Mod. Rewit.* **2014**, *3*, 43–46. (In Polish)
21. Laskowski, J.; Dankowski, M. Przykład utraty stateczności ściany frontowej pozostawionej po rozebraniu kamienicy. *Bud. Arch.* **2018**, *17*, 119–124. [CrossRef]
22. Eschenasy, D. Development Along Old Party Walls. *Structure* **2017**, 12–15. Available online: https://www.structuremag.org/wp-content/uploads/2017/05/C-StrucPractices-Eschensay-Jun17.pdf (accessed on 10 October 2020).
23. Jermołowicz, P. Roboty Ziemne Przy Realizacji Obiektów Budowlanych i Instalacji Podziemnych—Zabezpieczenie Wykopów. 2014. Available online: https://docplayer.pl/46189190-Roboty-ziemne-przy-realizacji-obiektow-budowlanych-i-instalacji-podziemnych-zabezpieczenie-wykopow.html (accessed on 15 September 2020).
24. Filip, R.K. Temporary Works Toolkit: Part10: Propping and needling. *Struct. Eng.* **2017**, *95*, 30–35.
25. Dachowski, R.; Gałek, K. Selection of the Best Method for Underpinning Foundations Using the Promethee II Method. *Sustainability* **2020**, *12*, 5373. [CrossRef]
26. Breymann, H.; Freiseder, M.; Schweiger, H.F. Deep excavations in soft ground, in situ measurements and numerical predictions. In Proceedings of the XIV International Conference on Soil Mechanics and Foundation Engineering, Hamburg, Germany, 6–12 September 1997; pp. 589–594.
27. Simpson, B. Retaining structures: Displacement and design. *Geotechnique* **1992**, *42*, 541–576. [CrossRef]
28. O'Rourke, T.D. Base stability and ground movement prediction for excavations in soft clay. In *Proceedings of the Conference Retaining Structures*; Thomas Telford: London, UK, 1993; pp. 131–139.

29. Burland, J.B.; Simpson, B.; St. John, H.D. Movements around excavations in London Clay. In Proceedings of the 7th European Conference on Soil Mechanics and Foundation Engineering, Brighton, UK, September 1979; Volume 1, pp. 13–19. Available online: https://www.researchgate.net/profile/John_Burland/publication/307234163_Movements_around_excavations_in_London_Clay/links/56b65bc808ae44bb3307a93d/Movements-around-excavations-in-London-Clay.pdf (accessed on 10 September 2020).
30. Michalak, H. Budynki głęboko posadowione a przemieszczenia podłoża i zabudowy w sąsiedztwie. *Geoinżynieria Drogi Mosty Tunele* **2008**, 66–76.
31. Long, M. Database for retaining wall and ground movement due to deep excavations. *J. Geotech. Geoenvironmental Eng.* **2001**, *127*, 203–234. [CrossRef]
32. Smoltczyk, U. *Geotechnical Engineering Handbook*; Ernst & Sohn, A Wiley Company: Berlin, Germany, 2002; Volume 1–3.
33. Mrowczynska, M.; Grochowska, E.; Gibowski, S. Monitoring Vertical Displacements of an Engineeriong Object with Masonry Walls. *J. Civil Eng. Environ. Arch.* **2018**, *XXXV, z. 65*, 53–62.
34. Yang, K.; Yan, L.; Huang, G.; Chen, C.; Wu, Z. Monitoring Building Deformation with InSAR: Experiments and Validation. *Sensors* **2016**, *16*, 2182. [CrossRef] [PubMed]
35. Meier, E.; Geiger, A.; Ingensand, H.; Licht, H.; Limpach, P.; Steiger, A.; Zwyssig, R. Hydrostatic levelling systems: Measuring at the system limits. *J. Appl. Geod.* **2010**, *4*, 91–102. [CrossRef]
36. Eschenasy, D.; Sirakis, G.; Gami, B. Protection of Existing Buildings During Construction. In Proceedings of the Build Safe, Life Safe Conference, New York, NY, USA, 21–25 September 2016.
37. Rosa, A.M.; Lehigh, S. Shoring up the Past New York City Style. *Structure* **2014**, 25–28.
38. RS and RSS Push-Pull Props. Instructions for Assembly and Use—Standard Configuration. 2020. Available online: http://www.peri.ltd.uk (accessed on 10 October 2020).
39. Abdallah, M.; Hussein, F. Interaction Deep Excavation—Adjacent Structure: Numerical Two and Three Dimensioning Modeling. *Adv. Mater. Res.* **2011**, *324*, 344–347. [CrossRef]
40. Zhang, D.D.; Zhang, W.G.; Zhou, J.F.; Zhu, C.B. The Finite Element Analysis of the Excavation on Adjacent Buildings Based on Mohr Coulomb Model. *Adv. Mater. Res.* **2011**, *374–377*, 2171–2175. [CrossRef]
41. Just, M. Analysis of a Deep Excavation in Diaphragm Walls, on Surrounding Buildings. *Arch. Inst. Civil Eng.* **2018**, 86–104. [CrossRef]
42. Kuchta, K. Stalowe stężenia tymczasowe w budynkach o konstrukcji murowej i żelbetowej. *Inż. Bud.* **2015**. Available online: https://inzynierbudownictwa.pl/stalowe-stezenia-tymczasowe-w-budynkach-o-konstrukcji-murowej-i-zelbetowej/ (accessed on 10 September 2020).

applied sciences

MDPI

Article

Selection of the Optimal Actions for Crashing Processes Duration to Increase the Robustness of Construction Schedules

Sławomir Biruk and Piotr Jaśkowski *

Faculty of Civil Engineering and Architecture, Lublin University of Technology, Nadbystrzycka 40, 20-618 Lublin, Poland; s.biruk@pollub.pl

* Correspondence: p.jaskowski@pollub.pl; Tel.: +48-81-538-4445

Received: 13 October 2020; Accepted: 12 November 2020; Published: 12 November 2020

Abstract: Both the construction clients and the contractors want their projects delivered on time. Construction schedules, usually tight from the beginning, tend to expire as the progress of works is disturbed by materializing risks. As consequence, the project's original milestones are delayed. To protect the due date and, at the same time, avoid changes to the logic of work, the manager needs to the project progress and, if delays occur, speed up processes not yet completed. The authors investigate the problem of selecting the optimal set of actions of responding to schedule delays. They put forward a simulation-based method of selecting schedule compression measures (speeding up processes) and determining the best moment to take such actions. The idea is explained using a simple case. The results confirm that it is possible to find an easily implementable schedule crashing mode to answer schedule disturbances. The proposed method enables minimizing the cost of schedule crashing actions and the cost of delays as well as increasing the robustness of the schedule by reducing differences between the actual and the as-planned process starts. It is intended as a decision support tool to help construction managers prepare better reactive schedules. The lowest costs are achieved if the acceleration measures are implemented with some time lag to the occurrence of delays.

Keywords: construction project; random conditions; online scheduling; simulation method; process mode selection

---

1. Introduction

1.1. Motivation

Risk and uncertainty are inherent in construction projects [1,2]. Construction is claimed to be more vulnerable to them than other types of economic activity [3]. The effects of a large number of project participants, long production cycles, and variable locations of work units add to the impacts of external environmental factors. The more technically complex a project is and the longer it takes, the greater the impact of risk, while the greater the probability of risk occurrence makes the scale of impact more difficult to assess [4].

In practice, the actual construction time is rarely in line with the initial schedules. This is due to the effects of random conditions. The recent literature presents two directions of non-deterministic construction scheduling: one based on stochastic methods, the other employing the fuzzy set theory. The latter treats the process duration not as a random variable, but as an imprecise (fuzzy) number [5]. However, both of them assume that an experienced planner can predict an approximate scenario of the occurrence of disturbing factors and their approximate effects.

Risk is understood as the possibility of an undesirable result (a loss). It is often determined on based on an estimated probability of the event occurring within a period of analysis [6]. Risk analysis is a process of assessing (estimating) the value of factors influencing the effects of a decision. Risk assessment is the process of determining the risk profile, allowing for the variability of the risk factors' impact. The risk profile is explicitly described by a probability density or distribution function of time, cost, or other parameters describing the loss. Apart from risk analysis and assessment, the risk management process includes the design and implementation of risk response measures. These affect the impact or likelihood of adverse events and are aimed at reducing or maintaining the risk at an acceptable level.

Network methods are commonly used in the construction industry to schedule non-repeatable projects that include non-cyclical, non-rhythmic, and heterogeneous processes. The methods that treat process durations as random variables (like PERT (Programm Evaluation and Review Technique), CYCLONE, Petri network) most often focus on estimating the probability of meeting a particular deadline or calculating the project duration at a given confidence level, rather than on creating expedient risk-aware schedules.

The scheduling process should focus on the reliability of the project due date, as exceeding it puts all parties of the contract in an uncomfortable situation: the client loses opportunities for operating the facility, while the contractor pays contractual penalties and loses reputation.

The project due date is not the only essential element of the plan: delays with milestones and even particular processes can become costly and propagate disturbance into the project's meticulously designed organization system of subcontractors and suppliers, all of them trying to synchronize multiple duties in many projects. Thus, moving a subcontractor's task to a later date may even be out of the question (it may result in breaking the contract), or at least lead to downtime and extra costs. Reliable process deadlines are also vital for managing the contractor's in-house resources, planning material supplies, and plant maintenance.

The scheduling methods themselves may hinder defining reliable dates—because of the way their models simplify reality. For instance, PERT and some other classic scheduling techniques assume that processes start immediately after the end of their predecessors, ignoring other process starting policies observed in practice. This way, it is not possible to model a constraint that a process cannot start earlier than on a predetermined date.

Imperfect scheduling techniques produce schedules that easily expire, and their target dates cannot be met. This naturally leads to the researchers' and practitioners' undertaking to develop proactive methods to create disturbance-proof schedules. However, there is a demand for reactive methods to support updating schedules to answer disruptions. They are expected to prompt the most economical measures to prevent the propagation of schedule delays.

1.2. Literature Review

The volatility of construction conditions results in the variability of construction process execution time and even in uncertainties of the scope of work. To allow for the non-deterministic character of projects and their environment, the schedulers frequently treat process durations as random variables. Beta [7], logarithmic-normal [8], Weibull [9], trapezoidal [10], or triangular [11] distributions are thus used. The triangular distribution is argued to be the most intuitive to interpret [12–14]. Nevertheless, randomizing the durations of processes opens up the possibility to use simulation techniques to analyze the possible development of projects.

The literature presents numerous simulation models of construction projects [15–17]. The model of Lu and AbouRitzk [18] combined the discrete event simulation with a simplified way of defining the critical path in the network. Shi [19] used activity based construction simulation modeling that uses only one graphic symbol to represent a process. Lee et al. [20] developed a stochastic simulation system (AS4), based on the CPM (Critical Path Method). Sadeghi et al. [21] presented an original method of planning projects in random conditions DESPEL (discrete event simulation with probabilistic event list); they accounted for resource availability constraints. Lee [22] developed the stochastic project scheduling simulation (SPSS) to estimate the probability of meeting the due dates and the criticality of processes. Aziz [23] constructed his repetitive-projects evaluation and review technique (RPERT) that combines PERT with the line of balance. Jaskowski and Biruk [24] used simulations to assess the process criticality according to the processs' impact on the project timelines. The model by Leu and Hung [25] enables resource leveling under uncertain activity durations and combines the Monte Carlo simulation method with genetic algorithms. Biruk and Rzepecki [26] compared the performance of several priority rules applied to scheduling a pipeline project; the simulation was to help select a priority rule for resource allocation most suitable from the point of timely project completion in random conditions.

Two distinct approaches to random scheduling are observable in the literature: the "offline" and the "online" [27,28]. The former, also referred to as proactive or predictive [29], consists of constructing a schedule that anticipates all possible future disruptions, with their identification resting upon the incomplete and uncertain information available before the start of the project. The latter is considered a continuous activity, with decisions on the processes' timing or scope, made and verified as the works progress and for a short planning horizon. The online scheduling uses the concepts of stochastic and reactive scheduling to create the schedule and update it.

The proactive approach consists of designing robust schedules to be immune to the future and uncertain disturbances. It uses robust optimization techniques that focus on keeping the schedule acceptable (i.e., meeting all constraints) for all realizations of the uncertain durations within an uncertainty set. These techniques can be applied even if the probability distribution of the parameters cannot be defined, provided that the ranges of their values are known [30]. However, this approach is considered conservative: it may produce solutions much worse than expected or even infeasible.

One of the ways to increase the schedule robustness is to allocate time buffers to processes using techniques based on contingency or redundancy. Some researchers put forward constructing a schedule in a traditional way (no risks considered), then adding time buffers at the end of all processes, and treating the buffers as an integral part of the expected process duration [31]. In this case, the time buffer does not prevent the propagation of disturbances throughout the schedule. For these reasons, other strategies of buffer allocation are recommended, for instance, placing buffers only before the particularly important processes [29,31,32].

The critical chain concept by Goldratt [33] is one of the first methods of designing disturbance-resilient schedules, where the completion date is protected by time buffers. Goldratt's critical chain does not allow the planner to define precise completion dates of individual processes—as with PERT, processes are modeled to start immediately on completion of their predecessors.

Buffer sizing and location are still unsolved issues [34,35]. Herroelen and Leus [32] constructed an optimization model to determine the size of time buffers in a schedule with discrete disturbances of a single process. However, as the level of complexity of real-life problems is much greater (many possible disruptions of numerous processes), it seems pointless to look for exact optimal solutions [29]. The heuristics of the adapted float factor, the resource-flow dependent float factor, the virtual activity duration extension, and the starting time criticality [32,34,36] belong to the best-known algorithms for buffer sizing.

Herroelen and Leus [32], as well as Van de Vonder et al. [34,36], maintain that a robust schedule with a fixed due date must minimize the instability cost function. The function is defined as the weighted sum of the expected deviations between the processes' as-scheduled start dates and the random variables of their durations.

The method of increasing the reliability of predictive schedules and minimizing the instability cost function was proposed by Jaśkowski [37]. An effective way to reduce the instability cost is to look for time-optimal baseline schedules to increase the total float to be distributed among the processes in the form of time buffers. For this purpose, a variety of schedule compressing methods can be applied.

With stochastic scheduling, no baseline schedule is needed. Consecutive activities are added to a previously built partial schedule according to a predefined scheduling policy (e.g., priority rule).

At each decision point, the policy determines which activity is to be incorporated into the schedule concerning all precedence relations and constraints [38].

The framework of simulation-based scheduling is widely used to select the best scheduling policy [39,40]. Wang et al. [41] used this method to compare the efficiency of the 20 most common priority rules of resource allocation. They conducted a full factorial experiment on a sample of 1260 projects combined into 420 project portfolios. Their results provide guidelines for selecting the most suitable priority rule according to both the schedule quality and the measures of robustness.

As for the reactive scheduling approach, it rests upon updating schedules whenever they expire due to disturbance. The scope of such updates is defined based on all available information on the project itself and its environment collected so far [42]. The rescheduling action can be planned at fixed intervals or in reaction to a substantial disturbance [39].

If the actual duration of some activity deviates from the baseline schedule, the common objective of rescheduling is to keep the discrepancies between the baseline and the updated schedule to a minimum. This typically consists of minimizing the weighted sum of the expected absolute deviations between the as-planned and actual activity start dates while maintaining the original objectives and constraints of the schedule.

As in the case of proactive scheduling, the reactive methods also use scheduling policies. For instance, Van de Vonder [43] proposed two new schemes of robust reactive schedule generation based on priority rules.

Pasławski [44] put forward a method to improve the performance of the reactive approach by increasing the flexibility of the initial schedules. He recommended preparing a set of acceptable variants of construction methods and organizational solutions for the processes. This was to facilitate schedule updating as disruptions occur.

The same general idea of selecting from a number of activity modes was used by Deblaere et al. [45]. In the course of the rescheduling process, they allowed changes in the mode of some activities while adhering to resource availability constraints. However, they focused their analysis on two types of schedule disruptions: variations in activity durations and resource disruptions, both of a discrete character and occurring at random moments. They also neglected the randomness of processes not yet completed.

Yang [46] argues that most practitioners are reluctant towards computationally complex schedule optimization procedures; they prefer simple scheduling rules. Therefore, this paper puts forward a simple method of selecting actions, reducing the duration of processes not yet completed and determining the moment of their implementation to reduce delays in starting processes or project stages. The proposed method of responding to schedule disruptions does not use any advanced optimization algorithms but helps to reduce the cost of increasing the robustness of the schedule.

2. Materials and Methods

2.1. Simulation Technique for Construction Project Planning

Simulation models have been used to describe, plan, and study complex construction projects for several decades. Simulation is a technique of solving problems that consists of tracking changes in the dynamic model of a system over time [47,48]. Simulation methods are used to analyze models too complex to be approached with analytical methods. Their main advantages are the lack of limitations on the model's structure and level of complexity, and the possibility to capture stochastic processes.

Simulation experiments on project network models with non-deterministic process durations help planners assess the impact of process duration variability on the project performance [49]. The probability distributions of process start times estimated in the course of simulation experiments may serve as the basis for contractual deadlines, such as subcontractors' commencement with work or the project finish, at a predefined level of confidence.

The first stage of a simulation experiment is the preparation of the model to study the impact of the system input parameters on the outputs. In the course of modeling the construction project, its scope is broken down into elements, work packages, processes, or even detailed construction operations—depending on the desired level of detail. Then, these components are combined into a network by introducing technological and organizational relationships. The next stage, collection and analysis of input, consists of determining the quantity of the work, the related workloads, and estimating the distribution type and parameters of process durations.

The next step is programming. The model can be coded using a general-purpose programming language (e.g., C++ or Python) or one of the dedicated simulation languages. The latter contains in-built mechanisms of system time-lapse and simulation control, random number generators, and procedures for collecting and presenting results. Moreover, they facilitate rapid modifications of the model, the input, and the constraints. The popular languages for discrete simulation are GPSS and Simscripti Simula. A set of convenient tools to analyze network models is offered by visual interactive simulation (VIS) systems that facilitates the modeling process. VIS packages (e.g., AnyLogic or Witness) enable the user with no programming skills to build a model, conduct simulation tests and analyze the results.

As the model is being programmed, and on completion of this process, verification is needed to confirm its correctness. This consists of checking its operation. Then, the model should be validated, which consists of assessing how exactly the model describes the real system [47]. Due to the one-time nature of construction projects, model validation is a difficult task. Most often, it consists of comparing the results generated by the model with results of other models, either analytical or other, verified, simulation models. The stage of planning experiments is to determine the values of input parameters. During the experiments, the observed values of the examined quantities are collected to determine—at the stage of analyzing the results—the confidence intervals for their means and standard errors. When designing simulation experiments, the aim is to minimize the length of the confidence intervals, which guarantees good quality of the results.

2.2. Modeling Process Duration with Risk

The credibility of the project model depends on the correctness of the input, in this case, the types and parameters of probability distributions of the process durations. The labor productivity benchmarks developed for the whole industry are estimated with the assumption of "average conditions". They do not account for unique conditions of a particular organization, project, location, actual composition and qualification of work gangs, and inevitable fluctuations of their performance, or weather. Most "standard labor productivity rates" are published as single values, with no hint on the scale of variability observed during data collection. Therefore, their use in the simulations is limited.

In practice, the types and parameters of process durations are assumed based on historical data or expert opinions. Due to the unique nature of construction projects, historical data are of limited use. Collecting productivity data in the course of a project and recording them together with data on particular conditions is time-consuming and expensive. The results become unreliable when the technical and organizational conditions change, and the use of statistical forecasting methods is risky, especially when the values of the forecasted parameters exceed the range of available data.

The quality of experts' estimates depends on their individual experience and subject to bias. Experts from the client's side tend to be over-optimistic, while the contractor prefers to be "on the safe side" and schedule processes to take longer. To balance opinions, group decision-making methods are applied [50].

The experience prompts that the probability density function of construction processes duration is right-skewed. According to Johnson [11], the triangular distribution—described using simple analytical dependencies understandable to practitioners—provides an adequate approximation of the beta distribution used in the PERT, and the results of the risk assessment do not differ significantly. Many authors (e.g., Johnson [11], Kotz and van Dorp [13]) recommend defining the parameters of triangular distribution based on the mode and quantiles $t_{a,p}$ and $t_{b,r}$ of order p and r (typically $p = 0.10$,

$q = 0.90$ or $p = 0.05, q = 0.95$). The method by Jaśkowski [24] may be used to determine the parameters of the triangular distribution of construction processes duration under various conditions.

2.3. Proposed Method to Improve Construction Schedule Robustness

The proposed method is intended to be applied in the course of the project to prompt reactions to schedule disturbances. It assumes that there exists a baseline schedule that defines the project completion date and the dates of key subcontracted processes. It assumes that a failure to meet these dates results in penalties. Their amounts are defined in the contract between the client and the general contractor and the contracts between the general contractor and the subcontractors. To mitigate delays and minimize penalties once delays occur, it is necessary to speed up processes not yet completed. The method helps select the most economical ways to do it.

The method encompasses the following steps:

1. Constructing the project's network model (breaking down the scope into processes, allocating resources (crews/subcontractors), defining relationships between processes, defining process durations).
2. Creating the baseline schedule and specifying processes whose start dates must be protected against delays, then determining the unit costs of delaying their start.
3. Designing variants of actions to reduce process durations and determining their costs.
4. Simulation studies of the project implementation model for various time reduction policies.
5. Finding the optimal time reduction modes.

The project network is represented by a directed acyclic unigraph $G = \langle V, E \rangle$ of a single start and a single end node. $V = \{0, 1, \ldots, n\}$ is the set of construction processes (schedule tasks). $E \subset V \times V$ is a two-argument relation describing the sequence of processes. A function $T : V \to R^+$ assigns durations $\boldsymbol{t}_i$ to processes $i \in V$; the durations are random values of predefined distribution types and parameters. The estimated costs of processes are expressed as c_i. The project's predefined due date sets the time for completion to T_{max}.

The baseline schedule is built using the expected values of process durations to meet the predefined due date. Alternatively, the baseline schedule may be based on process durations corresponding to a particular quantile of the duration distribution.

To improve the schedule's robustness against disruption, it is advisable to distribute the free float as time buffers located before the processes whose start dates need to be protected (such as subcontracted processes with start dates contractually fixed or processes that involve an expensive hired plant). The set of processes that need to be protected is V_d.

It is assumed that the processes from the set of V_d can start no earlier than on the date set for them in the baseline schedule; the literature refers to this scheduling policy as the *railway policy*. The unit cost of delaying the process that belongs to V_d is c_i^d and represents contractual penalties or other costs attributable to such delay.

The baseline schedule's start dates of any process of the project is marked as s_i. In the course of the project, the processes are going to start as their predecessors are actually completed. In the case of processes $i \in V_d$, they are not allowed to start earlier than on the date set for them in the baseline schedule. Due to the stochastic nature of process duration, their starts can be delayed. Thus, the actual start of a process s_i^r may come later than the as-scheduled start ($s_i^r > s_i$).

As the delays are detected, the manager needs to decide on actions that prevent the propagation of disruptions on the processes to follow. These actions aim at reducing the execution time of delayed processes that are not yet started and consist in adding resources (reinforcing crews, using a more efficient plant), change of construction methods, working overtime, incentivizing the crews to work harder, etc. Inevitably, they result in an extra cost. The viable options of ways to compress the time of any process $i \in V$ (also their combinations) form a set W_i. The options differ in the resulting process time and cost. Let us assume that the option-related process duration $\boldsymbol{t}_{ij}$ is a random variable of known distribution type and parameters, and the option-related process cost c_{ij} is deterministic. Therefore, the expected value of the duration of process i, if decided to be delivered using option $j \in W_i$, is $\Delta_{ij} = t_i - t_{ij}$, where t_{ij} is the expected value of the random variable $\boldsymbol{t}_{ij}$. Let us put the options in the ascending order according to the values of Δ_{ij} and number them accordingly ($j = 1, 2, \ldots$).

Selecting actions of reacting to the process start delays consists of determining the best option and the best time of its deployment. The latter is defined by a lag, marked as λ, between the baseline start of the process and the moment when acceleration measures begin.

It is assumed that the first option of the time compressing actions (i.e., the option that offers the highest acceleration) is selected if $s_i^r - s_i \geq \lambda$. If $s_i^r - s_i \geq \Delta_{i,j+1} + \lambda$, then the next option is to be selected.

To facilitate operative management, the lag λ is constant and equal for all processes. Its value is subject to optimization: λ is defined in a way that minimizes the sum of the cost of process start delays, C_d, and the cost of the duration compression measures, C_w:

$$\min C(\lambda) : C(\lambda) = C_d + C_w = \sum_{i \in V_d} c_i^d \cdot E(\boldsymbol{s}_i - s_i) + \sum_{i \in V} \sum_{j \in W_i} \left(c_{ij} - c_i\right) \cdot E\left(\boldsymbol{x}_{ij}(\lambda)\right), \tag{1}$$

where $\boldsymbol{s}_i$ is the random variable representing the start of process $i \in V_d$, and $\boldsymbol{x}_{ij}(\lambda)$ is an auxiliary random variable; it equals 1 if option j is decided for the delivery of process i, and 0 otherwise; the value depends on the value of λ. The expected values of $\boldsymbol{s}_i$ and $\boldsymbol{x}_{ij}(\lambda)$ are estimated based on simulation experiments for different λ.

3. Results

The application of the method is presented in an example. It is based on the schedule of a project to build a single block of flats (rainforced concrete frame filled with masonry, monolithic floor slabs—a structure typical for Polish housing). The network model (Figure 1) presents relationships between processes entrusted to specialized crews/subcontractors. The project scope is broken down into 14 processes plus two dummy nodes: project start and project finish.

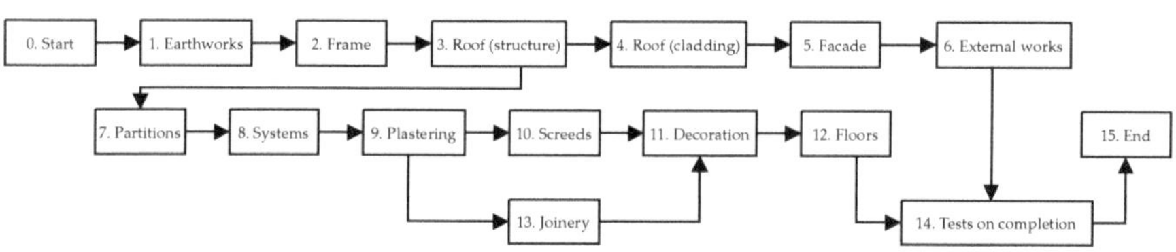

Figure 1. Project network (example).

The process durations are defined as random values of triangular distribution, and their costs are deterministic. The values of costs were derived from a real-life cost plan, and the parameters of random variables of process durations were the construction superintendent's estimates gathered during an interview. Table 1 lists the values of process parameters.

The predefined time for completion is T_{max} = 200 days. Figure 2 presents the baseline schedule. Processes marked black are the processes whose start days must be protected against disruption (so the processes that belong to the set V_d). The unit costs of their delays are set to 1% of their total value—for each day of delay (they can be understood as penalties agreed with the subcontractors hired to deliver them). These processes are to be started according to the *railway policy*. The unit cost of delaying the whole project is also 1% of the total project cost (the sum of costs shown in Table 1), so 270,045 EUR/day (delay penalty of the main contract).

Table 1. List of processes and their parameters.

i	Process Name	Minimal Duration (Days)	Maximal Duration (Days)	Mode (Days)	Expected Value (Days)	Cost (EUR)
		t_{ai}	t_{bi}	t_{mi}	t_i	c_i
1	Earthworks	5	13	6	8	200,000
2	Frame	33	48	39	40	24,000,000
3	Roof (structure)	19	32	24	25	140,000
4	Roof (cladding)	20	37	33	30	175,000
5	Facade	42	62	49	51	165,000
6	External works	39	49	44	44	125,000
7	Partitions	19	32	24	25	340,000
8	Systems	20	26	32	26	600,000
9	Plastering	14	27	22	21	170,000
10	Screeds	21	35	28	28	320,000
11	Decoration	21	31	26	26	230,000
12	Floors	16	26	18	20	80,000
13	Joinery	26	33	28	29	500,000
14	Tests on completion	1	1	1	1	0

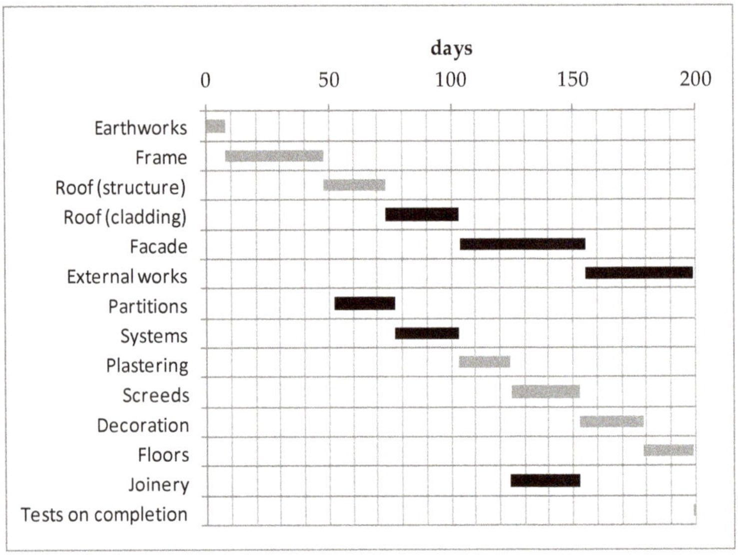

Figure 2. Baseline schedule (example).

It was assumed that the project starts at moment zero (no delay). The "Earthworks" are to be carried out using the baseline methods (no time reduction measures—no options available). The same holds for the "Tests on completion".

As for the remaining processes, each is assigned two options of time-reducing measures; their parameters are presented in separate tables (Tables 2 and 3) for clarity. The options with longer expected values of process durations are grouped in Table 2. Following the convention of numbering options in ascending order according to the scale of duration reduction (described in the previous chapter), this table presents the set of options with index $j = 1$. Table 3 groups options that compress durations more strongly, thus $j = 2$. Please note that stronger compression was assumed to be more costly (last column in Tables 2 and 3).

Table 2. Parameters of the first group of options of process duration compression measures.

i	Process Name	Minimal Duration (Days)	Maximal Duration (Days)	Mode (Days)	Expected Value (Days)	Cost (EUR)
		t_{ai1}	t_{bi1}	t_{mi1}	t_{i1}	c_{i1}
2	Frame	32	44	35	37	25,000,000
3	Roof (structure)	17	27	22	22	150,000
4	Roof (cladding)	21	36	27	28	185,000
5	Facade	39	57	48	48	195,000
6	External works	36	43	38	39	160,000
7	Partitions	16	29	21	22	355,000
8	Systems	20	24	28	24	720,000
9	Plastering	12	25	20	19	180,000
10	Screeds	21	33	24	26	340,000
11	Decoration	19	29	24	24	250,000
12	Floors	14	24	16	18	95,000
13	Joinery	24	31	26	27	620,000

Table 3. Parameters of the second group of options of process duration compression measures.

i	Process Name	Minimal Duration (Days)	Maximal Duration (Days)	Mode (Days)	Expected Value (Days)	Cost (EUR)
		t_{ai2}	t_{bi2}	t_{mi2}	t_{i2}	c_{i2}
2	Frame	30	39	33	34	26,500,000
3	Roof (structure)	16	21	20	19	165,000
4	Roof (cladding)	16	30	20	22	200,000
5	Facade	36	51	39	42	220,000
6	External works	31	40	37	36	180,000
7	Partitions	15	25	20	20	380,000
8	Systems	15	22	26	21	900,000
9	Plastering	10	23	15	16	200,000
10	Screeds	17	28	21	22	355,000
11	Decoration	17	25	21	21	285,000
12	Floors	12	20	16	16	100,000
13	Joinery	20	29	23	24	705,000

The simulation model was coded using GPSS language, and the simulations were conducted in GPSS Word Minuteman Software. The experiment was repeated with different lags (λ) for introducing the time compression measures. Each experiment involved 10,000 simulation runs.

Table 4 lists the expected values of delayed start dates of processes, juxtaposing three cases. Case I allows no time-reduction measures—all processes are delivered using methods assumed for the baseline. Case II offers a choice between the baseline methods and options coming only from the first group (Table 2). Case III makes it possible to choose from the baseline option and both options of time-reducing measures.

Table 4. Expected values of delayed start dates of processes.

Case	λ	$E(s_4-73)$	$E(s_5-104)$	$E(s_6-155)$	$E(s_7-50)$	$E(s_8-77)$	$E(s_{13}-124)$	$E(s_{15}-200)$
I	0	2.0170	2.4767	3.5912	0.8064	0.8705	4.1804	6.8009
II	0	0.5744	1.1926	1.7892	0.3075	0.567	1.6492	1.5647
	1	0.8067	1.3093	1.9607	0.4225	0.6224	1.899	1.9133
	2	1.0812	1.4799	2.1809	0.5515	0.6938	2.2433	2.3916
	3	1.3551	1.6813	2.4390	0.6728	0.7593	2.6359	2.9381
	4	1.5962	1.8885	2.7090	0.7656	0.8186	3.0373	3.5479
	6	1.8647	2.1944	3.1113	0.8064	0.866	3.6283	4.6561
	8	1.9758	2.3692	3.3577	0.8064	0.8705	3.9299	5.5422
	10	2.0126	2.4507	3.4987	0.8064	0.8705	4.0787	6.1302
	12	2.0170	2.4695	3.5582	0.8064	0.8705	4.1432	6.4640
	14	2.0170	2.4752	3.5813	0.8064	0.8705	4.1678	6.6369

Table 4. *Cont.*

Case	λ	$E(s_4-73)$	$E(s_5-104)$	$E(s_6-155)$	$E(s_7-50)$	$E(s_8-77)$	$E(s_{13}-124)$	$E(s_{15}-200)$
III	0	0.3996	1.0807	1.5231	0.2531	0.5611	0.7877	1.0931
	1	0.6393	1.1681	1.7651	0.4023	0.6183	1.1684	1.4434
	2	0.9419	1.3128	2.0323	0.5515	0.6934	1.6552	1.9456
	3	1.2521	1.5042	2.3074	0.6728	0.7593	2.181	2.5129
	4	1.5239	1.7265	2.5748	0.7656	0.8185	2.6871	3.1614
	6	1.8480	2.0809	3.0018	0.8064	0.866	3.4279	4.3833
	8	1.9747	2.3255	3.3029	0.8064	0.8705	3.8276	5.3736
	10	2.0126	2.4296	3.4705	0.8064	0.8705	4.0335	6.0459
	12	2.0170	2.4647	3.5501	0.8064	0.8705	4.1266	6.4243
	14	2.0170	2.4752	3.5789	0.8064	0.8705	4.1635	6.6190

Figures 3 and 4 show the relationships between the value of the lag (λ) for introducing duration compression measures and the expected value of the total cost of delays (C_d) and time reduction measures (C_w) for cases II and III, respectively.

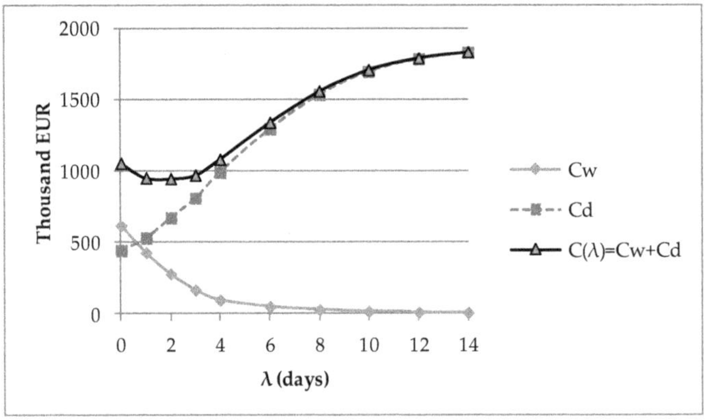

Figure 3. Relationship between the value of the lag (λ) for introducing duration compression measures and the expected value of the total cost of delays (C_d) and time reduction measures (C_w) for case II.

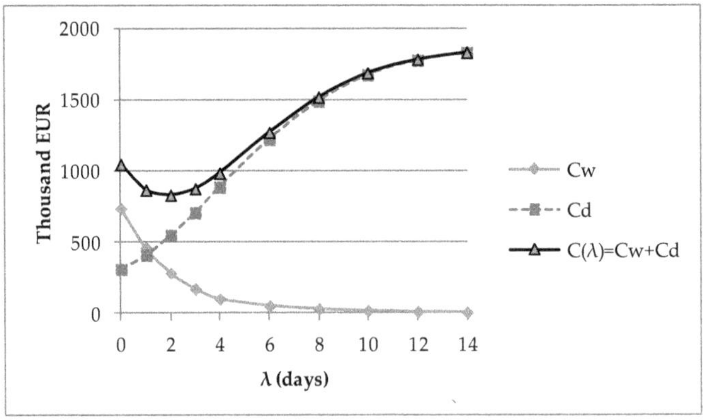

Figure 4. Relationship between the value of the lag (λ) for introducing duration compression measures and the expected value of the total cost of delays (C_d) and time reduction measures (C_w) for case III.

Figure 5 shows how the optimal lag (λ) and the total cost $C(\lambda)$ (penalties plus cost of schedule compression measures) depend on the penalty rate. Please note that the penalty per unit of time is calculated as a percentage (i.e., penalty rate defined in the contract) of the value assigned to a process.

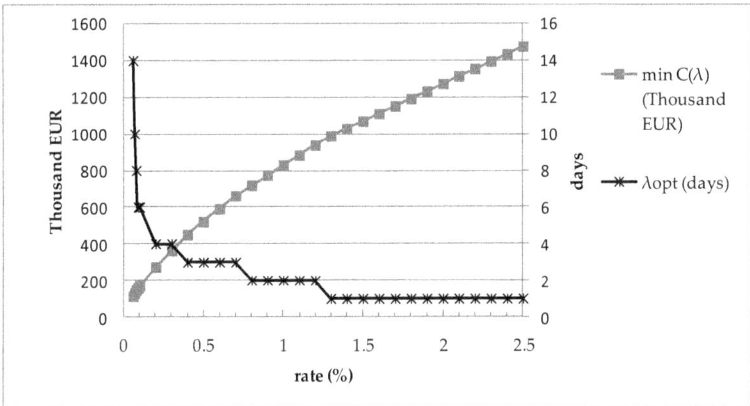

Figure 5. Effect of penalty rate on the optimal lag (λ) and the total cost of delays and duration compression measures for case III.

4. Discussion

The results of the simulation experiment made it possible to determine the optimal value of lag λ between the occurrence of a delay and the moment of implementing the duration compression measures.

For both cases analyzed in the example, this lag was the same (two days). Let us consider process four (roof cladding): its baseline start was scheduled for day 73 ($s_4 = 73$). The expected value of its duration is $t_4 = 30$ (baseline), and, if accelerated by switching to the first option, $t_{41} = 28$. Therefore, $\Delta_{41} = 2$. If all predecessors of process four are completed by day 70 (earlier than scheduled in the baseline), so $s_4^r = 70$, the process must start on day 73 anyway—because of the *railway policy* that rules this process. As $\Delta_{41} + \lambda = 4$, if the actual start of process four happened between day 75 and 77: $75 \leq s_4^r \leq 77$, this process should be conducted according to the first option of duration compressing measures. If process four was observed to start later than on day 77 ($s_4^r > 77$), switching to the second option of duration compressing measures was advised.

Oddly enough, a lower total cost $C(\lambda)$ was obtainable by introducing more expensive duration reduction measures from the second group of options: $C(\lambda) = 828{,}882.84$ EUR in case III, whereas $C(\lambda) = 946{,}833.96$ EUR in case II. However, this fact is attributable to the possibilities of stronger compression offered in case III—stronger compression means a reduction in delay penalties (C_p). The expected values of duration compression measures (C_w) proved similar: (case II—276,897 EUR, case III—282,029 EUR).

The results of the sensitivity analysis of the optimal lag to changes in penalty rate (Figure 5) indicate that the lower the penalty rates, the less stable the optimal solutions. Thus, even a small change in the contract regarding the penalty rate, may call for repeating the optimization procedure. A smaller penalty rate gives the construction manager more time to implement the schedule compression measures.

The results can be compared with the effects of actions taken intuitively by construction managers. In practice, to avoid contractual penalties, these actions are undertaken each time a delay occurs, and implemented immediately. Such actions are consistent with the proposed approach with $\lambda = 0$. The total cost corresponding to such a rule is 1,056,214.98 EUR for case II and 1,045,415.87 EUR for case III. Therefore, the application of the proposed method enables a reduction in the cost and financial penalties on average by 109,381.02 EUR (case II) and 216,533.03 EUR (case III).

5. Conclusions

The performance of construction projects depends largely on the efficiency of the operative management and decisions taken in the course of the project. The proposed method was intended to support decisions at the stage of designing the implementation of construction projects. It helps assess the chances of meeting the project due date and considers the effects of corrective measures (working overtime, hiring additional resources, etc.) in terms of both cost and time.

The proposed approach is innovative: the existing methods of reactive scheduling respond to schedule disturbances by relocating resources and rescheduling the processes that have not started yet—in a way that minimizes the weighted sum of the differences between the updated and the baseline process starts [43]. The authors found only one study [45] that, in addition to the above, considers the problem of selecting the process acceleration measures.

The authors assume that, due to risk, the process durations can be modeled as random values. In contrast to other approaches, the aim is not to build an optimized schedule after each disturbance. Instead, an optimal decision rule to pick the schedule acceleration measures is desirable. It is therefore not necessary to carry out an optimization procedure every time a disruption occurs. The quality of the decision rule is evaluated in the course of the simulations: its outcomes are assessed based on the distribution of results obtained in many simulation runs. Therefore, as the assumptions are different, the results obtained using the proposed method are not directly comparable with the results generated employing the methods proposed in the literature.

The proposed method is intuitive. It does not involve complex optimization calculus which the site engineers might be not familiar with. The simulation model needs to be developed only once, and the simulation tests are repeated only when the parameters of the model are changed—the lag time. They can be performed using practically any simulation package available on the market. The data for the model are obtained based on expert opinions, as in PERT, which is widely established in the construction industry.

Construction activity is considered particularly exposed to risk and uncertainty. Nevertheless, assumptions of full knowledge of work organization parameters and on the influence of disturbing factors are frequently made by construction schedulers. The reason may be the availability of software that supports only deterministic planning, or a natural human preference for exact numbers that define project dates. However, deterministic schedules tend to expire. The approach of subsequent updates in reaction to changes (incremental design strategy) is usually less efficient than the strategy of searching for an optimal solution in given conditions (proactive approach), yet it is commonly used in practice. Therefore, the direction of further research is to develop pro-active scheduling methods that account for the possibility to switch from one mode of operation to some other, selected out of a set of options differing in duration, cost, and even resources.

Author Contributions: Conceptualization, methodology, P.J. and S.B.; description of the problem, P.J.; simulation model programming, S.B.; validation, P.J. and S.B.; simulation experiments, S.B.; formal analysis, P.J. and S.B.; data collection, S.B.; writing—original draft preparation, writing—review and editing, P.J. and S.B. Both authors have read and agreed to the published version of the manuscript.

Funding: This research was financially supported by the Ministry of Science and Higher Education in Poland within the statutory research number FN/63.

Conflicts of Interest: The authors declare no conflict of interest.

References

1. Majid, M.Z.A.; McCaffer, R. Factors on non-excusable delays that influence contractors' performance. *J. Manag. Eng.* **1998**, *14*, 42–49. [CrossRef]
2. Skorupka, D. Identification and Initial Risk Assessment of Construction Projects in Poland. *J. Manag. Eng.* **2008**, *24*, 120–127. [CrossRef]
3. Flanagan, R.; Norman, G. *Risk Management and Construction*; Blackwell Publishing: Oxford, UK, 1993.

4. Zou, P.X.; Zhang, G.; Wang, J. Understanding the key risks in construction projects in China. *Int. J. Proj. Manag.* **2007**, *25*, 601–614. [CrossRef]
5. Ibadov, N.; Kulejewski, J. Construction projects planning using network model with the fuzzy decision node. *Int. J. Environ. Sci. Technol.* **2019**, *16*, 4347–4354. [CrossRef]
6. Edwards, P.J.; Bowen, P.A. Risk and risk management in construction projects: Concepts, terms and risk categories re-defined. *J. Constr. Procure.* **2000**, *5*, 42–57.
7. Golenko-Ginzburg, D. *Stochastic Network Models in Innovative Projecting*; Science Book Publishing House: Voronezh, Russia, 2011.
8. Mohan, S.; Gopalakrishnan, M.; Balasubramanian, H.; Chandrashekar, A. A lognormal approximation of activity duration in PERT using two time estimates. *J. Oper. Res. Soc.* **2007**, *58*, 827–831. [CrossRef]
9. McCombs, E.L.; Elam, M.E.; Pratt, D.B. Estimating Task Duration in PERT using the Weibull Probability Distribution. *J. Mod. Appl. Stat. Methods* **2009**, *8*, 282–288. [CrossRef]
10. Dawood, N. Estimating project and activity duration: A risk management approach using network analysis. *Constr. Manag. Econ.* **1998**, *16*, 41–48. [CrossRef]
11. Johnson, D. The triangular distribution as a proxy for the beta distribution in risk analysis. *J. R. Stat. Soc. Ser. D Stat.* **1997**, *46*, 387–398. [CrossRef]
12. Williams, T.M. Practical use of distributions in network analysis. *J. Oper. Res. Soc.* **1992**, *43*, 265–270. [CrossRef]
13. Kotz, S.; Van Dorp, J.R. *Beyond Beta—Other Continuous Families of Distributions with Bounded Support and Applications*; World Scientific Pub Co Pte Ltd.: Singapore, 2004.
14. Vose, D. *Risk Analysis: A Quantitative Guide*; John Wiley and Sons Inc.: Hoboken, NJ, USA, 2008.
15. Abourizk, S.; Wales, R. Combined Discrete-Event/Continuous Simulation for Project Planning. *J. Constr. Eng. Manag.* **1997**, *123*, 11–20. [CrossRef]
16. Lee, H.-S.; Shin, J.-W.; Park, M.; Ryu, H.-G. Probabilistic Duration Estimation Model for High-Rise Structural Work. *J. Constr. Eng. Manag.* **2009**, *135*, 1289–1298. [CrossRef]
17. Shahin, A.; Abourizk, S.; Mohamed, Y.; Fernando, S. Simulation modeling of weather-sensitive tunnelling construction activities subject to cold weather. *Can. J. Civ. Eng.* **2014**, *41*, 48–55. [CrossRef]
18. Lu, M.; Abourizk, S. Simplified CPM/PERT Simulation Model. *J. Constr. Eng. Manag.* **2000**, *126*, 219–226. [CrossRef]
19. Shi, J.J. Activity-Based Construction (ABC) Modeling and Simulation Method. *J. Constr. Eng. Manag.* **1999**, *125*, 354–360. [CrossRef]
20. Lee, D.-E.; Bae, T.-H.; Arditi, D. Advanced Stochastic Schedule Simulation System. *Civ. Eng. Environ. Syst.* **2011**, *29*, 1–18. [CrossRef]
21. Sadeghi, N.; Fayek, A.R.; Ingolfsson, A. Simulation-Based Approach for Estimating Project Completion Time of Stochastic Resource–Constrained Project Networks. *J. Comput. Civ. Eng.* **2012**, *26*, 558–560. [CrossRef]
22. Lee, D.-E. Probability of Project Completion Using Stochastic Project Scheduling Simulation. *J. Constr. Eng. Manag.* **2005**, *131*, 310–318. [CrossRef]
23. Aziz, R.F. RPERT: Repetitive-Projects Evaluation and Review Technique. *Alex. Eng. J.* **2014**, *53*, 81–93. [CrossRef]
24. Jaskowski, P.; Biruk, S. The method for improving stability of construction project schedules through buffer allocation/statybos vykdymo grafiko stabilumo užtikrinimas paskirstant laiko rezervus. *Technol. Econ. Dev. Econ.* **2011**, *17*, 429–444. [CrossRef]
25. Leu, S.-S.; Hung, T.-H. An optimal construction resource leveling scheduling simulation model. *Can. J. Civ. Eng.* **2002**, *29*, 267–275. [CrossRef]
26. Biruk, S.; Rzepecki, Ł. Simulation model for resource-constrained construction project. *Open Eng.* **2019**, *9*, 211–217. [CrossRef]
27. Gupta, D.; Maravelias, C.T. On the design of online production scheduling algorithms. *Comput. Chem. Eng.* **2019**, *129*, 106517. [CrossRef]
28. Gupta, D.; Maravelias, C.T. Framework for studying online production scheduling under endogenous uncertainty. *Comput. Chem. Eng.* **2020**, *135*, 106670. [CrossRef]
29. Van De Vonder, S.; Demeulemeester, E.; Herroelen, W. Proactive heuristic procedures for robust project scheduling: An experimental analysis. *Eur. J. Oper. Res.* **2008**, *189*, 723–733. [CrossRef]

30. Balouka, N.; Cohen, I. A robust optimization approach for the multi-mode resource-constrained project scheduling problem. *Eur. J. Oper. Res.* 2019. [CrossRef]
31. Moonseo, P.; Feniosky, P.-M. Reliability buffering for construction projects. *J. Constr. Eng. Manag.* **2004**, *130*, 626–637.
32. Herroelen, W.; Leus, R. The construction of stable project baseline schedules. *Eur. J. Oper. Res.* **2004**, *156*, 550–565. [CrossRef]
33. Goldratt, E.M. *Critical Chain*; The North River Press: Great Barrington, MA, USA, 1997.
34. Van De Vonder, S.; Demeulemeester, E.; Herroelen, W.; Leus, R. The use of buffers in project management: The trade-off between stability and makespan. *Int. J. Prod. Econ.* **2005**, *97*, 227–240. [CrossRef]
35. Połoński, M.; Pruszyński, K. Impact of Baseline Terms on the Course of Critical Paths and Time Buffers in the Modified Goldratt's Method. *Arch. Civ. Eng.* **2013**, *59*, 313–320. [CrossRef]
36. Van De Vonder, S.; Demeulemeester, E.; Leus, R. The trade-off between stability and makespan in resource-constrained project scheduling. *Int. J. Prod. Res.* **2006**, *44*, 215–236. [CrossRef]
37. Jaskowski, P. Methodology for enhancing reliability of predictive project schedules in construction. *Ekspolatacja i Niezawodn. Maint. Reliab.* **2015**, *17*, 470–479. [CrossRef]
38. Rostami, S.; Creemers, S.; Leus, R. New strategies for stochastic resource-constrained project scheduling. *J. Sched.* **2018**, *21*, 349–365. [CrossRef]
39. Suwa, H. A new when-to-schedule policy in online scheduling based on cumulative task delays. *Int. J. Prod. Econ.* **2007**, *110*, 175–186. [CrossRef]
40. Chakrabortty, R.K.; Rahman, H.F.; Ryan, M.J. Efficient priority rules for project scheduling under dynamic environments: A heuristic approach. *Comput. Ind. Eng.* **2020**, *140*, 106287. [CrossRef]
41. Wang, Y.; He, Z.; Kerkhove, L.-P.; Vanhoucke, M. On the performance of priority rules for the stochastic resource constrained multi-project scheduling problem. *Comput. Ind. Eng.* **2017**, *114*, 223–234. [CrossRef]
42. Janczura, M.; Kuchta, D. Proactive and reactive scheduling in practice. *Res. Pap. Wrocław Univ. Econ.* **2011**, *238*, 34–51.
43. Van De Vonder, S.; Ballestín, F.; Demeulemeester, E.; Herroelen, W. Heuristic procedures for reactive project scheduling. *Comput. Ind. Eng.* **2007**, *52*, 11–28. [CrossRef]
44. Paslawski, J. Flexibility Approach in Construction Process Engineering/Lankstumas Statybos Proceso Inžinerijoje. *Technol. Econ. Dev. Econ.* **2008**, *14*, 518–530. [CrossRef]
45. Deblaere, F.; Demeulemeester, E.; Herroelen, W. Reactive scheduling in the multi-mode RCPSP. *Comput. Oper. Res.* **2011**, *38*, 63–74. [CrossRef]
46. Yang, K.-K. A comparison of dispatching rules for executing a resource-constrained project with estimated activity durations. *Omega* **1998**, *26*, 729–738. [CrossRef]
47. Law, A.; Kelton, W.D. *Simulation Modeling and Analysis*, 3rd ed.; McGraw-Hill Education: Boston, MA, USA, 2000.
48. Miłosz, M.; Pastuszak, Z.; Miłosz, E. Decision-making simulation games as a tool for verifying enterprise logistic strategies. In Proceedings of the INTED2015: 9th International Technology, Education and Development Conference, Madrid, Spain, 2–4 March 2015; pp. 3558–3567.
49. Elmaghraby, S.E. On criticality and sensitivity in activity networks. *Eur. J. Oper. Res.* **2000**, *127*, 220–238. [CrossRef]
50. Demeulemeester, E.L.; Herroelen, W.S. *Project Scheduling, A Research Handbook*; Kluwer Academic Publishers: Boston, MA, USA, 2002.

applied sciences

MDPI

Article

Development of Alfa Fiber-Based Mortar with Improved Thermo-Mechanical Properties

Siham Sakami [1], Lahcen Boukhattem [2,3,*], Mustapha Boumhaout [2] and Brahim Benhamou [2,4]

1 L3G, Faculty of Sciences and Techniques, Cadi Ayyad University, Marrakesh 40000, Morocco; s.sakami@uca.ma

2 EnR2E Laboratory, National Center of Studies and Research on Water and Energy, CNEREE, Cadi Ayyad University, Marrakech 40000, Morocco; mustapha.boumhaout@edu.uca.ac.ma (M.B.); BBenhamou@uca.ma (B.B.)

3 LMPEQ, National School of Applied Sciences, Cadi Ayyad University, Safi, Marrakech 46000, Morocco

4 ProcEDE Laboratory, Cadi Ayyad University, Marrakech 40000, Morocco

* Correspondence: l.boukhattem@uca.ma

Received: 10 October 2020; Accepted: 6 November 2020; Published: 12 November 2020

Abstract: This work deals with the development of a new composite based on mortar reinforced with optimally sized alfa fiber (AF). Experimental investigations of physical and thermo-mechanical properties of the new AF mortar composite are performed for AF weight fraction varying from 0% to 5%. This simple material preparation process is described and scanning electron microscopy (SEM) is undertaken to analyze the morphology of this composite. It shows a random dispersion of the AF into the mortar matrix. Physical properties such as open porosity, water absorption, and bulk density fluctuations with AF mass content are measured. Measured thermal conductivity is compared to the values generated by different prediction models. Good agreement, within 9.6%, is obtained with data predicted by Woodside–Messmer's method. It is demonstrated that this simple blending of AF into mortar improves significantly the thermo-mechanical behavior of the new composite. An addition of 5% of AF weight content makes composite material lighter by about 15%, enhances its insulating thermal capabilities by about 57% and increases its heat diffusion damping rate by about 49%. Moreover, the composite mechanical (flexural and compressive) strength increases by up to 10% for an AF weight content of 1%.

Keywords: alfa fiber; cement; composite; density; mechanical properties; thermal properties

1. Introduction

Recently, the use of natural plant fibers as reinforcement to concrete has become very attractive [1]. This is due to the increase of environmental awareness and to the advantages of these fibers such as availability, recyclability, low cost, environmentally friendly, harmlessness, abrasion free, biodegradability, and better thermo-mechanical performance when compared to conventional fibers. Different natural fibers (date palm, cotton, sisal, flax, hemp, jute, ramie, kenaf, bamboo, banana, coir, coconut and wheat straws) could be good candidates for blending with other compounds (cement, clay, sand, lime, gypsum, mortar, concrete, etc.) in order to develop new thermal insulating composite materials. Therefore, several studies have been conducted on various vegetable fibers from sisal [2], coconut [3], hemp [4], kenaf [5], banana [6], date palm [7], bamboo [8], jute [9] to alfa fiber [10].

The alfa plant, also called esparto grass stems (*Stipa tenacissima*), grows abundantly in the Mediterranean basin, especially in countries such as Spain, Italy, Libya, Tunisia, Algeria, and Morocco. This fiber is a permanent plant that does not disappear during winter and grows in independent shrubs where it can reach a height between 1 to 1.20 m [11]. It covers many million hectares in this area, with about 4,000,000 ha in Algeria, 3,186,000 ha in Morocco, 600,000 ha in Tunisia, 350,000 ha in Libya, and 300,000 ha in Spain [12]. The annual production is approximately 250,000 t in Algeria, 125,000 t in Morocco, and 75,000 t in Tunisia [13].

Alfa fiber is selected in this study thanks to its high availability in dry region and its interesting thermo-mechanical properties. Its thermal conductivity can be as low as 0.042 $W{\cdot}m^{-1}{\cdot}K^{-1}$ [14] which is in the same order of magnitude as hemp and flax [15]. Moreover, it has a better mechanical strengths as reported in the previous studies [16,17]. These properties make alfa fiber (AF) a reinforcement candidate for mineral materials. In building construction field, several researchers have dealt with the manufacturing of AF cementitious composite materials.

Krobba et al. [18] performed an experimental study on a repair mortar based on dune sand and microfibers with various volume percentages, lengths (3 to 5 mm) and diameters (150–200 μm). The results showed that the use of 0.75% of alfa micro-fiber increases mechanical properties of mortar.

Jabali et al. [19] studied the influence of different volume fractions of long alfa fibers (150 mm) on thermomechanical performance of mortar-based composite. This procedure is very labor extensive since the AF is distributed as a single layer within the mortar slab at a specific depth. They measured a reduction in thermal conductivity of 37% at 1.5% of alfa fiber content and an improvement of flexural strength by up to 27% for 1% of AF.

Elhamdouni et al. [20] presented a comparative study of thermo-mechanical properties of two materials: clay/alfa and clay/straw. They showed that for a weight fraction of fiber above 2% the adhesion between fiber and clay becomes very low which in turn affects the mechanical performance of these two composite materials.

On the other hand, interest in these new composite materials has drawn researcher's attention to developing new prediction tools to estimate the new composite thermal conductivity. Braiek et al. [21] have compared experimental results and theoretical equivalent thermal conductivity of materials with two phases, continuous gypsum phase and dispersed date palm fibers phase. Taoukil et al. [22] have carried out a comparison of experimental and theoretical approaches of the material's thermal conductivity with two-phase medium made up of mortar matrix and wood wool insulator.

Despite the extensive research work on alfa fiber in construction materials, this fiber is still not heavily used in building. The objective of this work is to present a procedure to build the AF mortar composite and to investigate the improvement of its thermo-mechanical properties in order to bring more attention to the potential of this abundant renewable resource of AF (which is mostly used in handcrafts only) in material building in the Mediterranean region.

A complete analysis of the effect of mortar blending with AF on physical and thermomechanical properties was executed for different weight fraction (ω) from 0% to 5%.

In the first part of this paper, the material preparation process is described; the elaborated new composites were scanned by the SEM to determine their morphology. In the second part, experimental measurements of the thermomechanical properties such as thermal conductivity, thermal diffusivity, flexural and compressive strengths were carried out. Moreover, mathematical models were used to compare predicted and measured thermal conductivities of the composite materials at dry state. These theoretical models have been developed/used by different researchers to calculate the equivalent thermal conductivity value of composite materials at dry state; they combine in different ways the thermal conductivity of two phases: solid phase (solid matrixes of mortar and fibers) and air phase in composites. These models were rearranged in this work to be applied for three phases: solid matrix of mortar, solid matrix of alfa fiber and air. The purpose of this comparison, between predicted and measured thermal conductivities as a function of volume percentage of AF's solid matrix (φ), is to demonstrate that the model that captures well the distribution of the three phases within the composite material is in good agreement with the experimental data.

2. Materials and Methods

2.1. Material Selection

The new composite material manufacturing consists of mixing cement, sand and alfa fiber (AF) with drinking tap water. The AF plant used to reinforce mortar was extracted from the Oujda region in the northeast of Morocco (Figure 1a). It was prepared in different steps. It was first prepared by submerging it in a salt water (35 g·L^{-1}) [23] at ambient temperature (about 30 °C) for 7 days and was washed with water jet. The goal of this soaking process was to remove sand and dust existing on the stem surface and to facilitate fiber separation. Subsequently, the plant stems were dried under the sun for one week. These low cost and environmental operations will eliminate most of the moisture in alfa stems and consequently make the grinding operation more efficient. The dried alfa stems were crushed (Figure 1b), separated, and cut into short fibers with a diameter between 0.7 and 1 mm, and a length between 1.5 and 2.5 cm (Figure 1c). These lengths were selected since [24] demonstrated that optimal mechanical performance was obtained at this length range.

Figure 1. Steps of alfa fiber (AF) preparation: (**a**) alfa plant, (**b**) crushed alfa stem after drying process and (**c**) used AF.

This natural treatment of AF will enhance the fibers adhesion with the cementitious matrix.

An alfa fiber diffractogram obtained from an X-ray diffractometry (XRD) test is shown in Figure 2. One can clearly notice that the spectrum contains two main peaks. The first one, wide and multiple, is obtained at Bragg angles of 15° and 16.54° corresponding to crystallographic planes of (1–10) and (110), respectively. The second one appears at 22.50° which is associated to a crystallographic plane of (200). These two peaks are characteristic of the presence of cellulose I. The cellulose component is considered as a bio-binder that can bond efficiently the fibers to mortar.

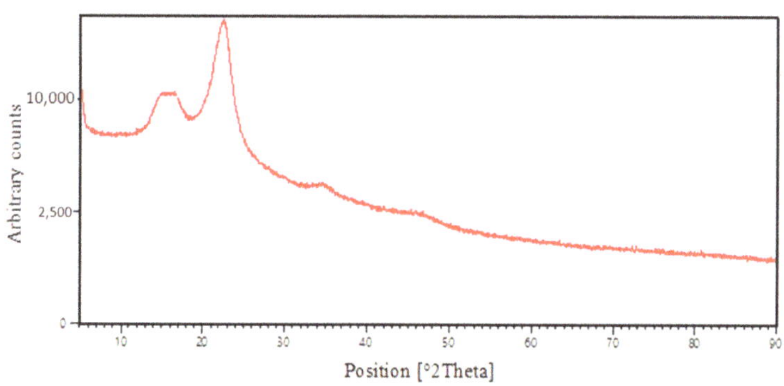

Figure 2. Alfa fiber X-ray diffractogram.

The Portland (CPJ 35- CEM II 22.5) cement is used in this study, and its technical characteristics are in agreement with the Moroccan norm (NM 10.01.004., 2003) [25]. The sand used is extracted from

Nafiss River (Loudaya, Morocco) with a maximum size of 2 mm and water mass content of 2.15%. The sand equivalent test indicates that the average value of its neatness is 87% (NF EN 933-8, 1999) [26]. Its chemical constitution is reported in Table 1.

Table 1. Chemical constitution of used sand.

Constituents	SiO_2	Al_2O_3	K_2O	Fe_2O_3	Na_2O	MgO	CaO	P_2O_5	TiO_2	SO_3	Cl
Content (%)	68.51	13.55	3.95	2.83	2.72	2.64	1.62	0.16	0.7	0.01	0.005

2.2. Preparation of the Alfa Fiber Composite Test Specimens

The preparation of the test samples consists of substitution of sand with different AF mass fractions to evaluate the impact on the thermo-mechanical properties compared to those of the reference mortar (RM). The latter was made from sand-cement with a mass ratio of 3 and a water-cement mass ratio of 0.6. Afterward, in addition to RM, five other new composite materials, labeled AFRM0.5%, AFRM1%, AFRM2%, AFRM3.5% and AFRM5%, were obtained by incorporating respectively the AF mass percentages of 0.5%, 1%, 2%, 3.5% and 5% into mortar. Prismatic molds of dimensions $4 \times 4 \times 16$ cm^3 were used for mechanical tests while parallelepiped molds with dimension of $26.4 \times 26.4 \times 4$ cm^3 were used for thermo-physical measurements (Figure 3a–c). All molds were filled with the mixed components and cured in the laboratory for 24 h under normal weather conditions: T = 20 ± 2 °C and relative humidity RH = 65 ± 5%. After demolding operation, the samples were placed in water at ambient temperature for 28 days, according to the norm (NF EN 196-1, 1995) [27]. Finally, for each AF mass percentage, three manufactured samples of AFRM composites were tested to determine the average values and uncertainties of thermo-mechanical properties. The manufactured samples are shown in Figure 3d–f.

Figure 3. Preparation and manufactured samples: (**a**) mixer, (**b**,**c**) prismatic and parallelepiped molds, (**d**,**e**) samples for mechanical and thermal tests and (**f**) AF binderless board.

2.3. Measurement Instruments and Experimental Methods Description

2.3.1. Scanning Electron Microscopy (SEM) Morphological Observation

A TESCAN ORSAY (VEGA3) scanning electron microscope (SEM) was used to observe the morphology of AFRM composite and examine the homogeneity and distribution levels of AF in mortar matrix. It is a versatile tungsten thermionic emission system intended for both high and low void processes. VEGA3 is fitted with an electron optics system based on a unique four-lens wide field optics design with a proprietary intermediate lens. The SEM pictures were taken with beam acceleration at 5 kV.

2.3.2. Open Porosity, Water Absorption, and Bulk Density

Open porosity P_0, water absorption W_a, and bulk density ρ_b measurements were conducted to determine the quality of mortar-AF composite materials. Their values were obtained by the Archimedes method of triple weighing, in agreement with the Standards Methods Designation: American society for testing and materials (ASTM C20–00) [28]. Immersion liquid used for these tests was clean water. Knowing dry weight (m_{dry}), soaked weight (m_{sat}) and hydrostatic weight (m_{hyd}) of the studied AFRM composites, the corresponding open porosity, water absorption, and bulk density are calculated respectively by Equations (1)–(3):

$$P_0 = \frac{m_{sat} - m_{dry}}{m_{sat} - m_{hyd}} \times 100 \quad (1)$$

$$W_a = \frac{m_{sat} - m_{dry}}{m_{dry}} \times 100 \quad (2)$$

$$\rho_b = \frac{m_{dry}}{m_{sat} - m_{hyd}} \times \rho_{water} \quad (3)$$

2.3.3. Thermo-Mechanical Characterization Methods

The EI700 device (two boxes method) was used to determine the thermal properties of insulating building materials [29]. It contains two boxes designed to take measurements in similar conditions. The first box was reserved for thermal conductivity k measurement and the second box was dedicated to thermal diffusivity α measurement Figure 4c. The process used to measure these two parameters was detailed in a previous study by Boumhaout et al. [30].

Figure 4. Devices used for thermo-mechanical tests: (**a**) air conditioned chamber, (**b**) two boxes method, (**c**) boxes of measured thermal conductivity and diffusivity, (**d**) flexural test, (**e**) two half prisms obtained from flexural test and (**f**) compressive test.

The accuracy of thermal conductivity measurement at steady state can be affected by ambient conditions variation. An air-conditioned chamber (see Figure 4a,b) was built to maintain the instrument surroundings at constant temperature during the experiment. Therefore, the experiment can be completed in a reduced duration between 6 and 10 h.

The device 270 kN capacity I20 201 press with velocity loading fixed at 500 $\pm$ 10 $N{\cdot}s^{-1}$ and 2400 $\pm$ 200 $N{\cdot}s^{-1}$ was used for mechanical testing. The measurements of flexural and compressive strengths were performed after 28 days of specimens curing process, according to the Moroccan norm (NM 10.1.005., 1994) [31] (Figure 4d,f). The flexural strength measurement method of prismatic specimen was determined by a three points bending technique. The two half prisms (Figure 4e) obtained from this test were used for compressive strength measurement (Figure 4f).

2.4. Thermal Conductivity Prediction Models

Many researchers have developed theoretical models to calculate the equivalent thermal conductivity value of composite materials. These models combine in different ways the thermal conductivity of two phases in composite materials at dry state: solid phase (solid matrixes of mortar and fibers) and air phase. These models were rearranged in this study to be applied for three phases: solid matrix of mortar, solid matrix of alfa fiber and air. Table 2 provides the rearrangement expressions of seven models presented in the literature (fourth column of Table 2). These models were applied to calculate the thermal conductivity of AFRM composites for different volume ratios of AF's solid matrix φ.

Table 2. Theoretical models.

Model	Equation		Comment	Reference
	$k_{//} = \varphi(1-P_t)k_{i-f} + (1-P_t)(1-\varphi)k_{i-M} + P_t k_a$	(4)		
Parallel	$\varphi = \frac{\rho_{abs\ RMAF} - \rho_{abs\ RM}}{\rho_{abs\ AF} - \rho_{abs\ RM}}$ $\rho_{abs\ RMAF} = \frac{1}{\omega \times \left(\frac{1}{\rho_{abs\ AF}} - \frac{1}{\rho_{abs\ RM}}\right) + \frac{1}{\rho_{abs\ RM}}}$	(5) (6)	φ: Volume ratio of solid matrix of AF per absolute volume of sample. $\rho_{abs\ RMAF}$: Absolute density of composite. P_t: Total porosity. k_{i-f} and k_{i-M}: Intrinsic thermal conductivities of AF and reference mortar (RM) respectively. k_a: Thermal conductivity of air. ω: Weight content of AF per weight of mixture. ρ_b: Bulk density of composite. $\rho_{abs\ RM}$ and $\rho_{abs\ AF}$: Absolute densities of reference mortar and AF.	
	$P_t = \left(1 - \frac{\rho_b}{\rho_{abs\ RMAF}}\right) \times 100$	(7)		
Serial	$k_{\perp} = \frac{1}{(\varphi/k_{i-f}) + (1-(\varphi+P_t))/k_{i-M} + P_t/k_a}$	(8)		
Beck's	$k = \sqrt{k_{\perp} k_{//}}$	(9)		
Woodside and Messmer	$k = k_{i-f}^{\varphi(1-P_t)} k_{i-M}^{(1-P_t)(1-\varphi)} k_a^{P_t}$	(10)		[32–34]
Auto-coherent	$k = k_{i-M}\left[1 + \frac{\tau}{(1-\tau)/3 + \frac{3+\delta(k_a/k_{i-M}-1)}{3(k_a/k_{i-M}-1) - \delta(k_a/k_{i-f}-1)(2k_{i-f}/k_{i-M}+1)}}\right]$	(11)	$\tau = (1-\varphi)P_t + \varphi$: Volume ratio of (air + AF) in the sample. $\delta = 1 - \frac{P_t}{\tau}$: Concentration of AF in relation to components (air + AF) in the sample.	[35–37]
Effective medium theory (EMT)	$\sum_j \phi_j\left(\frac{k-k_j}{k+k_j}\right) = 0$ $\phi_f\left(\frac{k-k_{i-f}}{k+k_{i-f}}\right) + \phi_M\left(\frac{k-k_{i-M}}{k+k_{i-M}}\right) + \phi_a\left(\frac{k-k_a}{k+k_a}\right) = 0$	(12) (13)	$\phi_f = \varphi(1-P_t)$: Volume ratio of solid matrix of AF per volume of sample. $\phi_M = (1-P_t)(1-\varphi)$: Volume ratio of solid matrix of mortar per volume of sample. $\phi_a = P_t$: Volume ratio of solid matrix of air per volume of sample.	[38,39]
Krischer–Kroll	$k = \frac{1}{(1-\varepsilon)/k_{//} + \varepsilon/k_{\perp}}$	(14)	ε: Volume fraction of the layers oriented perpendicular to the heat flux direction	[40]

The goal is to determine the effect of the three components (mortar, air and AF) distribution on the predicted thermal conductivity and to compare them to the measurement data.

3. Results and Discussions

3.1. SEM Morphological Analysis of the Alfa Fiber Reference Mortar (AFRM) Composites

Scanning electron microscopy was used to analyze the morphology of AFRM composites. The analysis of the SEM pictures displayed in Figure 5 indicates clearly that alfa fibers are randomly dispersed, separated, and embedded into mortar matrix. Moreover, all pictures show that the fiber surfaces are impregnated with mortar paste, which could indicate good bonds between AF and the surrounding solid matrix. The voids observed in Figure 5e,f between alfa fibers and mortar matrix could be generated due to the drying operation, worse miscibility, and lower fiber dispersion in the mortar.

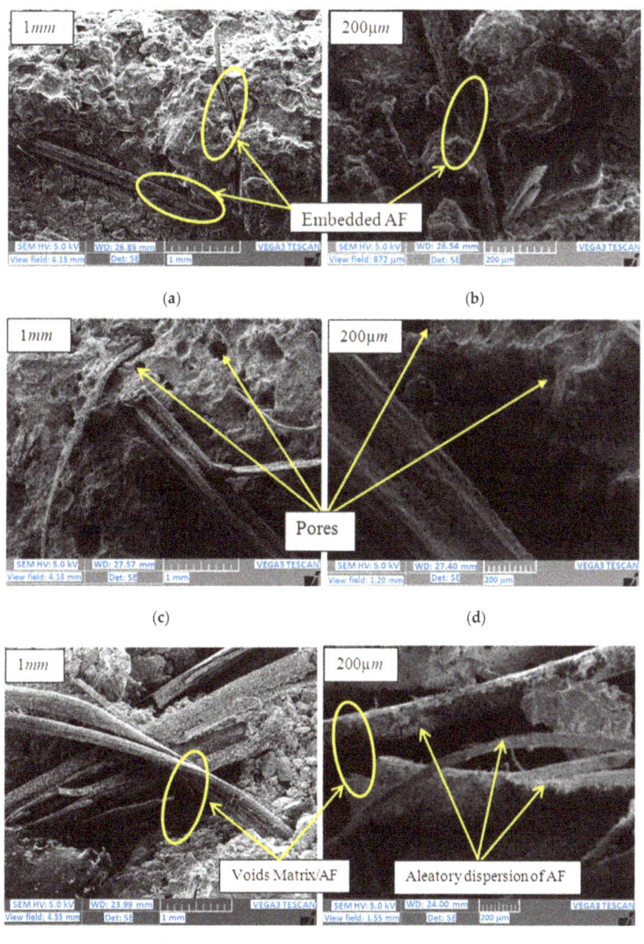

Figure 5. Scanning electron microscope (SEM) pictures of composites at two different scale lengths 1 mm and 200 μm: (**a**,**b**) AFRM 1%, (**c**,**d**) AFRM 2%, (**e**,**f**) AFRM 5%.

3.2. *Thermophyical Performance Analysis*

3.2.1. Open Porosity, Water Absorption, and Bulk Density

The fluctuations of open porosity, water absorption, and bulk density of composites as a function of AF are presented respectively in Figure 6a,b and Figure 7. It can be seen that the addition of AF leads to an increase in porosity by a factor up to 48.71%, an increase of water absorption by about 74% due to the hydrophilic character of AF, and a decrease of the density by about 14.68% for 5% of AF mass content. This is expected since the AF density (247 kg·m^{-3}) is lower than that of mortar (1748 kg·m^{-3}). These results can also be explained by the formation of voids at the inter-facial areas between AF and solid matrix due to the air bubbles trapping by fibers during the mixing and drying processes.

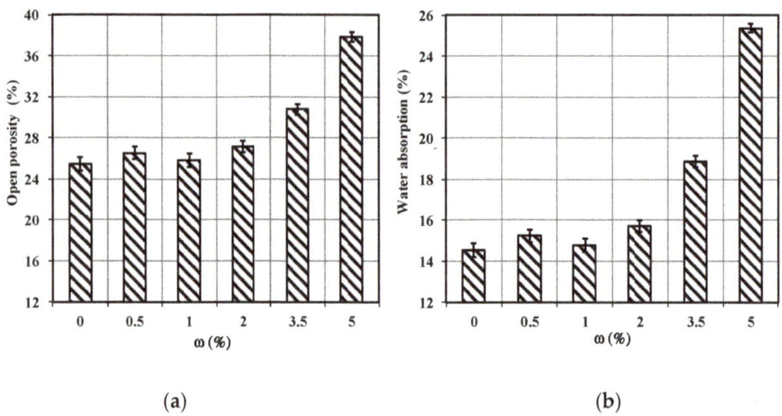

Figure 6. Effect of AF mass content at day 28 on open porosity (**a**) and on water absorption (**b**). Measurements uncertainties are respectively below 0.7% and 0.33%.

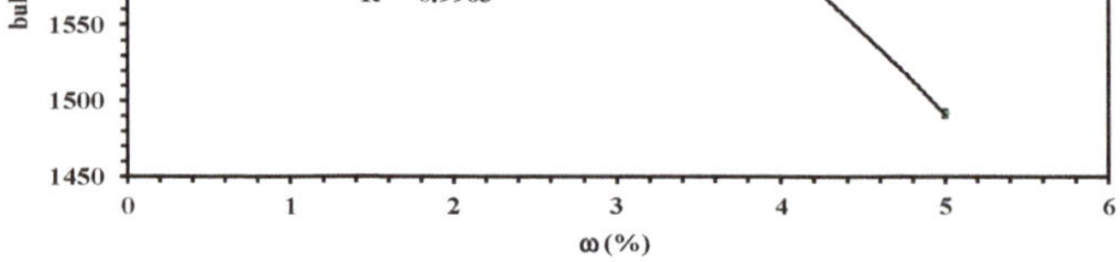

Figure 7. Effect of AF mass ratio on bulk density at day 28. Measurements uncertainty is below 0.2%.

Moreover, porosity and water absorption of AFRM1% slightly decrease respectively by 2.4% and 3% compared to AFRM0.5% while the bulk density remains nearly constant from 0% to 1% of AF inclusion. Furthermore, mortar reinforced with 1% of AF yields to good homogeneity and a better consolidated structure of the composite as indicated in the SEM pictures (Figure 5a).

In order to determine the thermal properties of composites at dry state, all prepared specimens were placed in a drying oven at 105 ± 1 °C with relative humidity of about 30% until mass stabilization within ±2 g for 24.

3.2.2. Thermal Conductivity Measurements

It can be observed from Figure 8 that the thermal conductivity of AFRM composite decreases with increasing fibers weight content, and consequently increases its thermal insulation capacity. The thermal conductivity of the reference mortar of about 0.809 $W{\cdot}m^{-1}{\cdot}K^{-1}$ drops to 0.347 $W{\cdot}m^{-1}{\cdot}K^{-1}$ for AFRM5%, corresponding to an improvement in insulation performance of around 57%. A similar result was reported by [19] who used a single layer fully made of AF with a length of 15 cm as reinforcement at a specific depth inside the cementitious matrix.

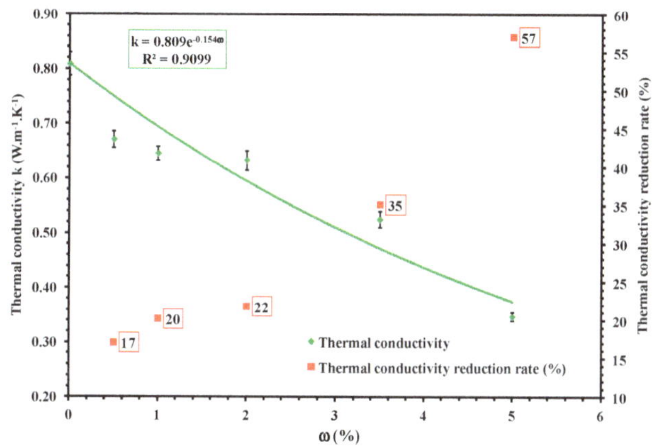

Figure 8. Thermal conductivity (primary axis) and its reduction rate (secondary axis) as function of AF mass content. Measurements uncertainty is below 2.9%.

This behavior is expected and can be explained by two arguments. First, the natural fibers are known to have lower thermal conductivities than those of mortar solid matrixes. The measured bulk thermal conductivity of AF binderless board and mortar are 0.056 ± 0.002 $W{\cdot}m^{-1}{\cdot}K^{-1}$ and 0.809 $W{\cdot}m^{-1}{\cdot}K^{-1}$ respectively. Second, the presence of fibers in solid matrix produces voids as shown in SEM pictures Figure 4e, causing growth in open porosity and a reduction in bulk density of AFRM composite. The latter becomes then more thermally insulating. This phenomenon is always taking place on composite with mineral matrix and vegetable fibers according to several works [41–43].

3.2.3. Intrinsic Thermal Conductivities Determination

The thermal conductivity prediction models shown in Table 2 require some parameters to be first determined. Solid matrix thermal conductivity k_{i-M} of RM is determined by solving a system of two unknowns (k_{i-M} and ε) ([40,44–46]):

$$\varepsilon_{dry}(k_{i-M}) = \frac{k_{\perp}\left(k_{//} - k_{dry}\right)}{k_{dry}\left(k_{//} - k_{\perp}\right)} \quad (15)$$

$$\varepsilon_{sat}(k_{i-M}) = \frac{k_{\perp}\left(k_{//} - k_{sat}\right)}{k_{sat}\left(k_{//} - k_{\perp}\right)} \quad (16)$$

$\varepsilon_{dry}(k_{i-M})$ and $\varepsilon_{sat}(k_{i-M})$ are respectively the values of ε at dry and saturated states of the sample. k_{dry} and k_{sat} are the thermal conductivities of the RM measured respectively at dry and saturated states.

The measured thermal conductivity of saturated RM sample is $k_{sat} = (1.06 \pm 0.02)$ W·m^{-1}·K^{-1}. The determination of k_{i-M} and ε is undertaken by using Newton's method, which consists in determining by successive approaches, in Scilab Software, the value of k_{i-M} for which:

$$\varepsilon_{dry}(k_{i-M}) = \varepsilon_{sat}(k_{i-M}) \tag{17}$$

where $k_{i-M} = 2.58$ W·m^{-1}·K^{-1}

$k_{i-f} = 0.21$ W·m^{-1}·K^{-1} is determined according to Beck's model Equation (9) at dry state of AF binderless board (the measured value is $k_{BB} = (0.056 \pm 0.02)$ W·m^{-1}·K^{-1}).

Total porosities and absolute densities of composites, and absolute density of fiber are given in Table 3.

Table 3. Required parameters for thermal conductivity prediction.

Composites	RM	AFRM 0.5%	AFRM 1%	AFRM 2%	AFRM 3.5%	AFRM 5%
P_t (%)	28.10	28.20	27.46	27.72	30.84	35.86
$\rho_{abs\ RMAF}$(kg·m^{-3})	2433	2422	2411	2389	2357	2326
$\rho_{abs\ AF}$(kg·m^{-3})			1263			

3.2.4. Predicted Thermal Conductivities Comparison

Experimental and theoretical data of thermal conductivity are displayed in Figure 9 as function of AF intrinsic volumetric fraction per absolute volume of sample, φ.

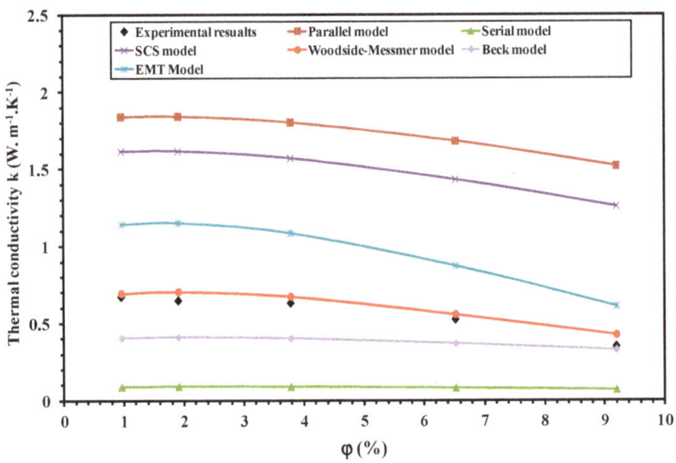

Figure 9. Comparison between measured and calculated thermal conductivities of AFRM composites as function of φ.

At first glance, one can remark that the measured thermal conductivities are between extreme values represented by parallel model (upper limit) and serial model (lower limit). They are respectively away from parallel model by about 221% and serial model by about 84.5%. The Auto-coherent model assumes all composite constituents to have a spherical shape, which is not the case for AF (see Figure 5). That explains the disagreement between the results obtained by this model and the experimental measurements which show a mean relative error of 175%. The results obtained by the effective medium theory (EMT) model are shifted from the experimental ones by about 73%. Beck's model prediction shows smaller difference with the experimental measurements with a mean relative error of 29%, because some alfa fibers have parallel and serial distributions.

The model that better predicts the measured thermal conductivities is Woodside and Messmer's model, with a mean relative error within 9.6%. This can be explained by the fact that the three phases (solid matrix, AF, and air) distribution as observed in SEM images of AFRM composites (Figure 5) is well captured by Woodside and Messmer's model. This finding can help in the future limit the need for additional thermal conductivity measurements if one has sufficient a priori information.

3.2.5. Thermal Diffusivity

The thermal diffusivity variation of AFRM composite as function of mass AF ratio is displayed in Figure 10. The thermal diffusivity is reduced by a factor of about 51% for 5% of AF mass content in the mixture. The heat transfer in the AFRM sample is damped due to the porosity growth within the sample with AF mass content increase, and to the alveolar arrangement of AF which resists the heat diffusion. It is observed from Figure 7 that the density, which was obtained from an independent measurement, decreases with AF content. Calculated specific heat (Figure 11) decreases as well for AF content below 1%. Thus, this decrease in thermal diffusivity is mainly driven by the thermal conductivity.

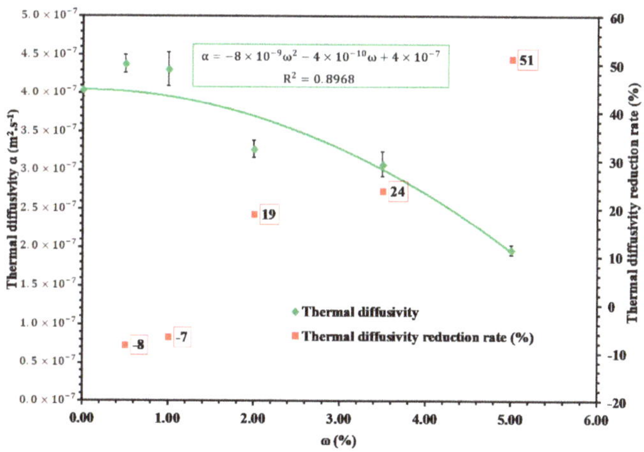

Figure 10. Thermal diffusivity (primary axis) and its reduction rate (secondary axis) as function of AF mass content. Measurements uncertainty is below 5.22%.

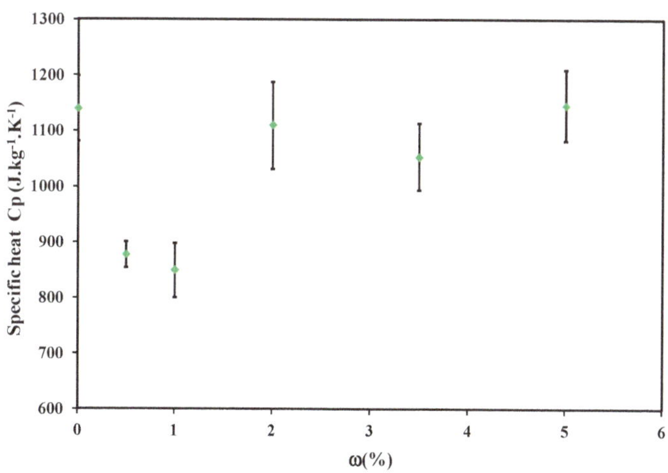

Figure 11. Specific heat variation with alfa fiber mass content. Measurements uncertainty is below 7%.

In addition, the thermal diffusivity of mortar, with no AF, ($4.04 \pm 0.0886\ 10^{-7}\ m^2 \cdot s^{-1}$) is greater by about four times than the measured one of AF ($0.98 \pm 0.04\ 10^{-7}\ m^2 \cdot s^{-1}$), which obviously impacts the thermal diffusivity of the composite material.

3.3. Time-Lag

Time-lag (TL) is defined as the time duration between two associated occurrences such as a cause and its effect. In our case, it is the time interval during which the heat flow peak passes through a flat slab from one side to another. The time-lag was calculated based on measured thermal diffusivity values using the following expression [47]:

$$TL = \frac{1}{2} e \sqrt{\frac{T_0}{\pi\alpha}} \tag{18}$$

where T_0 is the periodic cycle of temperature variation (h) ($T_0 = 24$ h) and e is the thickness of the sample (m).

The calculated time-lag as function of AF mass content is displayed in Figure 12 for samples with a thickness of 4 cm. As expected, one can notice that the time-lag increases with AF mass content increase. It reaches 2.21 h for an AF content of 5%. This means that the reinforcement of mortar with AF enhances the insulation performance of mortar.

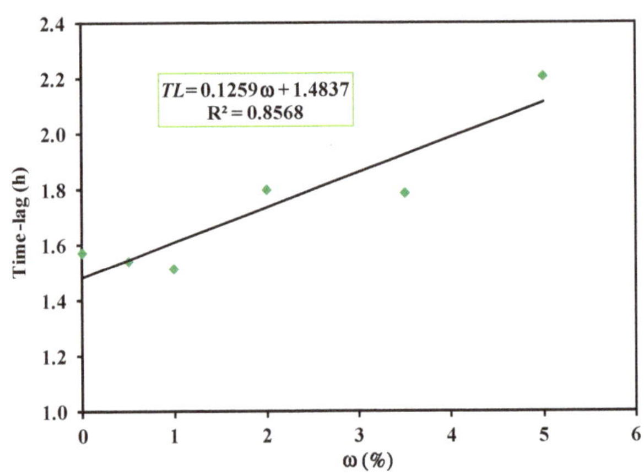

Figure 12. Time-lag variation as function of AF mass content.

Furthermore, Table 4 compares the time-lag of this AF-based composite with 5% to those of different materials used in the building construction sector. All compared materials have the same wall thickness of 20 cm. It can be seen that AFRM5% has a good time-lag. It is greater than that of hemp concrete (9.6 h) followed by cellular concrete (8.6 h). This result indicates that the composite AFRM material developed in this study can reduce outdoor weather conditions' propagation through the building envelope by about 10.37 h for a composite thickness of 20 cm. That is, the midday heat does not reach the indoor building until night and vice-versa for the mid-night cold.

Table 4. Comparison between the time-lag of AFRM 5% material and those of other materials presented in the literature.

Material (e = 20 cm)	ρ (kg·m^{-3})	R (W^{-1}·m^2·K)	α (m^2·s^{-1})	Time-Lag (h)
AFRM	1492	0.58	1.97×10^7	10.37
Hemp concrete	440	1.81	2.27×10^7	9.6
Cellular concrete	600	1.43	2.74×10^7	8.6
Earth block	1850	0.20	5.24×10^7	6.2
Solid brick	1950	0.23	5.41×10^7	6.3
Concrete	2300	0.11	7.93×10^7	5.1

3.4. Mechanical Performance Analysis

Figures 13 and 14 display the flexural and compressive strengths variation of composite with AF mass content.

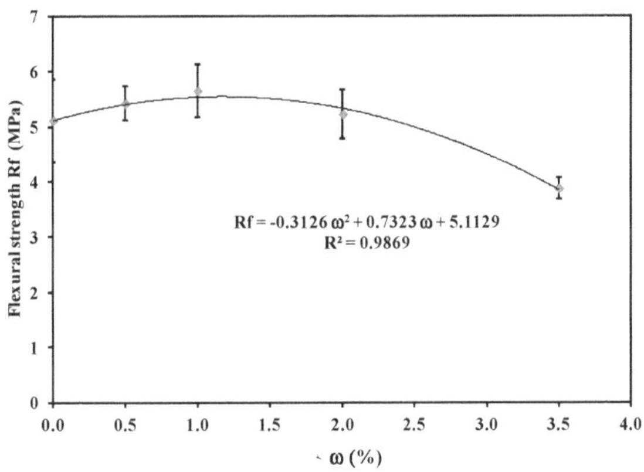

Figure 13. Flexural strength variation as function of AF mass content at day 28. Measurements uncertainty is below 8.5%.

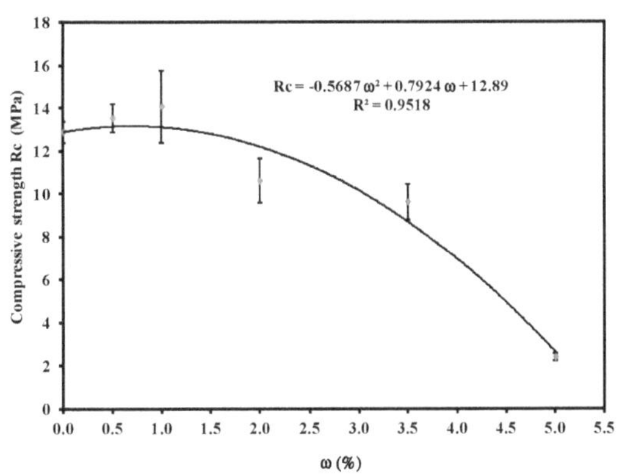

Figure 14. Variation of compressive strength with AF mass content at day 28. Measurements uncertainty is below 12%.

Concerning flexural strength (Figure 13), it is worth highlighting that it increases with the growth of the AF mass ratio from 0% to 1% where a maximum value of about 5.65 MPa is achieved. This improvement in flexural strength of about 10.45% is attributed to the good AF–mortar matrix adhesion as explained in Section 3.2, to the enhancement of mechanical bonds between solid matrix (mortar) and reinforced AF, and to the high mechanical characteristics of AF, including high tensile strength and stiffness. However, for an AF ratio higher than 1%, it decreases to reach 3.87 MPa value for an AF mass ratio of 3.5%. This is due to the composite's porosity increase due to the excessive content of fibers in the mixture, resulting in a reduction of compactness and cohesion within the AF composite. This result is in agreement with previous research that dealt with mortar reinforcement with AF [48,49].

Compressive strength as a function of AF mass content is shown in Figure 14 with a measurement error of less than 12% for all specimens. It can be noticed that the compressive strength behavior is compatible with the flexural strength behavior as the AF mass content is increased. The maximum is obtained at a weight percentage of 1%. This maximum represents an increase of 9% in the compressive strength. Beyond this percentage, it decreases by around 82.5% at a weight fraction of 5%. This diminution is due to the growth in porosity which affects negatively the compressive strength.

3.5. Material Classification of the Developed Composite Based on Compressive Strength and Thermal Conductivity

The lightweight material classification can be made based on the compressive strength (Rc) and thermal conductivity (k). These two quantities are considered as crucial parameters in choosing suitable material types to be used in thermal insulation buildings. The diagram reported in Figure 15 represents a relationship between these two variables for each AF mass percentage. It can be noted that as AF mass content increases, the thermal conductivity of AFRM composite decreases considerably along with compressive strength, except AFRM1% for which Rc reaches a maximum value of about 14.07 MPa.

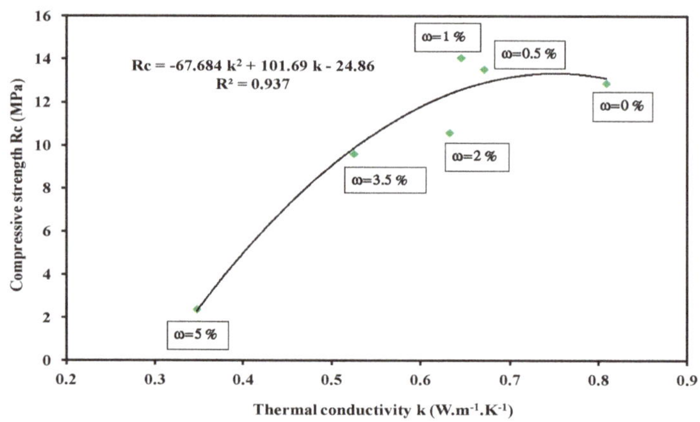

Figure 15. Correlation between compressive strength and thermal conductivity.

Thereafter, based on the functional classification of lightweight concrete of RILEM [50] (International Union of Laboratories and Experts in Construction Materials, Systems and Structures) and the results of Figure 15, this developed composite can be classified in two main classes as a function of AF mass content. Composites with Rc greater than 3.5 MPa and k less than 0.75 $W{\cdot}m^{-1}{\cdot}K^{-1}$ belong to Class II, while those with Rc greater than 0.5 MPa and k less than 0.3 $W{\cdot}m^{-1}{\cdot}K^{-1}$ are in Class III.

Using the transfer functions obtained from Figures 8 and 14, the composites built with a weight fiber percentage less than 4.8% can be classified among structural and thermal insulating materials of Class II while those built with higher fiber percentage can be good thermal insulating candidates and thus belong to Class III.

Finally, it has been shown from the results above that mortar reinforced with AF can be classified as a suitable candidate for exterior walls, partitions, ceiling, roofs and other building structural elements.

4. Conclusions

In this paper, mortar reinforced with alfa fiber for thermal insulating building construction was developed. The physical and thermal-mechanical properties of this new composite were investigated for AF mass ratio ranging from 0% to 5%. The analysis of the SEM pictures demonstrates that alfa fibers are randomly dispersed, separated, and well embedded in the mortar matrix which could indicate good bonds between AF and the surrounding solid matrix.

The addition of AF into mortar makes the new composite lighter by reducing its density by up to 15%, increasing its porosity by around 51% and improving its thermal conductivity by up to 57%. Comparison between the measured thermal conductivity and that predicted by Woodside–Messmer's model shows a good agreement, within 9.6%. This finding can help in the future to limit the need for thermal conductivity measurements if one has sufficient a priori information. Moreover, good heat damping properties were reached as the thermal diffusivity of the mortar-AF composite is decreased by up to 49% compared to the reference material. The incorporation of 0.5% to 1% of AF mass content leads to a better homogeneity and more consolidated structure within the composite. An enhancement of its mechanical properties by about 10.45% for the flexural strength and 9% for the compressive strength was observed. The composite samples with AF mass content less than 4.8% could be classified as structural and thermal insulating materials. Above 4.8%, it could be used as an insulating material.

In conclusion, this AF mortar composite, thanks to its simple manufacturing process, can be a very attractive material to be used in building in the Mediterranean region.

Author Contributions: Experiment planning, S.S.; experiment measurements and data analysis, S.S., L.B. and M.B.; theoretical models, L.B., M.B. and S.S.; writing-original draft preparation, S.S. and L.B.; writing-review and editing, L.B. and B.B. All authors have read and agreed to the published version of the manuscript.

Funding: This research received no external funding.

Conflicts of Interest: The authors declare no conflict of interest.

References

1. Pacheco, T.F.; Jalali, S. Cementitious building materials reinforced with vegetable fibres: A review. *Constr. Build. Mater.* **2011**, *25*, 575–581. [CrossRef]
2. De Carvalho Bello, C.B.; Boem, I.; Cecchi, A.; Gattesco, N.; Oliveira, D.V. Experimental tests for the characterization of sisal fiber reinforced cementitious matrix for strengthening masonry structures. *Constr. Build. Mater.* **2019**, *219*, 44–55. [CrossRef]
3. Ali, M.; Liu, A.; Sou, H.; Chouw, N. Mechanical and dynamic properties of coconut fibre reinforced concrete. *Constr. Build. Mater.* **2012**, *30*, 814–825. [CrossRef]
4. Seng, B.; Magniont, C.; Lorente, S. Characterization of a precast hemp concrete. Part I: Physical and thermal properties. *J. Build. Eng.* **2019**, *24*, 1–46. [CrossRef]
5. Elsaid, A.; Dawood, M.; Seracino, R.; Bobko, C. Mechanical properties of kenaf fiber reinforced concrete. *Constr. Build. Mater.* **2011**, *25*, 1991–2001. [CrossRef]
6. Nadhari, W.N.A.W.; Danish, M.; Nasir, M.S.R.M.; Geng, B.J. Mechanical properties and dimensional stability of particleboard fabricated from steam pre-treated banana trunk waste particles. *J. Build. Eng.* **2019**, *26*, 22–25. [CrossRef]
7. Belakroum, R.; Gherfi, A.; Bouchema, K.; Gharbi, A.; Kerboua, Y.; Kadja, M.; Maalouf, C.; Mai, T.H.; El Wakil, N.; Lachi, M. Hygric buffer and acoustic absorption of new building insulation materials based on date palm fibers. *J. Build. Eng.* **2017**, *12*, 132–139. [CrossRef]
8. Muhtar; Dewi, S.M.; Wisnumurti; Munawir, A. Enhancing bamboo reinforcement using a hose-clamp to increase bond-stress and slip resistance. *J. Build. Eng.* **2019**, *26*, 1–13. [CrossRef]
9. Kesikidou, F.; Stefanidou, M. Natural fiber-reinforced mortars. *J. Build. Eng.* **2019**, *25*, 1–6. [CrossRef]
10. Elhamdouni, Y.; Khabbazi, A.; Benayad, C.; Dadi, A.; Ahmid, O.I. Effect of fiber alfa on thermophysical characteristics of a material based on clay. *Energy Procedia* **2015**, *74*, 718–727. [CrossRef]

11. Rogge, E. Extraction et Etude des Propriétés Physiques et Mécaniques des Fibres d'Alfa (*Esparto grass*) en vue d'Applications Textiles. Master's Thesis, Ecole Nationale Supérieure d'Ingénieurs Sud-Alsace (UHA), Mulhouse, France, 2010.
12. L'alfa. Importance écologique et socio-économique. Portail de l'agriculture marocaine. *Terre Vie* **2002**, *61–62*, 1–3.
13. Couderc, R. Une «cueillette industrielle»: L'alfa en Algérie. *Méditerranée* **2016**, *19*, 3–16. [CrossRef]
14. Dallel, M. Evaluation du Potentiel Textile des Fibres d'Alfa (*Stipa Tenacissima* L.): Caractérisation Physico-Chimique de la Fibre au Fil. Ph.D. Thesis, Haute Alsace University, Mulhouse, France, 2012.
15. Korjenic, A.; Petránek, V.; Zach, J.; Hroudová, J. Development and performance evaluation of natural thermal-insulation materials composed of renewable resources. *J. Build. Eng.* **2011**, *43*, 2518–2523. [CrossRef]
16. Paiva, M.C. Alfa fibres: Mechanical, morphological and interfacial characterization. *Compos. Sci. Technol.* **2007**, *67*, 1132–1138. [CrossRef]
17. Mounir, J.; Slah, W.B.M.; Mohamed, B. Characterization of mechanical extracted alfa fibres. *Int. J. Fiber Text. Res.* **2014**, *4*, 1–4.
18. Krobba, B.; Bouhicha, M.; Zaidi, A.; Lakhdari, M. Formulation of a repair mortar based on dune sand and natural microfibers. *Concr. Solut.* **2014**, 91–95.
19. Jaballi, H.; Miraoui, S.; Hassis, I. Optimization of flexural strength and thermal conductivity of mortar reinforced with alfa fibers. *Appl. Mech. Mater.* **2015**, *799–800*, 794–799. [CrossRef]
20. El Hamdouni, Y.; Khabbazi, A.; Benayad, C.; Dadi, A.; Idriss Ahmid, O. Contribution to the thermal and mechanical behavior of the two materials at the base of clay reinforced by fibers ALFA and of straw fibers. *Res. J. Appl. Sci. Eng. Technol.* **2016**, *12*, 490–497. [CrossRef]
21. Braiek, A.; Karkri, M.; Adili, A.; Ibos, L.; Ben Nasrallah, S. Estimation of the thermophysical properties of date palm fibers/gypsum composite for use as insulating materials in building. *Energy Build.* **2017**, *140*, 268–279. [CrossRef]
22. Taoukil, D.; El Bouardi, A.; Ajzoul, T.; Ezbakhe, H. Effect of the incorporation of wood wool on thermo physical proprieties of sand mortars. *KSCE J. Civ. Eng.* **2012**, *16*, 1003–1010. [CrossRef]
23. Maghchiche, A.; Haouam, A.; Immirzi, B. Extraction and characterization of algerian alfa grass short fibers (stipa tenacissima). *Chem. Chem. Technol.* **2013**, *7*, 339–344. [CrossRef]
24. Kriker, A.; Debicki, G.; Bali, A.; Khenfer, M.M.; Chabannet, M. Mechanical properties of date palm fibres and concrete reinforced with date palm fibres in hot-dry climate. *Cem. Concr. Compos.* **2005**, *27*, 554–564. [CrossRef]
25. IMANOR. *Liants Hydraulique-Ciments Composition, Spécifications et Critères de Conformité*; NM 10.01.004; IMANOR: Rabat, Morocco, 2003.
26. AFNOR. *Essais Pour Déterminer les Caractéristiques Géométriques des Granulats—Partie 8: Évaluation des Fines—Équivalent de Sable*; NF EN 933-8 (NF P 18-622-8); AFNOR: Paris, France, 1999.
27. AFNOR. *Méthodes D'essai des Ciments, Partie I. Détermination des Résistances Mécaniques*; NF EN 196-1; AFNOR: Paris, France, 1995.
28. ASTM International. *Standard Test Methods for Apparent Porosity, Water Absorption, Apparent Specific Gravity, and Bulk Density of Burned Refractory Brick and Shapes by Boiling Water*; ASTM C20-00; ASTM Int.: West Conshohocken, PA, USA, 2000; Volume 3, PA 19428-2959.
29. Menguy, G.; Laurent, M.; Moutarda, A.; Leveau, J. *Cellule de Mesure des Caractéristiques Thermophysiques des Matériaux E1700. Société Weber Broutin*; Bulletin Technique, Deltalab: Carcassonne, France, 1986.
30. Boumhaout, M.; Boukhattem, L.; Hamdi, H.; Benhamou, B. Thermomechanical characterization of a bio-composite building material: Mortar reinforced with date palm fibers mesh. *Constr. Build. Mater.* **2017**, *135*, 241–250. [CrossRef]
31. AFNOR. *Techniques des Essais sur les Liants Hydrauliques*; NM 10.1.005; AFNOR: Paris, France, 1994.
32. Cabaleiro, D.; Nimo, J.; Pastoriza-Gallego, M.J.; Piñeiro, M.M.; Legido, J.L.; Lugo, L. Thermal conductivity of dry anatase and rutile nano-powders and ethylene and propylene glycol-based TiO_2 nanofluids. *J. Chem. Thermodyn.* **2015**, *83*, 67–76. [CrossRef]
33. Beck, J.M.; Beck, A.E. Computing thermal conductivities of rocks from chips and conventional specimens. *J. Geophys. Res.* **1965**, *70*, 5227–5239. [CrossRef]
34. Woodside, W.; Messmer, J.H. Thermal conductivity of porous media. I. Unconsolidated Sands. *J. Appl. Phys.* **1961**, *32*, 1688. [CrossRef]

35. Boutin, C. Conductivité thermique du béton cellulaire autoclavé: Modélisation par méthode autocohérente. *Mater. Struct.* **1996**, *29*, 609–615. [CrossRef]
36. Benazzouk, A.; Douzane, O.; Mezreb, K.; Laidoudi, B.; Quéneudec, M. Thermal conductivity of cement composites containing rubber waste particles: Experimental study and modelling. *Constr. Build. Mater.* **2008**, *22*, 573–579. [CrossRef]
37. Bederina, M.; Marmoret, L.; Mezreb, K.; Khenfer, M.M.; Bali, A.; Quéneudec, M. Effect of the addition of wood shavings on thermal conductivity of sand concretes: Experimental study and modelling. *Constr. Build. Mater.* **2007**, *21*, 662–668. [CrossRef]
38. Mark, C.; Burns, P.J.; Sommerfeld, R.A. Quasi-steady temperature gradient metamorphism in idealized, dry snow. *Numer. Heat Transf. Part. A Appl.* **1994**, *25*, 259–278.
39. Maxwell, D.C. *A Treatise of Electricity and Magnetism*, 3rd ed.; Dover Publications Inc.: New York, NY, USA, 1954; Volume 1–2.
40. Krischer, O. *Die Wissenschaftlichen Grundlagen der Trocknungstechnik*; Springer: Berlin/Heidlberg, Germany, 1963; Volume 1.
41. Dawood, E.T.; Ramli, M. Properties of high-strength flowable mortar reinforced with palm fibers. *Int. Sch. Res. Netw.* **2012**, 1–5. [CrossRef]
42. Belhadj, B.; Bederina, M.; Makhloufi, Z.; Goullieux, A.; Quéneudec, M. Study of the thermal performances of an exterior wall of barley straw sand concrete in an arid environment. *Energy Build.* **2015**, *87*, 166–175. [CrossRef]
43. Taallah, B.; Guettala, A. The mechanical and physical properties of compressed earth block stabilized with lime and filled with untreated and alkali-treated date palm fibers. *Constr. Build. Mater.* **2016**, *104*, 52–62. [CrossRef]
44. Bal, H.; Jannot, Y.; Gaye, S.; Demeurie, F. Measurement and modelisation of the thermal conductivity of a wet composite porous medium: Laterite based bricks with millet waste additive. *Constr. Build. Mater.* **2013**, *41*, 586–593. [CrossRef]
45. Rahim, M.; Douzane, O.; Tran Le, A.D.; Langlet, T. Effect of moisture and temperature on thermal properties of three bio-based materials. *Constr. Build. Mater.* **2016**, *111*, 119–127. [CrossRef]
46. Boukhattem, L.; Boumhaout, M.; Hamdi, H.; Benhamou, B. Moisture content influence on the thermal conductivity of insulating building materials made from date palm fibers mesh. *Constr. Build. Mater.* **2017**, *148*, 811–823. [CrossRef]
47. Chikhi, M. Young's modulus and thermophysical performances of bio-sourced materials based on date palm fibers. *Energy Build.* **2016**, *129*, 589–597. [CrossRef]
48. Krobba, B.; Bouhicha, M.; Kenai, S.; Courard, L. Formulation of low cost eco-repair mortar based on dune sand and Stipa tenacissima microfibers plant. *Constr. Build. Mater.* **2018**, *171*, 950–959. [CrossRef]
49. Miraoui, I.; Jaballi, S.; Hassis, H. Analysis of the mechanical properties of mortar reinforced with long unidirectional alfa fibers in different curing conditions. *Mech. Compos. Mater.* **2016**, *52*, 545–554. [CrossRef]
50. Rilem, L. Functional classification of lightweight concrete. *Mater. Struct.* **1978**, *11*, 281–283.

Publisher's Note: MDPI stays neutral with regard to jurisdictional claims in published maps and institutional affiliations.

applied sciences

MDPI

Article

Modeling the Drying of Capillary-Porous Materials in a Thin Layer: Application to the Estimation of Moisture Content in Thin-Walled Building Blocks

Gennadiy Kolesnikov * and Timmo Gavrilov

Institute of Forestry, Mining and Construction Sciences, Petrozavodsk State University, Lenin pr., 33, 185910 Petrozavodsk, Russia; gtimmo@mail.ru
* Correspondence: kolesnikovgn@ya.ru

Received: 25 August 2020; Accepted: 28 September 2020; Published: 4 October 2020

Abstract: Drying, as a process of changing the moisture content and temperature of capillary-porous materials, is a necessary step in many technologies. When predicting moisture changes, it is necessary to find a balance between the complexity of a model and the accuracy of the simulation results. The purpose of this work was the development of a mathematical model for drying a capillary-porous material with direct consideration of its initial moisture content and drying temperature. Methods of mathematical modeling were used in the work. Using the developed model, an analysis of the features of the drying process of materials with high and low initial moisture content has been carried out. The analytical relationship for determining the time at which the extremum of the drying rate is reached has been substantiated. A model has been developed to directly take into account the influence of the initial material moisture content and drying temperature. The simulation results are consistent with the experiments on drying ceramic blocks for construction which are described in the literature. The obtained results can be taken into account in studies of the effect of drying modes on the energy consumption of a drying process.

Keywords: thin layer drying; moisture; drying rate; sustainable development

---

1. Introduction

Drying is a necessary step in many technological processes for processing capillary-porous organic and inorganic materials. From a physical point of view, drying is a process of heat and mass transfer, the result of which is a decrease in the liquid content of a material. This liquid can be organic solvents, but most often it is water. There are many known drying methods. However, in any case, a significant input of heat and time is required to carry out the drying. With an increase in the drying temperature, the time spent decreases, but the energy consumption for heat production increases. If drying is carried out under natural conditions, then the duration of this process may be unacceptably long. On the other hand, excessively high drying temperatures can lead to a decrease in product quality. In order to reduce the time and energy consumption, methods of drying the material in a thin layer are often used, which has been studied in a large number of works [1–3]. This drying method is discussed in this paper.

Note that in addition to forced drying in artificial conditions, drying in natural conditions is also used [4]. The elements of wooden roofs function in conditions of cyclical effects of natural drying and humidification [5–7], for the diagnosis of which the considered models can be used. The same models can be adapted as a tool for studying the life cycle of building materials [8,9], thermal insulation materials and wall structures [10], as a contribution to achieving the main goal of sustainable development.

The problem is that drying of capillary-porous materials is a complex phenomenon [11–16], but practice requires simple methods to predict and optimize drying time and temperature quite accurately. Tests of real materials and structures are expensive. Therefore, it is recommended to predict the drying process using mathematical models [1,12,13].

Reviews of drying models in a thin layer are given in [17] (analysis of five models published from 1921 to 1978), [14] (review of 14 models), [15] (six models considered), [1] (67 models from 2003 to 2013), [16] (10 models from 2013 to 2019). Models of material drying in a thin layer allow predicting the moisture content of the material depending on the drying time, while not requiring complex calculations. For example [15], a Page model $M_R = \exp(-kt^n)$ with two parameters, in which t—drying time, k and n—empirical constants (parameters), is often used as one of the best in terms of accuracy and analytical simplicity [1,17]. However, this model does not formally reproduce the explicit (separate) effect of the initial moisture content and the drying temperature of the material in the above calculation formula. Initial moisture content and temperature are taken into account indirectly using the parameters k and n. In other words, the influence of the initial moisture content and drying temperature is indistinguishable and aggregated in the parameters k and n along with the influence of the structural features of the capillary-porous medium and other factors. For example, the values k and n in modeling the drying of a capillary-porous material with a thickness of 25 mm, considered in [15] (Table 3), were taken equal to 0.0474 and 1.67 (T = 343 K), respectively; 0.217 and 1.50 (T = 373 K); 0.503 and 1.40 (T = 393 K). Obviously, the decomposition of parameters k and n, i.e., separate consideration of the influence of the initial moisture content of the material and the drying temperature, will expand the predictive capabilities of the model, which will make it possible to better understand the characteristics of drying various materials and increase their competitiveness. The implementation of this predictive capability is investigated in the article using the example of Newton's model.

In logical connection with the above brief analysis, the purpose of this work is determined: to develop a mathematical model for drying a capillary-porous material with direct consideration of its initial moisture content and drying temperature, including indirect consideration of other factors.

2. Materials and Methods

In accordance with the above purpose, the development of the model was carried out using mathematical modeling methods. For the theoretical substantiation of the drying model, an approach was modified, a simplified version of which is known from [18]. Let us consider the basic concepts and definitions necessary for building a model from a methodological point of view.

2.1. Main Concepts and Definitions

Let a capillary-porous material of mass M contain a dry matter of mass M_{dry} and water of mass M_{water}:

$$M = M_{dry} + M_{water} \tag{1}$$

The ratio of M_{dry} and M_{water} depends on the processes of interaction with the environment. Obviously $M_{dry}/M + M_{water}/M = 1$. According to the physical meaning, M_{dry}/M—the concentration of dry matter, M_{water}/M—the concentration of water in the capillary-porous material (by weight).

Based on the fact that the test material with mass M contains water, it is logical to call the ratio M_{water}/M—the moisture content of the material, determined on a wet basis (w.b.). Accordingly, the ratio M_{dry}/M determines the dry matter concentration on the same basis (w.b.). Let us denote:

$$C_{water}^{(w.b.)} = \frac{M_{water}}{M} \tag{2}$$

$$C_{dry}^{(w.b.)} = \frac{M_{dry}}{M} \tag{3}$$

In the literature, dry basis moisture content (or d.b.) $C_{water}^{(d.b.)}$ is often used as dimensionless characteristics of the state of a capillary-porous material. By analogy with [19–21], we write:

$$C_{water}^{(d.b.)} = \frac{M_{water}}{M_{dry}} \tag{4}$$

Characteristics (2), (3) and (4) are interval variables:

$$0 < C_{water}^{(w.b.)} < 1,\ 0 < C_{dry}^{(w.b.)} < 1 \tag{5}$$

$$0 < C_{water}^{(d.b.)} < \infty \tag{6}$$

The relation between $C_{water}^{(w.b.)}$ and $C_{water}^{(d.b.)}$ is found using (2) and (4). For this, taking into account Equation (1), we write the equality $C_{water}^{(w.b.)}M = C_{water}^{(w.b.)}\left(M_{dry} + M_{water}\right)$ and divide both of its parts by M_{dry}. Taking into account (4), we get:

$$C_{water}^{(w.b.)} = \frac{C_{water}^{(d.b.)}}{1 + C_{water}^{(d.b.)}} \tag{7}$$

Equation (7) implies that $C_{water}^{(w.b.)} \le C_{water}^{(d.b.)}$.
Using (7), we find

$$C_{water}^{(d.b.)} = \frac{C_{water}^{(w.b.)}}{1 - C_{water}^{(w.b.)}} \tag{8}$$

Figure 1 illustrates the relationship between $C_{water}^{(w.b.)}$ and $C_{water}^{(d.b.)}$. Note that if the moisture content of the material is low (≤0.2), then $C_{water}^{(w.b.)} \approx C_{water}^{(d.b.)}$.

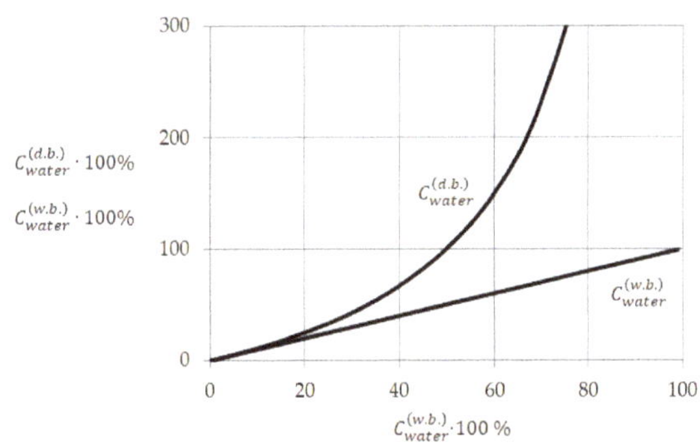

Figure 1. The relationship between moisture values determined on the wet basis and on the dry basis.

Relations (5) and (6), as well as Figure 1 suggest that it is easier to use $C_{water}^{(w.b.)}$ from Equation (2) as an indicator of moisture in a theoretical study.

2.2. Substantiation of the Model

Let us consider a sample of a capillary-porous material of mass M (1). Drying is performed at a temperature T °C. Over time t, due to a decrease in the moisture content of the material, the values M_{water} and M decrease by the same value ΔM_{water}. Then at the moments of time t and $\tilde{t} = t + \Delta t$, the mass of moisture is M_{water} and ($M_{water} - \Delta M_{water}$), respectively; the mass of a sample at the same times is equal to M and $(M - \Delta M_{water})$. At the same moments of time, relative moisture content (2) is determined by relations (9) and (10), respectively:

$$C_{water}^{(w.b.)} = \frac{M_{water}}{M} \tag{9}$$

$$\tilde{C}_{water}^{(w.b.)} = \frac{M_{water} - \Delta M_{water}}{M - \Delta M_{water}} \tag{10}$$

For a sufficiently small value of Δt, we can assume that the value of ΔM_{water} is proportional to Δt. In addition, the value of ΔM_{water} is proportional to the amount of water M_{water}. The total influence of other technological factors will be taken into account by the coefficient τ. Thus:

$$\Delta M_{water} = \frac{\Delta t}{\tau} M_{water} \tag{11}$$

The coefficient τ has the dimension of time; its value is determined experimentally and remains constant, but only within the framework of solving a specific problem.

Let's move on to dimensionless parameters. Let's write:

$$\theta = \frac{t}{\tau} \tag{12}$$

$$\Delta\theta = \frac{\Delta t}{\tau} \tag{13}$$

Then instead of (11) we get

$$\Delta M_{water} = \Delta\theta M_{water} \tag{14}$$

Using (2) and (14), we transform (10) to the form:

$$\tilde{C}_{water}^{(w.b.)} = \frac{C_{water}^{(w.b.)} - \Delta\theta C_{water}^{(w.b.)}}{1 - \Delta\theta C_{water}^{(w.b.)}} \tag{15}$$

Taking into account (9) and (15), we determine the change in relative moisture content over the time interval Δt, that is $\Delta C_{water}^{(w.b.)} = \tilde{C}_{water}^{(w.b.)} - C_{water}^{(w.b.)}$. After transforming this equality, taking into account relations (1) and (4), neglecting the second-order value $\Delta\theta\Delta C_{water}^{(w.b.)} \approx 0$, we obtain:

$$\tilde{C}_{water}^{(w.b.)} = \frac{C_{water}^{(w.b.)} - \Delta\theta C_{water}^{(w.b.)}}{1 - \Delta\theta C_{water}^{(w.b.)}} \tag{16}$$

With $\Delta\theta \rightarrow 0$, instead of (16), we obtain a differential equation, which can be written in the form:

$$\frac{dC_{water}^{(w.b.)}}{C_{water}^{(w.b.)}(1 - C_{water}^{(w.b.)})} = -d\theta \tag{17}$$

Integrating both sides of equality (17), we get $\ln \frac{C_{water}^{(w.b.)}}{C_{water}^{(w.b.)} - 1} = -\theta + A$. The constant of integration A is found from the condition that the initial moisture content of the material $C_{water,start}^{(w.b.)}$ is known, i.e., if θ

= 0, then $C^{(w.b.)}_{water} = C^{(w.b.)}_{water,start}$. After transformations, we will find $e^{\theta} = \frac{C^{(w.b.)}_{water}\left(1-C^{(w.b.)}_{water,start}\right)}{\left(1-C^{(w.b.)}_{water}\right)C^{(w.b)}_{water,start}}$. From here we express $C^{(w.b.)}_{water}$ as follows:

$$C^{(w.b.)}_{water} = \frac{C^{(w.b.)}_{water,start}}{C^{(w.b.)}_{water,start}(1-e^{\theta}) + e^{\theta}} \tag{18}$$

From (18) it follows that the normalized moisture ($C^{(w.b.)}_{water}/C^{(w.b.)}_{water,start}$) directly depends on the initial moisture $C^{(w.b.)}_{water,start}$. Relation (18) can be converted to the form:

$$C^{(w.b.)}_{water} = \frac{e^{-\theta}}{\left(C^{(w.b.)}_{water,start}\right)^{-1} + e^{-\theta} - 1} \tag{18a}$$

From (18a) it follows: if $\theta \to \infty$, then $C^{(w.b.)}_{water} \to 0$.

2.3. Newton's Model

From a practical point of view, it is important to get an answer to two questions:

(1) Do the moisture values calculated by formula (8) differ from the values determined using other models, for example, Newton's model $M_R = e^{-kt}$ [17]?

(2) To what basis (w.b. or d.b.) do the simulation results correspond?

The answer to the second question is especially important because, as shown above, the estimates of moisture content are almost the same only at low material moisture, but the discrepancy between the estimates quickly increases with increasing moisture content (Figure 1).

Regarding the choice of Newton's model, which is also called the Lewis model, we note that this model is the simplest and is often used by researchers [1,16].

To get answers to the questions formulated above in an analytical form, we will perform the following transformations. We transform Equation (8) to the form (19):

$$C^{(d.b.)}_{water} = \frac{C^{(w.b.)}_{water}}{1 - C^{(w.b.)}_{water}} = \frac{1}{\left(C^{(w.b.)}_{water,start}\right)^{-1} - 1} \tag{19}$$

Substitute (18a) into (19). We get after the transformations:

$$C^{(d.b.)}_{water} = \frac{C^{(w.b.)}_{water,start}e^{-\theta}}{1 - C^{(w.b.)}_{water,start}} = \frac{MC^{(w.b.)}_{water,start}e^{-\theta}}{M\left(1 - C^{(w.b.)}_{water,start}\right)} \tag{20}$$

Here $MC^{(w.b.)}_{water,start} = M_{water,start}$ (which follows from (2)). In addition, taking into account Equations (2), (3) and (1), we write: $M\left(1 - C^{(w.b.)}_{water,start}\right) = M - M_{water,start} = M_{dry,start}$. Then, taking into account Equation (4), we get:

$$C^{(d.b.)}_{water} = \frac{M_{water,start}}{M_{dry,start}}e^{-\theta} = C^{(d.b.)}_{water,start}e^{-\theta}. \tag{21}$$

Taking into account relation (12), we come to the conclusion that model (21) coincides with Newton's model [19], which is often written in normalized form [14]:

$$\frac{C^{(d.b.)}_{water}}{C^{(d.b.)}_{water,start}} = e^{-kt}. \tag{22}$$

Here $k = \tau^{-1}$. The value $\tau = k^{-1}$ can be determined using known techniques [17,19].

Normalized moisture (22) does not directly depend on the initial moisture (but the initial water content taken into account indirectly by the coefficient k).

Summarizing, we state that relations (21) and (22) are equivalent to Newton's model and determine the moisture content on a dry basis (d.b.). Models (18) and (18a) determine the moisture content on a wet basis (w.b.). In addition, it should be noted that relations (18) and (18a) explicitly take into account the effect of initial moisture content, which is especially important in predicting the drying time of materials with high initial moisture content. As noted above, explicit consideration of the influence of the initial moisture content of the material, presumably, expands the predictive capabilities of the model. The validity of this assumption is confirmed in the following presentation of the material of the study.

3. Results and Discussion

3.1. *Influence of Initial Material Moisture Content (Wet Basis and Dry Basis)*

Using the above relations (18), (18a), (20), and (21), we perform model calculations, the results of which illustrate the effect of the initial moisture content on the drying process. The simulation results are shown in Figure 2.

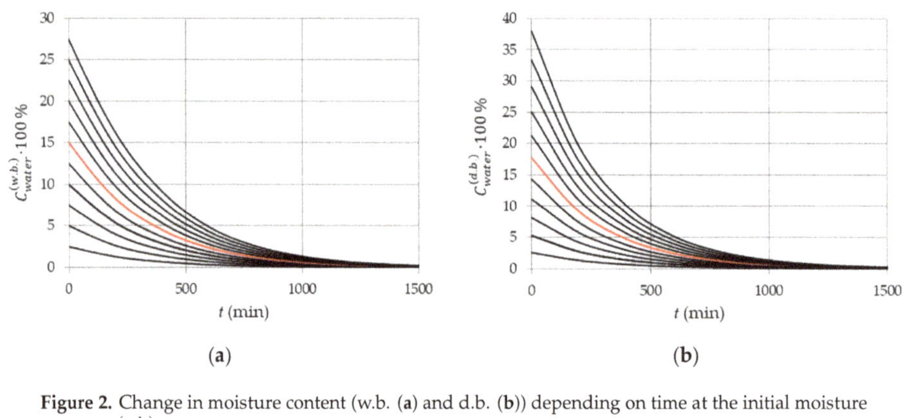

Figure 2. Change in moisture content (w.b. (**a**) and d.b. (**b**)) depending on time at the initial moisture content $C^{(w.b.)}_{water,start} = 0.025, 0.05, \ldots, 0.15, \ldots, 0.275$.

The following initial data were used in the calculations: initial moisture $C^{(w.b.)}_{water,start} = 0.025, \ldots,$ 0.275 (Figure 2a); equivalent initial moisture content $C^{(d.b.)}_{water,start} = 0.026, \ldots, 0.379$, calculated by the Equation (8) (Figure 2b); parameter $\tau = 300$ min (11) in all variants. Calculations are performed according to Equations (18a) (Figure 2a) and (21) (Figure 2b).

The normalized moisture content (Figure 3), depending on the basis (w.b. or d.b.), is defined as the ratio $C^{(w.b.)}_{water}/C^{(w.b.)}_{water,start}$ or $C^{(d.b.)}_{water}/C^{(d.b.)}_{water,start}$, respectively.

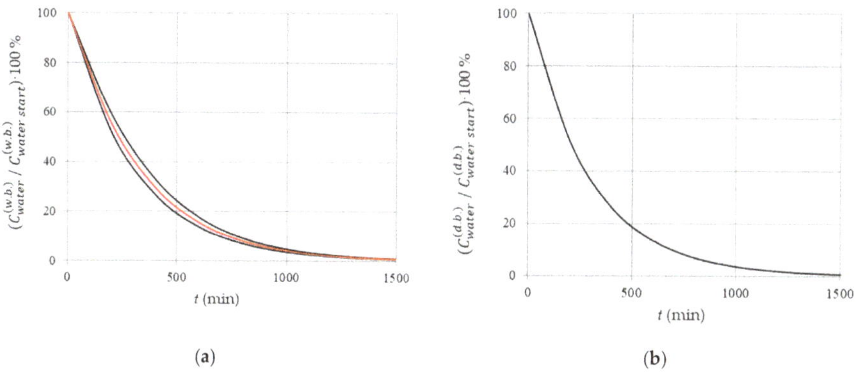

Figure 3. Normalized moisture content (w.b. (**a**) and d.b. (**b**)) depending on time at the initial moisture content by Figure 2. Ratio $C_{water}^{(d.b.)}/C_{water,start}^{(d.b.)}$ is indifferent to initial moisture.

Figure 2 shows that if the initial moisture content is low enough (no more than 0.1), then the predicted moisture content is almost independent of the choice of the basis (w.b. or d.b.). However, with an increase in the initial moisture content, the influence of the choice of the basis also increases.

Figure 3b and Equation (22) show that the normalized initial moisture content (d.b.) in this model does not depend on the initial moisture content, in other words, the model may not be informative enough. At the same time, the normalized initial moisture content (w.b.) determined using Equation (18) reflects the influence of the initial moisture content:

$$\frac{C_{water}^{(w.b.)}}{C_{water,start}^{(w.b.)}} = \frac{1}{C_{water,start}^{(w.b.)}(1 - e^{\theta}) + e^{\theta}} \tag{22a}$$

Continuing the discussion, it is important to pay attention to Figure 1, from which it follows that the influence of the basis (w.b. or d.b.) will increase with increasing initial moisture content. This circumstance, as noted above, is not taken into account in the right-hand side of Equation (22), in contrast to Equation (22a). Therefore, the use of normalized moisture content (w.b.) (22a) may be more appropriate when simulating the drying of materials with high initial moisture content. To check this assumption, we perform calculations at a sufficiently high initial moisture content of the material (0.70, 0.725, … , 0.95 (w.b.), which is equivalent to the initial moisture interval (2.33, … , 19.00 (d.b.)). The simulation results are shown in Figures 4 and 5. Calculations were performed using formulas (18a) (Figure 4a), (21) (Figure 4b), (22a) (Figure 5a), and (22) (Figure 5b).

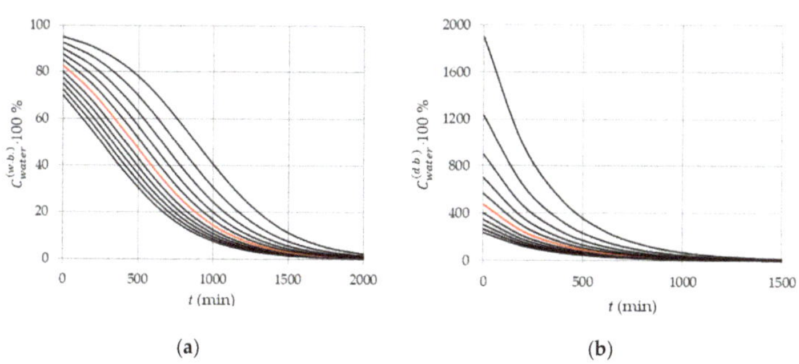

Figure 4. Change in moisture content (w.b. (**a**) and d.b. (**b**)) depending on time at the initial moisture content 0.70, 0.725, … , 0.825, … , 0.95 (w.b.).

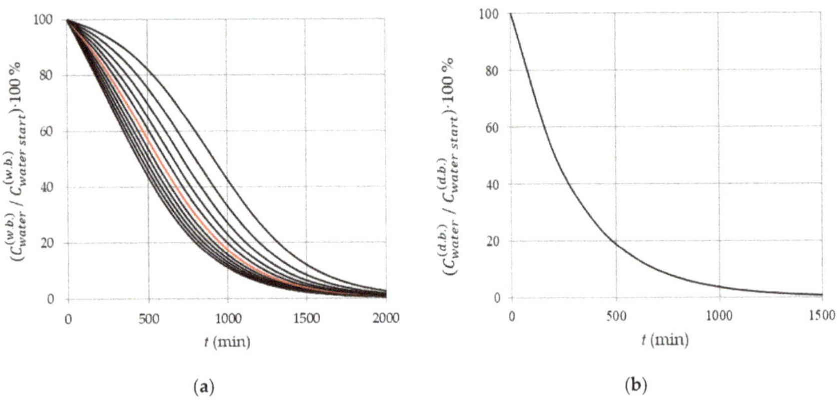

Figure 5. Normalized moisture content (w.b. (**a**) and d.b. (**b**)) depending on time at the initial moisture content (w.b.) by Figure 4. The ratio $C^{(d.b.)}_{water}/C^{(d.b.)}_{water,start}$ is independent of the initial moisture content (22).

Analysis of other features of the curves in Figures 4 and 5 is performed in Section 3.2.

3.2. Inflection Point on the Drying Curve and the Rate of the Drying Process

Figures 4 and 5 show that as the initial moisture content of the material increases, an inflection point appears on the curve in Figures 4a and 5a, in contrast to the curve in Figures 4b and 5b. Formally, this means that at the inflection point, the second derivative of the moisture function with respect to time is zero. Let t^* be the abscissa of the inflection point; P—the right side of relation (22a). Then from the equation $\frac{d^2P}{dt^2} = 0$, we find after transformations:

$$t^* = \tau \ln \frac{C^{(w.b.)}_{water}}{1 - C^{(w.b.)}_{water}} \tag{22b}$$

In the problem under consideration $t^* \geq 0$, $\tau > 0$. The physical meaning of the problem corresponds to such values $C^{(w.b.)}_{water}$, for which $\frac{C^{(w.b.)}_{water}}{1-C^{(w.b.)}_{water}} \geq 1$, i.e., $0.5 \leq C^{(w.b.)}_{water} < 1$. For example, if $C^{(w.b.)}_{water} = 0.85$ and $\tau = 300$ min, then $t^* = 520.4$ min.

The question then arises: What features of the drying process is simulated by the inflection point?

Answer: If an inflection point exists (22b), then the drying process rate increases up to the inflection point $0 < t < t^*$. At a point $t = t^*$, the rate of the drying process reaches an extreme. If $t > t^*$, then the rate of the drying process decreases. If $T^* = 0$, then the drying rate only decreases from start to finish of this technological process.

Figure 6 illustrates the noted features of the drying process for material with high initial moisture content.

The initial moisture content $C^{(w.b.)}_{water,start}$ of the material and the parameter τ are equal to 0.85 and 300 min, respectively. Point 1 is the inflection point on the curve simulating the dependence of on time. At point 3, as noted above, $\frac{d^2P}{dt^2} = 0$, $t = t^*$ 520.4 min. At point 2, the drying process rate is extreme at the same value of t.

Experimental data that confirm the existence of the above features of the drying process can be found in the literature, for example, in the graphs in Figure 2c from [20]. However, it was not possible to find a theoretical justification for these features, including analogs of relation (22b).

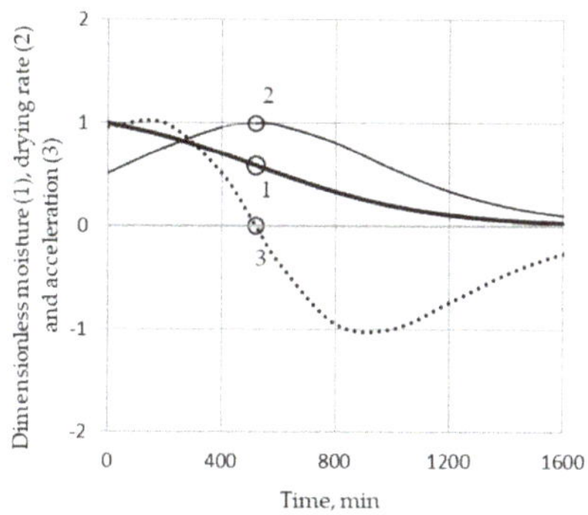

Figure 6. Normalized dimensionless characteristics: moisture content (w.b.), drying rate module and drying process acceleration.

3.3. *Influence of Drying Temperature*

The dependence of moisture content on drying temperature can be taken into account indirectly, for example, by changing the coefficient k in Newton's model. This issue was studied in more detail, for example, in [19], where, in particular, for modeling the drying of one of the materials at temperatures of 35, 45 and 55 °C, the values of coefficient k 0.34, 0.049 and 0.016, respectively, were obtained. A similar approach is used in [15]. Thus, the coefficient k summarizes the effect of temperature and other technological features of drying. Obviously, it is important for practice to know the influence of each factor on the drying process [1,22–25]. However, this is a difficult task. Let us consider a simple model in which the drying temperature and initial moisture content are separately taken into account; the influence of other factors is modeled in total.

Let's use the results presented above to study one of the possible approaches to model building, taking into account the drying temperature in a thin layer.

Restricting ourselves to the option of drying at a positive temperature, we can assume that in relation (11), the value of ΔM_{water} is proportional to the amount of water M_{water} and the drying temperature T °C (in this work, the drying temperature is assumed to be a constant value; to take into account the temperature, a dimensionless coefficient $\phi = T$ °C$/T_{ref}$ °C is used, where T_{ref} is the reference temperature equal to 100 °C. Thus, instead of (11), we can write:

$$\Delta M_{water} = \frac{\Delta t}{\tau} M_{water}\phi \tag{23}$$

For substantiating relation (23), one can additionally refer to the work [26], (p. 11), according to which the change in the moisture content of the material is directly proportional to the temperature of the air for drying.

Regarding the coefficient ϕ, note that the value of T_{ref} = 100 °C is chosen to obtain the coefficient in dimensionless form. In the particular case under consideration, the choice of T_{ref} is not critical, i.e., another suitable value may be used, since possible deviations will be compensated by the above parameter τ. However, all temperature values in the considered model are assumed to be positive.

Using (23), following the logic of obtaining relations (14)–(18a), we write down relations (24)–(28):

$$\Delta M_{water} = \Delta\theta M_{water}\phi \tag{24}$$

$$\tilde{C}^{(w.b.)}_{water} = \frac{C^{(w.b.)}_{water} - \Delta\theta\phi C^{(w.b.)}_{water}}{1 - \Delta\theta\phi C^{(w.b.)}_{water}} \tag{25}$$

$$\Delta C_{water}^{(w.b.)} = \Delta\theta\phi C_{water}^{(w.b.)}\left(1 - C_{water}^{(w.b.)}\right) \tag{26}$$

$$\frac{dC_{water}^{(w.b.)}}{C_{water}^{(w.b.)}(1 - C_{water}^{(w.b.)})} = -d\phi\theta \tag{27}$$

$$C_{water}^{(w.b.)} = \frac{C_{water,start}^{(w.b.)}}{C_{water,start}^{(w.b.)}(1 - e^{\theta\phi}) + e^{\theta\phi}} \tag{28}$$

By analogy with (18a), we transform (28) to the form (28a):

$$C_{water}^{(w.b.)} = \frac{e^{-\theta\phi}}{\left(C_{water,start}^{(w.b.)}\right)^{-1} + e^{-\theta\phi} - 1} \tag{28a}$$

Using (28a), we find the moisture content $C_{water}^{(d.b.)}$ by using Equation (8).

By analogy with (22), let us determine the normalized values of moisture content in the form of ratios $C_{water}^{(w.b.)}/C_{water,start}^{(w.b.)}$, $C_{water}^{(d.b.)}/C_{water,start}^{(d.b.)}$.

Note that the models considered in this paper belong to the class of models of drying in a thin layer [1]. In models of drying a material in a thin layer, it is assumed that the temperature and moisture content of the material almost does not change over its thickness. If these changes are significant, then more complex models based on differential equations in partial derivatives are used [8,13,22–25] and numerical modeling [26,27]. To get an idea of the adequacy of model (28) from a physical point of view, it is necessary to compare it with experimental data.

3.4. Comparison with the Experimental Data on Drying Ceramic Blocks for Construction, Known from the Literature

We will use the results of an experimental study of drying ceramic blocks for construction, known from the literature [26]. In this case, the use of models for drying a capillary-porous material in a thin layer is acceptable due to the fact that the thickness of the walls of the blocks is rather small (0.63, ... , 0.94 cm) [26,27].

The initial moisture content of the material is ~0.17 (d.b.) (Table 1 in [26]), which according to (7) corresponds to $C_{water}^{(w.b.)}$ = 0.17/(1 + 0.17) ≈ 0.15. In the cited work [26], experimental data on the change in the moisture content of the same material over time are given in graphical form for a number of experiments in which the drying temperature was 50, 60, 70, 80, 90, and 100 °C. The simulation results obtained using relations (24)–(28a) are shown in Figures 7 and 8. $\tau = 140$min (see Section 3.4). The consistency of the experimental and calculated data is illustrated in Figure 9.

Due to the fact that the data presented in Figure 8 confirm the adequacy of the model and the reliability of the simulation results. The characteristic of the process rate can be the slope ratio of the tangent to the curves in Figures 7 and 8. If the drying temperature increases, then the rate of this process also increases, and the time to completion of the process decreases, which corresponds to the data known from the literature.

Figure 7 and calculations by Equation (28a) show that when drying to a moisture content of 0.04 (w.b.) at a temperature of 50, 60, ... , 100 °C, each successive temperature step (10 °C) corresponds to decreasing time intervals. This feature can be explained by the fact that the relative value of the temperature increment decreases with each step. In other words, an increase in temperature from 50 to 60 °C means an increase in temperature by 10/50 = 20%; from 60 to 70 °C—by 10/60 ≈ 17%; from 90 to 100 °C—by 10/90 ≈ 11%. These results do not contradict the experimental studies known from the literature [26], according to which the rate of change in the moisture content of the material is directly proportional to the drying temperature.

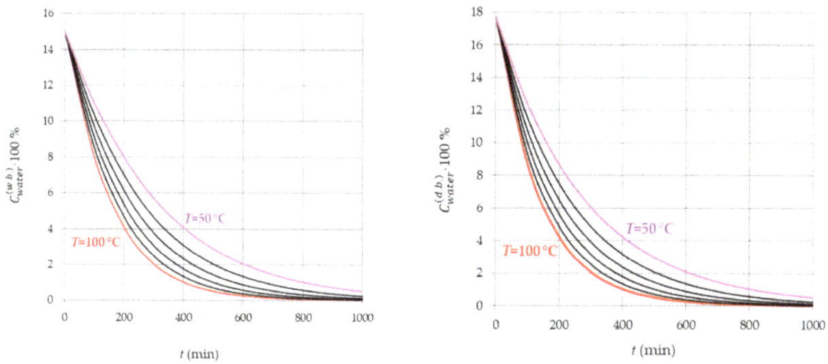

(a) (b)

Figure 7. Dependence of moisture content (w.b. (**a**) and d.b. (**b**)) on time and drying temperature 50, 60, . . . , 100 °C at the initial material moisture of 0.15 (w.b.).

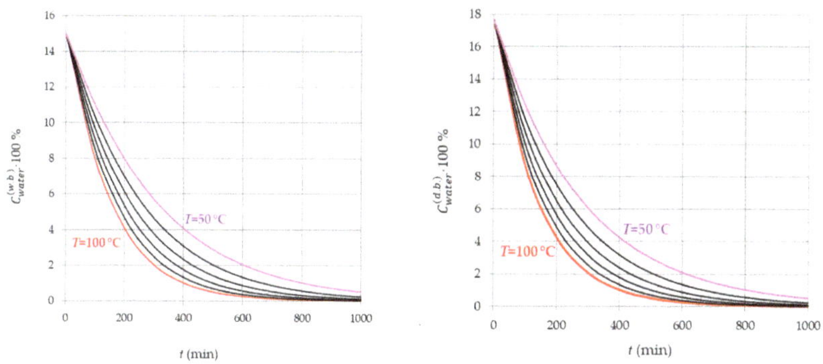

(a) (b)

Figure 8. Normalized moisture content (w.b. (**a**) and d.b. (**b**)) depending on the time and drying temperature 50, 60, . . . , 100 °C at the initial material moisture of 0.15 (w.b.).

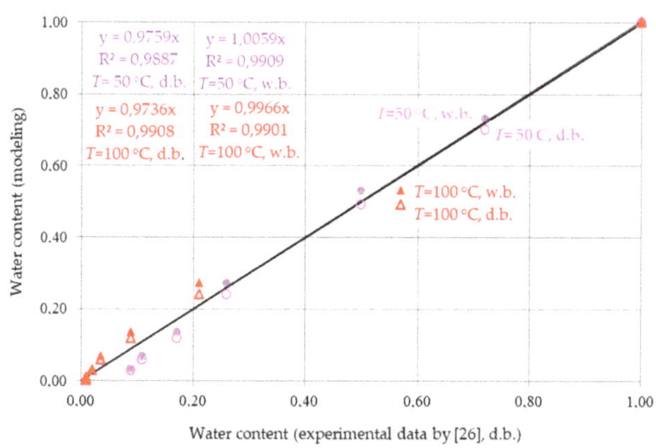

Figure 9. Comparison of experimental and calculated values of normalized moisture content (w.b. and d.b.). Drying temperature: 50 and 100 °C. Initial material moisture: 0.15 (w.b.) ≈ 0.17 (d.b.). The experimental data were obtained after processing the plots according to Figure 8 in paper [26].

3.5. Analysis of the Results: Methodological Aspects

The results presented above allow us to answer a question that is important from a practical point of view: what is the physical meaning of the coefficient τ in Equations (11), (12), (22), etc.? Also, it is methodologically important to know how to calculate the coefficient τ? To search for answers, we will use relations (22).

Let's write: $C_{water}^{(d.b.)}(t) = C_{water.,start}^{(d.b.)} e^{\frac{t}{\tau}}$. Let $\tau = t$. Then $C_{water}^{(d.b.)} = C_{water,start}^{(d.b.)} e^{-1} \approx C_{water,start}^{(d.b.)} / 2.718$ Consequently, the value τ is equal to the drying time t at which the moisture content decreases by a factor $e \approx 2.718$.

To calculate the coefficient τ, it is necessary to carry out a test drying and build a plot of the dependence $C_{water}^{(d.b.)}(t)$. The point on the plot for which $C_{water}^{(d.b.)} \approx C_{water,start}^{(d.b.)} / 2.718$ corresponds to $t = \tau$.

If we use formula (28a), then the coefficient τ is calculated in the same way, but in this case the test drying is carried out only at a temperature of $T = T_{ref}$, i.e., for $\phi = 1$ in Equation (23). This is how the value $\tau = 140$ min was found for the above modeling results (Figures 7 and 8).

The presented method for determining coefficient τ is recommended for calculations on wet and dry basis. The transition from one basis to another is carried out according to Equation (7).

With regard to calculations on a wet and dry basis, it is methodologically important to note the following. Figures 2 and 3 show that calculations results on a wet and dry basis are practically equivalent at low moisture. However, as moisture increases, the equivalence of the calculated data on a wet and dry basis is lost (Figures 4 and 5). Taking this circumstance into account, it is possible to formulate restrictions on the field of application of the calculation formulas, for which, however, further experimental and theoretical studies are required. From a methodological point of view, it is important to note that in this work a phenomenological approach was used to modeling changes in the moisture content of a capillary-porous material in a thin layer. This approach is used in applied sciences, for example, [28,29]. As is known, phenomenological models take into account only the observed (external) properties of objects and do not take into account the internal mechanisms of phenomena, for example, the change in the moisture content of a capillary-porous material is studied without detailed analysis of moisture transfer. A detailed consideration of the transfer of mass and heat would lead to a model that belongs to a different class [8,25]. Analysis of models that do not belong to the class of phenomenological models is beyond the scope of our work.

4. Conclusions

(1) A mathematical model has been developed for thin-layer drying of a capillary-porous material with direct consideration of its initial moisture content and drying temperature, along with indirect consideration of the influence of other factors.

(2) Using the developed model, the analysis of the features of the drying process of materials with high and low initial moisture content has been carried out. The dependence of the ratios of the normalized moisture content on the choice of the basis (w.b. or d.b.) has been proved.

(3) It is shown that the function of normalized moisture content (w.b.) directly depends on the initial moisture content and, as a consequence, is more informative than the function of normalized moisture content (d.b.), in which the initial moisture content is indirectly and indiscernibly taken into account in combination with the temperature and other technological factors of drying. It was found that if the initial moisture content (w.b.) is in the range (0.5; 1.0), then the rate of the drying process reaches an extreme in this range. An analytical relationship for determining the time at which the drying rate is extreme has been substantiated.

(4) The developed mathematical model of thin-layer drying makes it possible to predict changes in the moisture content of the material and the duration of its drying, depending on the initial moisture content of the material and the drying temperature. Thus, the number of indirectly taken into account factors has been reduced, which, accordingly, increases the predictive capabilities of

the model when justifying recommendations for improving drying technologies in the interests of sustainable development.

(5) The adequacy of the model and the assessment of the reliability of the results of model calculations are confirmed by their agreement with the experimental data related to the drying of ceramic blocks for construction, known from the literature.

Author Contributions: Conceptualization, G.K. and T.G.; methodology, G.K.; validation, G.K. and T.G.; formal analysis, G.K.; investigation, G.K.; writing—original draft preparation, T.G.; writing—review and editing, G.K.; project administration, G.K. All authors have read and agreed to the published version of the manuscript.

Funding: This research received no external funding.

Conflicts of Interest: The authors declare no conflict of interest.

References

1. Kucuk, H.; Midilli, A.; Kilic, A.; Dincer, I. A Review on Thin-Layer Drying-Curve Equations. *Dry. Technol.* **2014**, *32*, 757–773. [CrossRef]
2. Jayas, D.S.; Cenkowski, S.; Pabis, S.; Muir, W.E. Review of thin-layer drying and wetting equations. *Dry. Technol.* **1991**, *9*, 551–588. [CrossRef]
3. Ostanek, J.; Ileleji, K. Conjugate heat and mass transfer model for predicting thin-layer drying uniformity in a compact, crossflow dehydrator. *Dry. Technol.* **2020**, *38*, 775–792. [CrossRef]
4. Visser, R.; Berkett, H.; Spinelli, R. Determining the effect of storage conditions on the natural drying of radiata pine logs for energy use. *N. Z. J. Sci.* **2014**, *44*. [CrossRef]
5. Asdrubali, F.; Ferracuti, B.; Lombardi, L.; Guattari, C.; Evangelisti, L.; Grazieschi, G. A review of structural, thermo-physical, acoustical, and environmental properties of wooden materials for building applications. *Build. Environ.* **2017**, *114*, 307–332. [CrossRef]
6. Sadłowska-Sałęga, A.; Wąs, K. Risk of Moisture in Diffusionally Open Roofs with Cross-Laminated Timber for Northern Coastal Climates. *Buildings* **2020**, *10*, 10. [CrossRef]
7. Teaca, C.A.; Roşu, D.; Mustaţă, F.; Rusu, T.; Roşu, L.; Roşca, I.; Varganici, C.D. Natural Bio-Based Products for Wood Coating and Protection against Degradation: A Review. *Bio. Resour.* **2019**, *14*, 4873–4901. [CrossRef]
8. Pecenko, R.; Challamel, N.; Colinart, T.; Picandet, V. Semi-analytical solution of Luikov equations for time-periodic boundary conditions. *Int. J. Heat Mass. Transfer.* **2018**, *124*, 533–542. [CrossRef]
9. Silva, W.P.; Silva, C.M.D.P.S.; Silva, L.D.; Farias, V.S.O. Drying of clay slabs: Experimental determination and prediction by two-dimensional diffusion models. *Ceram. Int.* **2013**, *39*, 7911–7919. [CrossRef]
10. Mitterpach, J.; Igaz, R.; Štefko, J. Environmental evaluation of alternative wood-based external wall assembly. *Acta Fac. Xylol. Zvolen Res. Publica Slovaca* **2020**, *62*, 133–149. [CrossRef]
11. Huang, S.; He, Q.; Miao, Z.; Wan, K.; Wan, Y. Multiphysics modeling of water transport in high-intensity lignite drying process on pore scale. *Energy Sources Part A Recovery Util. Environ. Effects* **2018**, *40*, 2580–2589. [CrossRef]
12. Golisz, E.; Jaros, M.; Głowacki, S. Modelling of biomass temperature in the drying process. *E3S Web Conf.* **2020**, *154*, 1004. [CrossRef]
13. Defraeye, T. Advanced Computational Modelling for Drying Processes—A Review. *Appl. Energy* **2014**, *131*, 323–344. [CrossRef]
14. Agbossou, K.; Napo, K.; Chakraverty, S. Mathematical Modelling and Solar Tunnel Drying Characteristics of Yellow Maize. *Am. J. Food Sci. Technol.* **2016**, *4*, 115–124. [CrossRef]
15. Cuevas, M.; Martínez-Cartas, M.L.; Pérez-Villarejo, L.; Hernández, L.; García-Martín, J.F.; Sánchez, S. Drying kinetics and effective water diffusivities in olive stone and olive-tree pruning. *Renew. Energy* **2019**, *132*, 911–920. [CrossRef]
16. Soodmand-Moghaddam, S.; Sharifi, M.; Zareiforoush, H.; Mobli, H. Mathematical modeling of lemon verbena leaves drying in a continuous flow dryer equipped with a solar pre-heating system. *Qual. Assur. Saf. Crop. Foods* **2020**, *12*, 57–66. [CrossRef]
17. Górnicki, K.; Kaleta, A.; Choińska, A. Suitable model for thin-layer drying of root vegetables and onion. *Int. Agrophys.* **2020**, *1*, 79–86. [CrossRef]

18. Kantyshev, A.V.; Zaitseva, M.I.; Kolesnikov, G.N. Model of wood impregnation after incomplete drying as an additional energy management tool. *J. Phys. Conf. Ser.* **2019**, *1333*, 032033. [CrossRef]
19. Ivanova, M.; Katrandzhiev, N.; Dospatliev, L. Using some mathematical models in modeling mushroom drying (agaricus bisporus). *Int. J. Appl. Math.* **2020**, *33*, 109–124. [CrossRef]
20. Nigay, N.A.; Kuznetsov, G.V.; Syrodoy, S.V.; Gutareva, N.Y. Estimation of energy consumption for drying of forest combustible materials during their preparation for incineration in the furnaces of steam and hot water boilers. *Energy Sources Part A Recovery Util. Environ. Effects* **2020**, *42*, 1997–2005. [CrossRef]
21. Oosawa, K.; Kanematsu, Y.; Kikuchi, Y. Forestry and Wood Industry. In *Energy Technology Roadmaps of Japan*; Kato, Y., Koyama, M., Fukushima, Y., Nakagaki, T., Eds.; Springer: Tokyo, Japan, 2016. [CrossRef]
22. Dong, W.; Chen, Y.; Bao, Y.; Fang, A. A validation of dynamic hygrothermal model with coupled heat and moisture transfer in porous building materials and envelopes. *J. Build. Eng.* **2020**, *101484*. [CrossRef]
23. Caccavale, P.; De Bonis, M.V.; Ruocco, G. Conjugate Heat and Mass Transfer in Drying: A Modeling Review. *J. Food Eng.* **2016**, *176*, 28–35. [CrossRef]
24. Koukouch, A.; Bakhattar, I.; Asbik, M.; Idlimam, A.; Zeghmati, B.; Aharoune, A. Analytical solution of coupled heat and mass transfer equations during convective drying of biomass: Experimental validation. *Heat Mass Transf.* **2020**, *56*, 1971–1983. [CrossRef]
25. Luikov, A.V. Systems of differential equations of heat and mass transfer in capillary-porous bodies. *Int. J. Heat Mass Transf.* **1975**, *18*, 1–14. [CrossRef]
26. Da Silva, A.M.V.; Delgado, J.M.P.Q.; Guimarães, A.S.; de Lima, W.M.P.B.; Gomez, R.S.; de Farias, R.P.; de Lima, E.S.; de Lima, A.G.B. Industrial Ceramic Blocks for Buildings: Clay Characterization and Drying Experimental Study. *Energies* **2020**, *13*, 2834. [CrossRef]
27. De Vasconcellos, M.A.; de Brito Correia, B.R.; Brandão, V.A.A.; de Oliveira, I.R.; Santos, R.S.; de Oliveira Neto, G.L.; de Lucena Silva, L.P.; de Lima, A.G.B. Convective Drying of Ceramic Bricks by CFD: Transport Phenomena and Process Parameters Analysis. *Energies* **2020**, *13*, 2073. [CrossRef]
28. Merlin, O.; Olivera-Guerra, L.; Hssaine, B.A.; Amazirh, A.; Rafi, Z.; Ezzahar, J.; Gentinec, P.; Khabbab, S.; Gascoina, S.; Er-Raki, S. A phenomenological model of soil evaporative efficiency using surface soil moisture and temperature data. *Agric. For. Meteorol.* **2018**, *256*, 501–515. [CrossRef]
29. Bisson, A.; Rigacci, A.; Lecomte, D.; Rodier, E.; Achard, P. Drying of silica gels to obtain aerogels: Phenomenology and basic techniques. *Dry. Technol.* **2003**, *21*, 593–628. [CrossRef]

applied sciences

MDPI

Article

Impact of Alcohol on Occupational Health and Safety in the Construction Industry at Workplaces with Scaffoldings

Marek Sawicki and Mariusz Szóstak *

Department of Building Engineering, Faculty of Civil Engineering, Wroclaw University of Science and Technology, 50-370 Wrocław, Poland; marek.sawicki@pwr.edu.pl

* Correspondence: mariusz.szostak@pwr.edu.pl

Received: 25 August 2020; Accepted: 22 September 2020; Published: 24 September 2020

Abstract: The value, care, and customs of workers are essential in terms of occupational health and safety. The abuse of alcohol is widely regarded as a serious threat to the lives, health, and safety of employees. The aim of the research was to identify the main problems that are associated with alcohol abuse and consumption at work among employees in the construction industry, with particular emphasis on workstations where work is carried out on construction scaffoldings. Data for the analysis were obtained from two different sources. The first one was post-accident documentation on occupational accidents. The second one was surveys collected during the research project. This study confirmed that excessive and disproportionate alcohol consumption can be the cause of an accident, and consequently death at workplaces with scaffolding. Of 219 accident reports, 17.4% indicated alcohol as a contributing factor. Analysis of accident documentations shows that in cases where alcohol was indicated as a contributing factor in an accident, the alcohol was consumed during the workday. The results obtained on the basis of the conducted research were able to constitute a justification for the directions of preventive actions carried out in order to reduce the number of occupational accidents in the construction industry caused by alcohol.

Keywords: health and safety; workplace; construction industry; scaffoldings; alcohol

1. Introduction

Alcohol abuse is widely considered a serious threat to the life, health, and safety of employees and bystanders [1]. The harmful use of alcohol is one of the leading risk factors for population health and cause of death worldwide [2]. According to a World Health Organization report, alcohol is in the third place among the risk factors for human health, and over 60 types of diseases and injuries are associated with it [3,4]. Alcohol consumption disorder (alcohol addiction) is a serious illness with far-reaching consequences for life and health [5].

It is well known that use of drugs (alcohol, cigarettes, psychoactive substances) influences the occurrence of an accident situation in various branches of the economy, as well as in the construction industry [6]. According to scientific research on alcohol abuse, construction workers are at the forefront of professions in which the percentage of alcoholics is significant [7,8].

A construction worker is a typically male teamwork profession that requires physical effort [9], is performed outdoors, and is often carried out in difficult weather conditions [10]. These conditions are conducive to drinking alcohol for relaxation [11]. Drinking alcohol is an accepted social and cultural habit in most Western countries [12]. An additional factor contributing to alcohol abuse among construction workers is the mental tension caused by working in the construction industry, which involves stress, time pressure, and superiors, as well as depression [13]. Alcohol abuse also has negative social consequences [14]. Alcohol consumption has a negative impact on human health [15].

Moreover, accidents caused by alcohol consumption translate into economic consequences [16], which cause, among other factors, a decline in the well-being of societies [17], as well as an increase in the direct cost of healthcare [18].

The phenomenon of alcohol abuse and consumption at work is a common problem in many sectors of the economy, especially in the construction industry [19]. Many construction workers have alcohol-related problems [20,21]. This problem also applies to workers at workplaces with scaffoldings [22]. Alcohol increases the risk of accident situations. Moreover, the most common events that cause accidents to workers on scaffolding are falls from height caused by reduced concentration, among other factors [23].

Statistical data published by the Statistical Office of the European Union (Eurostat) are the basis for the statement that the serious problem of alcohol consumption exists [24]. Only 16.4% of men in 2014 declared that they had "never or not in the last 12 months" consumed alcohol. This value is the frequency of alcohol consumption for 27 countries of the European Union (EU27). On the other hand, consumption of alcohol "every day" and "every week" were declared by 14.7% and 35.5%, respectively, of men in the 27 countries of the EU. The highest daily frequency of alcohol consumption was in Portugal (38.6%), and the highest weekly frequency in the United Kingdom (51.6%).

Furthermore, statistical data published by the Central Statistical Office of Poland [25] regarding the average annual alcohol consumption per capita of a statistical citizen is worrying. According to these data, the level of alcohol consumption increased from approximately 6 L per person per year in 2002 to 10 L in 2018. The rise in the amount of alcohol consumption resulted in a decrease in the rate of life expectancy, and from 2013 halted the tendency towards increased life expectancy. In addition, according to data in Poland, alcoholics constitute about 2% of the population, i.e., 600,000–800,000 people, and alcohol abusers constitute about 12% of the population, i.e., 3.6–4.8 million people.

There were two aims in this study. First, the overarching goal of the research was to identify the main problems associated with the consumption of alcohol at work among employees in the construction industry, with particular emphasis on jobs related to work on scaffolding. The second goal of the study was to determine patterns of alcohol consumption among construction workers in Poland.

This research is relevant to the development of interventions for reduction of occupational accidents and of interest to public health. The results obtained on the basis of the conducted research may be a justification for the directions of preventive actions carried out in order to reduce the number of occupational accidents in the construction industry caused by alcohol. This will significantly contribute to an increase of the level of occupational safety in the construction industry.

2. Literature Review

The results of studies conducted by Benner's team concerning the relationship between the amount of consumed alcohol and the general cause of death of construction workers are important with regards to the analyzed issue [26]. The study was conducted on 8043 German construction workers aged 25–64 who were employed as the following professions: plumber, carpenter, painter, plasterer, bricklayer, unskilled construction worker, office worker, engineer, and architect. Information regarding smoking, the number of people consuming alcohol, and the amount of alcohol consumed daily was obtained on the basis of medical investigations carried out by medical staff. Thus, only 7.4% of all people participating in the study declared abstinence, while the remaining people consumed alcohol daily in an amount from 1 g/day to over 100 g/day. The research assumed that 50 g of alcohol per day corresponds to the consumption of 1 L of beer or 0.5 L of wine, while 100 g of alcohol per day corresponds to 2 L of beer or 1 L of wine. Occasional drinkers (from 1–49 g/day) were the most numerous group, constituting over half of the total drinkers (53.3%). A total of 26.4% of employees declared alcohol consumption at an amount of 50–99 g/day, while 12.9% declared the consumption of over 100 g/day. In addition, research showed that alcohol consumption is highly correlated with smoking. Over 65.0% of respondents smoked cigarettes. Hypertension was the most common issue among workers who smoked and drank. During the study, there were 172 deaths—mainly in the group of employees declaring consumption of an amount of 1–49 g/day. Studies showed that the highest number of deaths occurred among people who declared alcohol consumption of an amount of over 50 g/day.

In turn, the Australian research team headed by Prof. H. Biggs, in cooperation with a large construction company and with the support of the Research Center of the National Fund for Sustainable Construction, implemented a research project that aimed to determine the impact of alcohol and other drugs on safety in the construction industry [27]. In total, 494 people were surveyed, with an average age of 35.7 (± 11.4). In the conducted surveys, employees were asked about the amount and frequency of alcohol consumption, drinking behavior that could indicate alcohol addiction, as well as the negative consequences of alcohol consumption. The obtained results showed that 58% of the surveyed people (i.e., 286 people) had contact with alcohol in their daily work. A high risk of alcohol-related problems was identified among 185 respondents, while alcohol addiction occurred in 43 respondents. The research was of great importance for Australia, and the results of the research were used to introduce harmonized standards in the construction industry in the field of occupational health and safety.

Subsequent studies conducted among young construction workers (the average age of people participating in the study was 21 years old) showed that about 65% of the respondents performed harmful and dangerous practices related to drinking alcohol. A total of 39% of respondents consumed alcohol 2–3 times a week, while 36% indicated that they consumed 10 or more alcoholic beverages at one time. The research also identified the positive correlation between harmful employee behavior and violence among drinkers (verbal violence, racial harassment, threats, attacks of aggression) [28].

According to the literature survey, the phenomenon of alcohol abuse and consumption at work is a common problem in many sectors of the economy [29]. This problem especially occurs in male-dominated industries [30]. On the basis of the Australian and New Zealand standard classification of economic sectors, industry is defined as male-dominated, with men accounting for around 70% of employees [31]. For example, in Australia, such sectors of the economy are agriculture (70%), construction (88%), mining (82%), production (74%), transport (77%) and municipal services (76%). In turn, in Poland, the 70% criterion is met by sectors of the economy such as mining and extraction (90%), construction (89%), energy production and supply (79%), transport and storage (77%), and municipal services (76%) [32]. Therefore, it can be concluded that the construction industry is a sector dominated by men who often have a problem with alcohol abuse and consumption at work.

The consumption of alcohol at a workstation leads to the deterioration of employee performance, which can in turn result in employee absenteeism at work, an occupational accident, other significant problems that are related to the occupational safety of employees, or interpersonal problems between workers [33]. It has also been reported that alcohol increases the risk of aggression and violence towards colleagues, as well as towards family members [34].

The increased risk of an accident under the influence of alcohol results from, among others, a reduced attention and cognitive ability, as well as reduced concentration and delay in making decisions and taking actions that can prevent the occurrence of an accident [35].

Alcohol influences, among others, psychomotor dysfunction (disorders of balance, speaking, thinking, and concentration; impaired motor coordination; reduced level of perception of threats; and others), and also damages many organs. The risk of getting sick and dying from alcohol abuse increases with an increasing alcohol consumption [36]. Alcohol consumption is associated with an increased risk of cardiovascular disease, elevated blood pressure, and liver disease [37]. Mental disorders such as anxiety and depression are common comorbidities in people with alcohol consumption disorders [38]. In addition, drinking large amounts of alcohol on an empty stomach may lead to hypoglycemia and a drop in blood sugar levels [39]. This condition can lead to a decrease in manual skills, fatigue, and irritability, which can in turn cause accidents. Often, people who consumed alcohol the previous day believe that they are already sober on the basis of their own judgment and well-being, which is often not true. Even the prevailing belief among many people that the decrease in blood alcohol content can be accelerated by a long sleep, a cold shower, or drinking water and coffee, is not true.

In addition, a very common consequence of consuming an excessive amount of alcoholic beverages, which usually occurs several hours, or the next day, after the consumption, is the phenomenon of malaise—the so-called alcohol hangover [40]. The most common ailments accompanying this phenomenon are headache, thirst, photophobia, hypersensitivity to noise, nausea, problems with concentration, weakness, irritability, and a decrease in the employee's ability to perceive threats and respond to them [41]. An employee under the influence of alcohol and psychoactive substances becomes both a threat to himself and to other people [42]. Alcohol also reduces the experience of pain. For example, the average pain threshold for employees without alcohol consumption is lower by 23% than that of occasional drinkers, and 42% lower than those who drink daily [43].

An investigation carried out by Liu et al. [44] determined that the average human body is able to burn 0.12–0.15‰ of alcohol per hour. The process of alcohol burning depends on many factors, among others, including sex, body weight, individual predispositions associated with metabolism, the amount and type of consumed food, and the state of health of the body. Small amounts of consumed alcohol are excreted with exhaled air and urine, only larger amounts attack the body and reach the brain and other organs. Excessive and prolonged alcohol consumption can lead to alcoholic liver damage (liver cirrhosis), myocardial damage, and brain damage (ischemic stroke).

In order to prevent excessive alcohol consumption by citizens in many countries around the world, guidelines are set for "safe" alcohol consumption. There are significant differences in alcohol policy between individual European countries and regions. National alcohol policies in the European Union are constantly changing [45]. Guidelines on alcohol policy and alcohol consumption are being developed in individual countries. Guidelines are usually developed by governments, e.g., the Ministry of Health, other Ministries, or government agencies that are responsible for alcohol policy. Information on alcohol consumption is based on the results of biomedical research and the relationship between the dose of alcohol consumed and the individual response of the body. Guidelines referring to alcohol consumption are presented in two forms: as grams of ethyl alcohol, or as the amount of "standard" drinks consumed during a day. For the quantity of "standard" drinks, the number of grams of ethyl alcohol per unit volume is determined, which enables an easy conversion to the level of pure alcohol. For example, the Portuguese National Food and Nutrition Council provides in its recommendations on "standard" units on the basis of wine consumption, while the Romanian Ministry of Health distinguishes in its guidelines the level of alcohol consumed in beer and wine. In turn, the guidelines of the Australian Council on Health and Medical Research on alcohol provide patterns of alcohol consumption, which are associated with long-term (chronic), as well as short-term (acute) harm to people consuming alcohol. These guidelines determine the appropriate level of the risk of health and social problems, including injury and death [46].

In turn, in Poland, a "standard" dose of alcohol is equal to 10 g or 12.5 mL of pure ethyl alcohol. This dose, calculated for the most commonly consumed types of alcoholic beverages, is equal to 200 g of 4.5% beer, 100 g of 10.0% wine, and 25 g of 40.0% vodka. When considering that alcoholic beverages are usually sold in volume measures, alcohol values in typical commercial units amount to 22.5 mL/18 g of alcohol for 500 mL of 4.5% beer (a "large beer"), 21 mL/16.8 g alcohol for 175 mL of 10.0% wine (a glass), and 20 mL/16 g of alcohol for 50 mL of 40.0% vodka. A low health risk for men is considered to be the consumption of up to four standard doses of alcohol per day (125 mL of vodka or 0.4 L of wine or two "large beers"), no more than five times a week. It is considered risky to drink more than six standard portions of alcohol per day [47,48].

According to the International Center for Alcohol Policies (ICAP) 14 Report called "International recommendations on alcohol consumption" [49], and in accordance with the recommendations of the World Health Organization (WHO), it is suggested, in Poland, to consume alcohol in an amount not exceeding two units per day (i.e., 20 g), a maximum of five times a week (no more than 100 g per week), with at least two days per week being without alcohol. According to the report, the smallest recommended amount of consumed alcohol, existing in Japan, is equal to one daily unit—19.75 g according to the Ministry of Health, Labor and Social Affairs. In contrast, the highest permitted amount of consumed alcohol occurs in France (up to five daily units, i.e., 60 g according to their National Medical Academy) and Spain (up to 70 g/day according to their Department of Health and Social Policy).

The particular goal of the study was to determine patterns of alcohol consumption among construction workers in Poland. In this study concerning alcohol consumption patterns, attention was paid to two important aspects that are related to the consumption of alcohol by employees: alcohol consumption in the workplace, i.e., during working hours or immediately before or after work, and also alcohol consumption after normal working hours. Although a relatively small number of people consume alcohol immediately before or during working hours, a significant proportion of employees consume alcohol in their free time after business hours [46]. Defining patterns is essential for the appropriate implementation of effective public health measures regarding prevention [50].

To date, there has not been any similar research carried out in Poland, and there is currently no information about alcohol consumption on Polish construction sites. Therefore, this paper is an attempt to fill this research gap.

3. Methodology of Research

The conducted research examined the consumption of alcohol by employees with regards to the daily amount of alcohol that was consumed by construction workers working at workplaces with the use of construction scaffolding. Data for the analysis were obtained from two sources. Figure 1 presents the map of the sites in Poland where the research was realized.

Figure 1. The map of the sites in Poland where the research was realized (own elaboration).

3.1. Accident Documentation

The first source of data concerning occupational accidents involved accident documentation (inspection reports) prepared by inspectors of the National Labor Inspectorate in Poland [23]. A total of 219 people who were injured in 2008–2017 in occupational accidents involving construction scaffolding were analyzed. On the basis of the Control Protocol, i.e., on the basis of a description of the circumstances and causes of the accident, it was possible to obtain information about the health status of an injured person during and after the accident. The report includes information, confirmed by a police officer or doctor admitting the injured person to the hospital ward, concerning blood alcohol content (in units: ‰ or mg/L); the consumed amount of alcoholic beverages; and also the statement "state after consumption", i.e., indicating alcohol consumption or "intoxication state". The amount of alcohol in the bloodstream is called blood alcohol concentration or BAC. BAC can be measured with a breathalyzer or by analyzing a sample of blood. It is measured by the number of grams of alcohol in 100 mL of blood. For example, a BAC of 0.08 means 0.08 g of alcohol in every 100 mL of blood.

It is worth noting that in Poland, as well as in many other European countries, alcohol consumption by employees is classified according to the level of ethyl alcohol in the blood [51]. Therefore, "after consumption" is understood as the content of ethyl alcohol in the blood from 0.20 to 0.50‰ or from 0.10 to 0.25 mg/L, while the "intoxication state" is considered to be the content of ethyl alcohol in the blood of above 0.50‰ or 0.25 mg/L [52].

On the basis of the control protocols, developed by inspectors of the National Labor Inspectorate, we were able to determine the exact time at which the accident occurred. On the basis of the time of the accident and the content of ethyl alcohol in the blood, we were able to estimate the amount of consumed alcohol. This alcohol was calculated per 500 mL of 4.5% beer consumed by an employee (one "large beer" is equivalent to the consumption of 175mL of 10.0% wine or 50 mL of 40.0% vodka). Figure 2 presents the Polish standard drink.

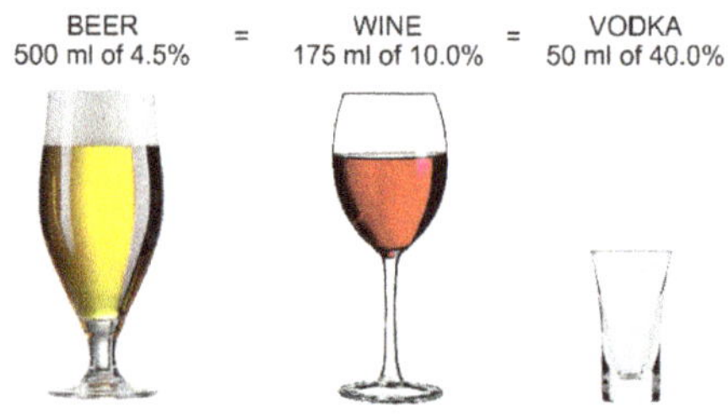

Figure 2. The Polish standard drink (own elaboration).

In order to determine:

- the maximum blood alcohol content (in ‰);
- the time after finishing drinking and reaching the maximum value (min);
- the time after finishing drinking and reaching the value of 0.00 ‰, i.e., total sobriety (min);

We used the ethyl alcohol content calculator [53]. It is widely known that the metabolism of alcohol involves individual differences [54]. The alcohol metabolism efficiency of different individuals will vary greatly [55]. Therefore, the reference unit was adopted for analysis. It was assumed that the reference unit is a man with a statistical height of a Pole equal to 1.80 m, a statistical body weight of 83 kg [11], and at the age that corresponds to the average age obtained in the analyses—40 years. This man has a normal body shape, normal alcohol consumption, and standard food consumption.

Figure 3 presents a graph of the change in the content of ethyl alcohol in the blood with regards to the dose of consumed alcohol (from 500 mL of 4.5% beer—1 beer, to 5500 mL of 4.5% beer—11 beers) and the time that elapsed since the end of drinking. For example, after consuming three beers (1500 mL of

4.5%) or three glasses of wine (525 mL of 10.0%) or three glasses of vodka (150 mL of 40.0%), the highest content of ethyl alcohol in blood (1.18‰)/blood alcohol concentration (BAC = 0.12 g/100 mL) content is 90 min after the end of drinking. The human body needs 480 min for content of ethyl alcohol in blood 0.0‰ (BAC = 0.00 g/10 0mL), which means the person is sober. In addition, Table 1 presents the characteristic data resulting from Figure 3. The Table 1 indicates the time when the content of ethyl alcohol in the blood is the highest and the time since the end of drinking indicating 0‰.

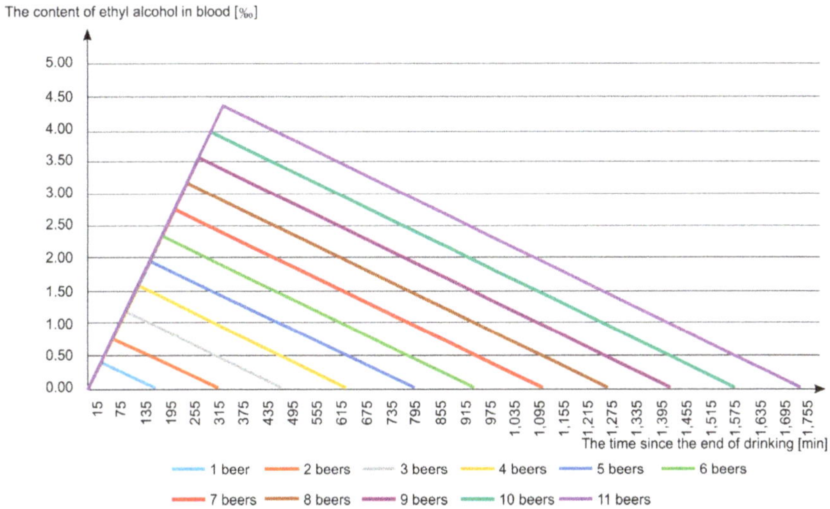

Figure 3. The content of ethyl alcohol in the blood with regards to the dose of alcohol consumed (from 1 beer to 11 beers) and the time since the end of drinking (own elaboration).

Table 1. Collective summary of the content of ethyl alcohol in the blood with regards to the dose of consumed alcohol (from 1 beer to 11 beers) and the time since the end of drinking (own elaboration).

Alcohol Dose			Content of Ethyl Alcohol in Blood	Blood Alcohol Concentration (BAC)	Time Since the End of Drinking
Number of 4.5% Beers/(mL)	Number of 10.0% Glasses of Wine/(mL)	Number of 40.0% Glasses of Vodka/(mL)	(‰)	(g/100 mL)	(min)
1/500	1/175	1/50	0.39 0.00	0.04 0.00	30 160
2/1000	2/350	2/100	0.78 0.00	0.08 0.00	60 320
3/1500	3/525	3/150	1.18 0.00	0.12 0.00	90 480
4/2000	4/700	4/200	1.56 0.00	0.16 0.00	120 640
5/2500	5/875	5/250	1.95 0.00	0.21 0.00	150 800
6/3000	6/1050	6/300	2.34 0.00	0.25 0.00	180 960
7/3500	7/1225	7/350	2.73 0.00	0.29 0.00	210 1120
8/4000	8/1400	8/400	3.12 0.00	0.34 0.00	240 1280
9/4500	9/1575	9/450	3.51 0.00	0.38 0.00	270 1440
10/5000	10/1750	10/500	3.90 0.00	0.42 0.00	300 1600
11/5500	11/1925	11/550	4.290.00	0.460.00	3301760

The following scenarios were adopted in the conducted analyses:

- Scenario 1: the injured person consumed alcohol on the day preceding the accident, while assuming that drinking finished at 21:00;
- Scenario 2: the injured person consumed alcohol during the work break preceding the accident. Three specific cases were assumed: (a) alcohol was consumed before starting work, i.e., 7:00; (b) alcohol was consumed during the breakfast break at 10:00; (c) alcohol was consumed during the lunch break at 13:00.

The main point of the scenarios above was to determine the amount consumed alcohol by the workers and the time of drinking. To determine the above hours, we used the Expert (Delphi) method. This method, by means of which obtained results were based upon, relied upon the opinions and assessments of competent experts [56]. The adopted assumptions resulted directly from the answers received from construction managers, work managers, and supervision inspectors about experience with working with construction workers. The experts were asked to, on the basis of their own experience and knowledge, present occurring situations or scenarios for workers under the influence of alcohol on their supervised construction sites. On the basis of the analysis of the information received from the expert, two different types of scenarios emerged: first—consumed alcohol on the day preceding the accident, and second—consumed alcohol during the work break preceding the accident.

3.2. Surveys

Surveys constituted the second source of data. The survey data were collected from January 2016 to December 2018 during the on-going research project called "Model of the assessment of risk of the occurrence of building catastrophes, accidents and dangerous events at workplaces with the use of scaffolding" ("ORKWIZ"). The surveys and questionnaires were prepared by a team from the Lublin University of Technology, under the leadership of K. Czarnocki [57]. From 1500 people working on the examined 120 construction sites (during the testing of 120 scaffoldings), 573 employees took part in the study (573 surveys were carried out among people working at the construction site of the examined scaffoldings).

Participation in the survey was voluntary and anonymous. Respondents had the right to refuse to participate without giving a reason. All the procedures performed in studies involving human participants were in accordance with the ethical standards and with the 1964 Helsinki Declaration and its later amendments [58]. According to the current guidelines of the Ethical Review Board at the Centre of Postgraduate Medical Education, Warsaw, Poland [59], an anonymous questionnaire-based cross-sectional study does not require separate consent.

On the basis of the information received from construction management, the researchers estimated that a total of 1500 people worked on the 120 examined construction scaffoldings. In 120 initial surveys, in the part concerning the scaffolding assembly team, the researchers asked employees about their age, seniority, and experience in scaffolding assembly. In contrast, in 573 personal surveys—covering 38.2% of the employees working on the scaffoldings in question—people were asked about drugs they used (alcohol, cigarettes, and other intoxicants).

The data collection method included questionnaires. The data obtained in this study were direct responses from individuals. As part of the research, the researchers developed standardized protocols for data collection. Furthermore, all study personnel were trained to conduct the research. It is well known that training of study personnel allows for the minimization of inter-observer variability [60]. The questions concerned private or sensitive topics, such as consumption of alcohol. Thus, self-reporting data may have been affected by an external bias caused by social desirability. The bias in this case can be referred to as social desirability bias [61]. Moreover, at the stage of validation, the data obtained from the questionnaire were analyzed using the methods of descriptive statistics in order to verify the variability of responses to individual questions. What is important is that the analysis of the collected data was performed after all the surveys had been completed. This approach was intended to reduce the interviewer's bias [62].

Respondents in the personal survey were first asked if they had ever consumed alcohol, and then whether they had consumed alcohol in the last 12 months. In the research, a person who did not drink alcohol was defined as a person who had never consumed alcohol, or a person who had consumed alcohol, but not in the last 12 months. If the respondents gave both positive answers, they were asked further questions about their frequency of drinking. An answer to the asked question was the amount of consumed "standard" alcoholic beverages, i.e., 4.5% beers with a capacity of 500 mL—"one large beer" during a day. The obtained data were subjected to detailed analysis, taking into account such parameters as age, marital status, and also the place of residence of the employees participating in the survey.

In the case of the age criterion, the following categories were applied: 18–19 years, 20–29 years, 30–39 years, 40–49 years, 50–59 years and >60 years. Marital status was classified as single, married, divorced, or widower. Permanent residence was classified as follows: village, a small town with up to 100,000 residents, and a city of over 100,000 residents.

The data were analyzed with Statistica v.13.3 (StatSoft Polska Sp z o.o.). Normality of distributions of continuous variables [63] was assessed by the Shapiro–Wilk test [64]. The distribution of categorical variables was shown by frequencies and proportions along with 95% confidence intervals [65]. Associations between personal characteristics (age, marital status, permanent residence) with drinking alcohol were conducted using the logistic regression analyses [66]. The socio-demographic characteristics were considered as independent variables [67]. In univariate logistic regression analyses [68], we considered all variables separately. Statistical inference was based on the criterion $p < 0.05$ [69].

4. Research Results

4.1. Analysis of Occupational Accidents

The control reports concerning 219 people injured in occupational accidents at workplaces that use building scaffoldings were analyzed with regards to the causes of the accidents. An interesting factor for the authors of this study was the human cause—consumption of alcohol, narcotic drugs, or psychotropic substances [70]. The cause related to alcohol consumption occurred in 38 injured people, which was 17.4% of all people injured in accidents on scaffolding. This means that every sixth occupational accident was caused by an abnormal sobriety of an employee.

In 22 control protocols, labor inspectors determined the exact value of the alcohol content in the blood of the injured person. In the remaining 16 cases, the protocol only provided information about the cause of the accident—alcohol consumption.

A detailed analysis of the alcohol content in a victim's blood showed that the lowest amount of ethyl alcohol in the blood was equal to 0.20‰ (state after consumption indicates alcohol consumption with typical symptoms of diffuse attention), while the highest identified value was equal to 4.16‰ (intoxication with typical symptoms such as balance disorder, speech disorder, drowsiness, decreased behavior, and movement control, and also impairment of hand–eye coordination). The average alcohol content value in a victim's blood was equal to 1.20 ± 1.10‰. Therefore, the state after consumption of alcohol was determined in the case of four people, while the remaining 18 people were in the state of intoxication.

The age structure of people for whom the accident was caused by alcohol is shown in Figure 4. The most numerous group was employees aged 40–49 (34.2% of all people with alcohol in their blood) and 30–39 (31.58%). It was in these age ranges (30–49) that the highest levels of ethyl alcohol were found in the blood. The average age of victims was 40 ± 8 years. It can be stated that the percentage of alcohol-consuming workers increases with age.

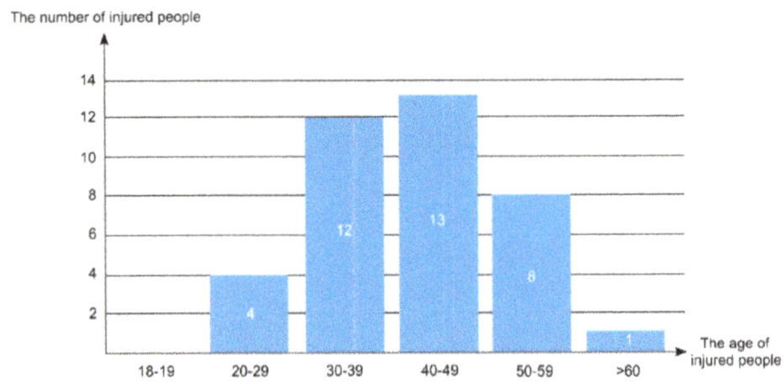

Figure 4. Age structure of injured people who were under the influence of alcohol based on control protocols (own elaboration).

The effect of the accident was also analyzed. Accidents were divided into fatal, severe, and light, and are shown in Figure 5. The age structure is as follows:

- Eight people had a fatal accident, which was 21% of the analyzed accidents. The alcohol content in five people for whom the blood alcohol content was known ranged from 0.53–4.16‰ (mean value of 1.97 ± 0.94‰). This means that all injured people were intoxicated (alcohol content greater than 0.50‰).
- Seventeen people had a severe accident, which represented 45% of all the analyzed cases. The alcohol content in 10 people for whom their blood alcohol content was known ranged from 0.22–3.40‰ (mean value of 1.47 ± 0.81‰). Two of them were in a state after the consumption of alcohol (alcohol content of less than 0.50‰), and the remaining eight people were drunk.
- Thirteen people suffered a light accident, which represented 34% of all the analyzed cases. The alcohol content in seven people for whom the blood alcohol value was known ranged from 0.44–3.00‰ (mean value 1.37 ± 0.86‰). Two of them were in a state after the consumption of alcohol (alcohol content of less than 0.50‰), and the remaining five people were drunk.

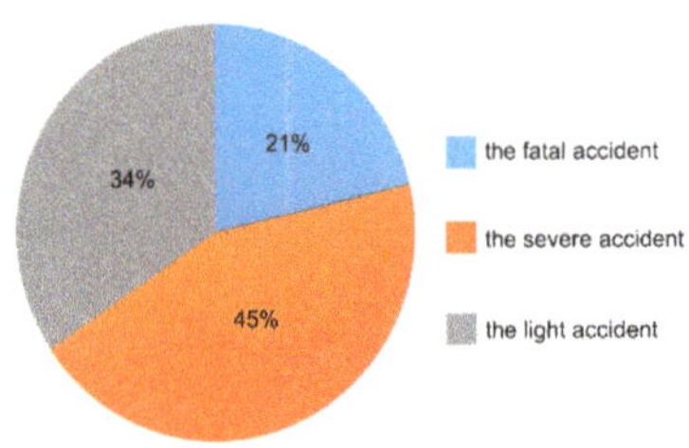

Figure 5. The result of an accident of people under the influence of alcohol on the basis of control protocols (own elaboration]).

On the basis of the control protocols, we were also able to determine the exact time at which the accident occurred. The analysis was carried out for 22 accidents for which the time of the accident and the content of alcohol in the victim's blood were known. Unreal situations were rejected for each accident, e.g., consumption of more than 5000 mL of 4.5% beer—10 beers (life-threatening situation). The most probable scenario was selected.

Table 2 presents the values of alcohol consumption with regards to the content of ethyl alcohol in the blood (‰), the time of the accident, the adopted scenario, and the estimated number of consumed beers. If the analysis resulted in an impossible situation (e.g., the time since the end of alcohol consumption and the blood alcohol content being impossible), the table indicates such a case with the abbreviation "*im*". The table highlights the assumed numbers of consumed alcoholic beverages in relation to 500 mL of 4.5% beer ("large beers").

Table 2. Estimated numbers of consumed alcoholic beverages ("large beers") (own elaboration).

No.	Severity	Content of Alcohol in Blood	Scenario 1		Scenario 2					
					(a)		(b)		(c)	
			Time	Number of Beers	Time	Number of Beers	Time	Number of Beers	Time	Number of Beers
(-)	(-)	(‰)	(min)	(pcs)	(min)	(pcs)	(min)	(pcs)	(min)	(pcs)
1	L	0.70	900	7	270	3	90	2	-	-
2	S	2.13	1080	10	480	7	300	6	120	*im*
3	L	0.53	720	5	120	2	-	-	-	-
4	S	1.23	680	7	80	*im*	-	-	-	-
5	S	1.85	1020	10	420	6	240	5	60	*im*
6	S	1.30	690	7	90	3	-	-	-	-
7	L	0.22	1070	7	470	3	290	2	110	1
8	F	3.00	810	>10	210	*im*	30	*im*	-	-
9	F	4.16	1075	>10	475	11	295	*im*	-	-
10	S	1.33	900	8	300	4	120	3	-	-
11	S	1.11	860	8	260	4	80	3	-	-
12	F	1.70	1070	10	470	6	290	5	110	*im*
13	L	1.00	840	7	240	3	60	2	-	-
14	L	0.46	1020	7	420	3	240	2	60	1
15	L	0.44	1020	7	420	3	240	2	60	1
16	S	1.30	885	8	285	4	105	3	-	-
17	L	0.20	780	5	180	2	-	-	-	-
18	F	2.74	990	>10	390	8	210	7	30	*im*
19	F	3.40	1110	>10	510	10	330	9	150	*im*
20	S	2.67	840	10	240	7	60	*im*	-	-
21	S	2.00	1110	>10	510	7	330	6	150	*im*
22	S	0.63	610	5	10	*im*	-	-	-	-

Information about the severity of the accident is present in Table 2 next to the number (No.) in the second column:

- "L" means light accident;
- "S" means severe accident;
- "F" means fatal accident.

For the authors of the study, nine cases (cases 2, 5, 8, 9, 12, 18, 19, 20, 21) aroused great concern and doubt—the obtained blood alcohol content was high enough (i.e., higher than 1.70‰ and amounting to 2.13‰, 1.85‰, 3.00‰, 4.16‰, 1.70‰, 2.74‰, 3.40‰, 2.67‰, 2.00‰, respectively) that allowing an employee to work in a life-threatening condition would be extremely irresponsible. Such an employee would have, and the authors had no doubts about this, come to work in a drunken state, which should have been immediately noticed by the construction supervisor (site manager, foreman). Further analysis of the causes of these accidents indicated the second cause of the accident, i.e., the lack of direct supervision over the operation that was performed by the injured party.

Other cases and the most probable scenarios indicate that

- injured parties were likely to drink alcohol during the work break immediately before the accident: scenario (2a)—cases: 6, 17 (1000–1500 mL of 4.5% beer: 2–3 beers), scenario (b)—cases: 1, 10, 11, 13, 16 (500–1500 mL of 4.5% beer: 1—3 beers), scenario (c)—cases: 7, 14, 15 (500 mL of 4.5% beer: 1 beer);
- injured parties most likely consumed alcohol the day before (scenario 1)—cases: 3, 4, 22 (2500–3500 mL of 4.5% beer: 5—7 beers).

4.2. *Analysis of Survey Data*

From 1500 people working on the examined 120 construction sites, 573 employees took part in the study. This was 38.3% of the people employed in the examined construction enterprises. The researchers tested 120 façade scaffoldings. The façade scaffoldings were divided into four groups with regards to the surface area: 30–300 m^2 (55 scaffolding), 300–600 m^2 (28 scaffolding), 600–900 m^2 (25 scaffolding), and 900—1500 m^2 (12 scaffolding). Figures 6–9 show examples of tested scaffoldings with regards to the surface area.

Figure 6. Example of façade scaffold with an area of 30–300 m^2 (authors' archive).

Figure 7. Example of façade scaffold with an area of 300–600 m^2 (authors' archive).

Figure 8. Example of façade scaffold with an area of 600–900 m^2 (authors' archive).

Figure 9. Example of façade scaffold with an area of 900–1500 m^2 (authors' archive).

Table 3 presents the obtained socio-demographic information on the respondents.

Table 3. Socio-demographic information (own elaboration).

Age of Employee			
Age Range	**Number of People**	**Percentage**	***p***
18–19	13	2.3%	<0.01
20–29	178	31.1%	
30–39	209	36.5%	
40–49	102	17.8%	
50–59	53	9.2%	
>60	18	3.1%	
Marital Status			
Marital Status	Number of People	Percentage	*p*
Single	251	43.8%	<0.01
Married	304	53.1%	
Divorced	11	1.9%	
Widower	7	1.2%	
Permanent Residence			
Place	Number of People	Percentage	*p*
A city of over 100,000 residents	179	31.2%	<0.01
A small town with up to 100,000 residents	204	35.6%	
Village	190	33.2%	

Figure 10 shows the age structure of the respondents. The employment structure is as follows:

- The largest group of respondents were employees aged 30–39, and they represented 36.5% of all the respondents. They were mainly employees with average experience who had worked for at least 10 years in their profession.
- The second group was made up of young employees aged 20–29 (31.1%), with a much shorter seniority, ranging from several days to a maximum of 10 years.
- Other respondents were experienced employees with more than 15 years of work experience at the age of 40–49 (17.8%) and 50–59 (9.2%), several employees over the age of 60 who were often at retirement age (3.1%), and adolescent employees (2.3%).

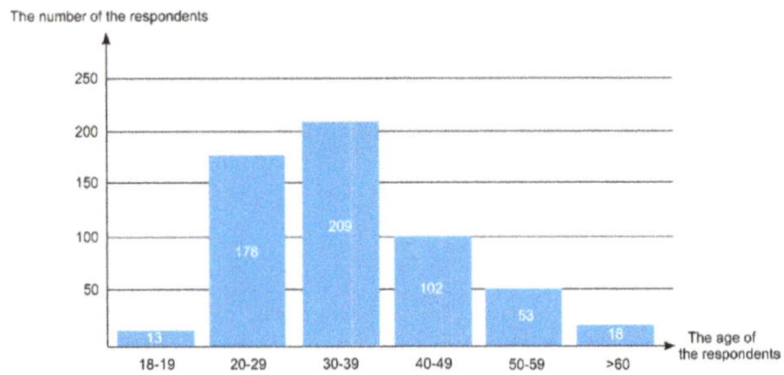

Figure 10. The age structure of the respondents (own elaboration).

The majority of the respondents were married (53.1% of all respondents), or bachelors (43.8%), 35.6% of whom indicated small towns as their place of permanent residence (up to 100,000 inhabitants), 33.2% of whom lived in villages, and 31.2% of whom lived in cities (over 100,000 inhabitants) ($p < 0.01$).

Figure 11 shows the obtained survey data concerning the consumption of alcohol by 573 construction employees who worked on the 120 assessed construction scaffoldings.

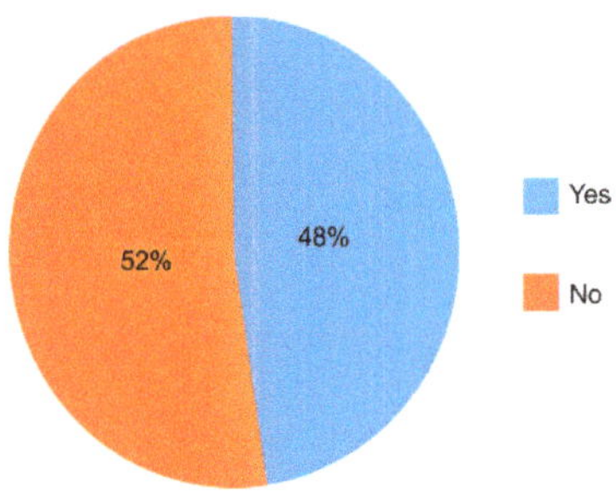

Figure 11. The consumption of alcohol by construction workers (own elaboration).

A total of 274 people (i.e., 47.8% of surveyed employees) declared that they consumed alcohol during the day, while the remaining 299 people (52.2%) declared that they had never consumed alcohol, or had consumed alcohol, but not during the last 12 months. During the research, no employee declared alcohol abuse, alcoholism, or alcohol consumption at work.

One of the questions that was asked in the survey concerned the number of consumed "standard" alcoholic beverages, i.e., 4.5% beers with a capacity of 500 mL—one "large beer". The obtained answers showed that the largest number of "large beers" consumed during a day (500 mL) was equal to 10 and was declared by nine people. The most common answer was one beer, and such information was given by 144 respondents, i.e., 25.1% of all the respondents. The remaining people declared the quantities that are shown in Figure 12. The average value for the studied population was 1000 mL ± 1000 mL of 4.5% beer (2 ± 2 beers).

The analysis of the place of residence showed that most often people declared living in the countryside—99 people, and in small towns (up to 100,000 inhabitants)—95 people. A total of 80 people declared that they permanently lived in cities (over 100,000 inhabitants).

The age of people who consumed alcohol, as well as the age of abstainers, was also analyzed. Table 4 presents the number of people with regards to their age.

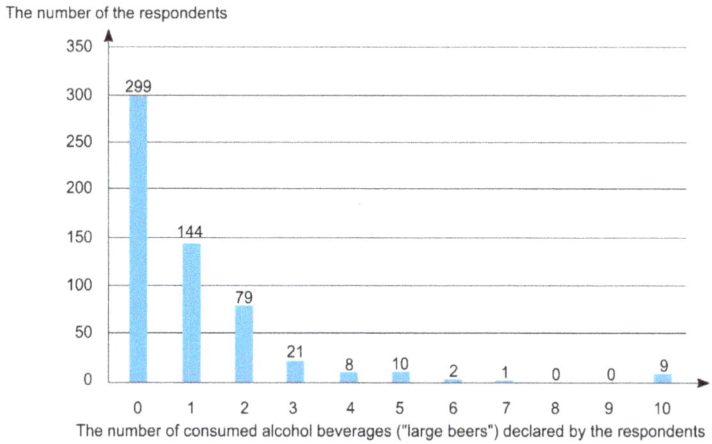

Figure 12. Number of consumed alcoholic beverages ("large beers") declared by construction workers during a day (own elaboration).

Table 4. The number of people consuming alcohol ("Yes"), and the number of abstainers ("No"), with regards to their age (own elaboration).

Age Range	Number of People	Answer	Number of People	Percentage
18–19	13	Yes No	3 10	23.1% 76.9%
20–29	178	Yes No	85 93	47.8% 52.2%
30–39	209	Yes No	100 109	47.8% 52.2%
40–49	102	Yes No	47 55	46.1% 53.9%
50–59	53	Yes No	29 24	54.7% 45.3%
>60	18	Yes No	10 8	55.6% 44.4%

The following conclusions can be drawn as a result of the obtained data (Table 4):

- The most numerous group who consumed alcohol during the day were employees aged 30–39 (100 people) and 20–29 (85 people);
- The highest percentage of people consuming alcohol occurred among employees aged over 60 (55.6%) and 50–59 (54.7%);
- The number of people consuming alcohol decreased with age (decreasing tendency) and the ratio of people consuming alcohol to the abstainers changed (the number of people consuming alcohol increased);
- Among younger employees (18–19, 20–29, 30–39), the percentage of people consuming alcohol was smaller than the percentage of abstainers.

During the tests, photographic documentation of the construction site was also carried out. It evidences that alcohol was consumed at the examined sites, e.g., Figure 13 shows an abandoned beer can at a litter site at the examined construction site, and Figure 14 shows a beer can in the assembly yard at the examined construction site.

Figure 13. An abandoned beer can at a litter site at the examined construction site (authors' archive).

Figure 14. Beer can in the assembly yard at the examined construction site (authors' archive).

4.3. Summary of Results

Analysis of 219 accident control protocols concerning occupational accidents in the construction industry involving construction scaffolding, which took place in Poland, allowed the following conclusions to be drawn:

- Alcohol is one of the causes that leads to occupational accidents. In the analyzed accidents on scaffoldings, the cause related to alcohol consumption occurred in 17.4% of all the victims. This means that every sixth occupational accident was caused by an abnormal sobriety of employees
- The largest group of people consuming alcohol were employees aged 40–49 and 30–39. The average age of victims was 40 ± 8 years. The proportion of drinkers increased with age.
- Analysis of the content of alcohol in blood with regards to the effect of an accident indicated that as the amount of alcohol in the blood rises, the severity of the accident increases. A high alcohol content contributes to the death of a construction worker.

An analysis of 573 construction employees working on the assessed 120 construction scaffoldings allowed the following conclusions to be formulated:

- 47.8% of surveyed employees declared that they drink alcohol during the day;
- the most common number of alcohol beverages consumed outside working hours was one "large beer" (500 mL of 4.5% beer);
- analysis of the place of residence showed that people most often declared living in the countryside and in small towns (up to 100,000 inhabitants);
- the proportion of people consuming alcohol increased with age (an increasing trend of consuming alcohol with age), and the proportion of people consuming alcohol to abstainers changed—the number of people consuming alcohol increased.

The place where an employee is working is constantly and dynamically changing [71]. With such dynamic environments, construction sites contain a significant quantity of unidentified or not well-assessed hazards that expose construction workers to additional safety risks during required operations [72]. Working on scaffolding carries a much higher risk of hazards than other types of construction works [73]. The main reason is the place of work itself—work at a height. Working from a height continues to be one of the major causes of fatality within the construction industry [74].

5. Discussion

The study identified the main problems associated with the abuse and consumption of alcohol at work among employees in the construction industry, with particular emphasis on jobs related to work on scaffolding. When the authors of the article examined the susceptibility of individual age groups to alcohol consumption, the pattern of alcohol consumption by construction workers in Poland emerged. It was found that every second person consumes alcohol during the day, and the number of people consuming alcohol increases with age.

Five limitations of our study should be mentioned. Firstly, it is worth noting that the accident documentation did not always contain all the necessary information about the content of ethyl alcohol in the blood of the injured person. Because the accident documentation (inspection reports) was prepared by inspectors of the National Labor Inspectorate, the authors were unable to interfere or supplement this archival data. It happened that such information was not provided/confirmed by the police officer or doctor admitting the injured person to the hospital ward. In the analyzed data, the cause related to alcohol consumption occurred in 38 injured people, which was 17.4% of all people injured in accidents, while the content of ethyl alcohol in the blood of the injured person was included only in 22 control protocols. Secondly, the content of ethyl alcohol in the blood was determined, assuming the reference unit. It should be remembered that the metabolism of alcohol involves individual differences [54]. The alcohol metabolism efficiency of different individuals will vary greatly [55]. Thirdly, the second source of the data analyzed was obtained on the basis of surveys. The proportion that took part in the survey was 38.2% of the target population. The data obtained in this study were direct responses from individuals. It may be assumed that the number of consumed "standard" alcoholic beverages might have been even higher than what the data revealed. No data were available regarding the comparisons of respondents and non-respondents. Since there was no officially recorded data, results obtained from the data collection method could be biased. Findings might have been subject to selection bias, although the data were collected with the greatest care. Fourth, since the second source of the data was collected via a survey and actively answered by working individuals, it did not contain data related to fatal accidents. Fifth, during the conducted research on the 120 examined construction scaffoldings, the authors did not collect any photographic documentation of some workers drinking an alcoholic beverage, for example, a can of beer. This was hard to do because the workers knew that the research team was conducting research and thus their behavior was close to "perfect". Moreover, while examining the scaffolding and conducting the survey, the authors did not register the situation that some workers were drinking alcoholic beverages. This may have been due to the fact that the construction workers consumed alcohol in hard-to-reach places on the construction site (in hiding).

The observations carried out also revealed that there was an additional, serious problem that had not yet been analyzed, which is the wide availability of alcoholic beverages. Unfortunately, a very unfavorable factor that affects the ease of drinking alcohol immediately before work or during breaks from work is the large, easy, and widespread availability of alcohol. It is worth noting that both in Poland and in other European countries, alcohol can be bought in different volumes and in every discount store or hypermarket. The assortment is very wide and available in various volumes, starting with the smallest bottles of 50 mL of 40.0% vodka (the shape of the bottle enables it to fit in a pocket), through to bottles of 200 mL of 40.0% vodka (measuring 7.8 × 17.0 cm with a flat shape that allows it to be stored, e.g., in a jacket pocket) and larger bottles (500 mL, 700 mL, and more). According to the authors, the availability of small-volume bottles should be limited, as this could help reduce the number of people who drink alcohol at work.

In addition, according to the authors, in order to improve the safety of employees in a workplace and to eliminate the problem of intoxicated construction workers, employers (including the construction manager and work managers) should be allowed to carry out sobriety checks of employees. Currently, an employer has the right to carry out an inspection only if two conditions are met: first—an employee has agreed to conduct such an examination (the examination is voluntary), and second—an employer has reasonable suspicion that an employee is under the influence of alcohol [75]. Therefore, at present, an employer cannot carry out such an examination if an employee does not agree to it. An employer has the right to request such an examination to be conducted by an authorized body, e.g., the police service. An employer is also not able to conduct routine sobriety checks on all or randomly selected employees.

6. Conclusions

The construction industry is recognized as one of the most dangerous industries. Much effort has been devoted to improving safety and to reduce hazards in the workplace, but less attention has been paid to the human factor, i.e., employees in the workplace. It should be remembered that the most important thing when considering safety at work is people. This study confirmed that alcohol consumption negatively affects the human body; reduces the ability to properly and safely, i.e., fault-free and accident-free, perform standard daily activities (such as driving, moving) and professional activities (e.g., work in an office, work on a construction site, work on scaffolding); and can also lead to death at a workstation.

Data for the analysis were obtained from two sources. The archival post-accident documentation, which was the first source of data from 2008–2017, allowed for the determination of the most probable scenario of alcohol consumption by employees during work. The advantage of these studies was the 10-year period of data collection, which allowed for the establishment of a certain trend in the accident situations. Unfortunately, the data were prepared by various inspectors of the National Labor Inspectorate and contained varying degrees of detail (from 38 post-accident documentation, only 16 protocols stated that the "injured person consumed alcohol" or was "under the influence of alcohol"). Therefore, when planning this type of research, it should be remembered that we may be dealing with incomplete data when using archival data. On the other hand, surveys (the second source of data) required the researchers to properly plan and conduct surveys. When preparing the questionnaire, it is important to remember that the questions should be logical, understandable to the respondents, and not be suggestive of answers. It is also particularly important to properly train the study personnel. The advantage of these studies was the testing of nearly 40% of people working at workplaces with scaffoldings.

This study confirmed that excessive and disproportionate alcohol consumption can be the cause of an accident, and consequently death (depending on the type of accident or physical ailment), at workplaces with scaffolding. Of 219 accident reports, 17.4% indicated alcohol as a contributing factor. Analysis of accident documentations shows that cases where alcohol was indicated as a contributing factor in an accident, and that the alcohol was consumed during the workday. Furthermore, the analysis of the blood alcohol content with regards to the effect of an accident indicates that with an increase in the amount of alcohol in the blood, the severity of the accident also increases. Comparative analysis of the results from the post-accident documentation and survey showed that the number of people consuming alcohol decreased with age, but the number of alcohol-related accidents did not decrease. Moreover, the percentage of people consuming alcohol slightly changed with age.

The research was conducted in five research areas, i.e., in five provinces (voivodeships) of Poland (Dolnośląskie, Lubelskie, Łódzkie, Mazowieckie, Wielkopolskie). The research of post-accident documentation showed that most of the accidents related to alcohol took place in central Poland, i.e., in the Łódzkie Voivodeship (12 accidents, i.e., that every four accidents were related to alcohol). On the other hand, the surveys showed that most of the people consuming alcohol worked on construction sites in the central and eastern part of Poland, i.e., in the Mazowieckie Voivodeship (59% of working people) and Lubelskie Voivodeship (57% of working people). Moreover, according to the answers provided by the respondents, the largest number of people who did not drink while working (i.e., abstinents) was in the Wielkopolskie Voivodeship (71%).

The use of alcohol is important a topic of occupational safety in the construction industry. This research is relevant to the development of interventions for reduction of occupational accidents and is of interest to public health. The results obtained on the basis of the conducted research are able to constitute a justification for the directions of preventive actions carried out in order to reduce the number of occupational accidents in the construction industry caused by alcohol. This will significantly contribute to an increase of the level of occupational safety in the construction industry.

Author Contributions: Conceptualization, M.S. (Marek Sawicki); data curation, M.S. (Marek Sawicki) and M.S. (Mariusz Szóstak); formal analysis, M.S. (Marek Sawicki) and M.S. (Mariusz Szóstak); investigation, M.S. (Marek Sawicki) and M.S. (Mariusz Szóstak); methodology, M.S. (Marek Sawicki) and M.S. (Mariusz Szóstak); visualization, M.S. (Mariusz Szóstak); writing—original draft preparation, M.S. (Marek Sawicki) and M.S. (Mariusz Szóstak); writing—review and editing, M.S. (Marek Sawicki) and M.S. (Mariusz Szóstak). All authors have read and agreed to the published version of the manuscript.

Funding: The article is the result of the implementation by the authors of the research project no. 244388 "Model of the assessment of risk of the occurrence of building catastrophes, accidents and dangerous events at workplaces with the use of scaffolding", financed by NCBiR within the framework of the Programme for Applied Research on the basis of contract no. PBS3/A2/19/2015.

Conflicts of Interest: The authors declare no conflict of interest.

References

1. Miller, T.R.; Zaloshnja, E.; Spicer, R.S. Effectiveness and benefit-cost of peer-based workplace substance abuse prevention coupled with random testing. *Accid. Anal. Prev.* **2007**, *39*, 565–573. [CrossRef]
2. Appleton, A.; James, R.; Larsen, J. The Association between mental wellbeing, levels of harmful drinking, and drinking motivations: A cross-sectional study of the UK adult population. *Int. J. Environ. Res. Public Health* **2018**, *15*, 1333. [CrossRef]
3. World Health Organization. *Global Status Report on Alcohol and Health 2018*; World Health Organization: Geneva, Switzerland, 2018; ISBN 978-92-4-156563-9. Available online: www.who.int/substance_abuse/publications/global_alcohol_report/en (accessed on 23 April 2020).
4. Anderson, P.; Baumberg, B. Alcohol in Europe—Public health perspective: Report summary. *Drugs* **2006**, *13*, 483–488. [CrossRef]
5. Brooks, A.T.; Raju, S.; Barb, J.J.; Kazmi, N.; Chakravorty, S.; Krumlauf, M.; Wallen, G.R. Sleep regularity index in patients with alcohol dependence: Daytime napping and mood disorders as aorrelates of interest. *Int. J. Environ. Res. Public Health* **2020**, *17*, 331. [CrossRef] [PubMed]
6. John, U.; Hanke, M.; Freyer-Adam, J. Health risk behavior patterns in a national adult population survey. *Int. J. Environ. Res. Public Health* **2018**, *15*, 873. [CrossRef] [PubMed]
7. Roche, A.; Chapman, J.; Duraisingam, V.; Phillips, B.; Finnane, J.; Pidd, K. Construction workers' alcohol use, knolwedge, perceptions of risk and workplace norms. *Drug Alcohol Rev.* **2020**. [CrossRef] [PubMed]

8. Strickland, J.; Wagan, S.; Dale, A.; Evanoff, A. Prevalence and perception of risky health behaviors among construction workers. *J. Occup. Environ. Med.* **2017**, *59*, 673–678. [CrossRef] [PubMed]
9. Hoła, B.; Szóstak, M. An occupational profile of people injured in accidents at work in the polish construcion industry. *Procedia Eng.* **2017**, *208*, 43–51. [CrossRef]
10. Hoła, B.; Nowobilski, T.; Szer, I.; Szer, J. Identification of factors affecting the accident rate in the construction industry. *Procedia Eng.* **2017**, *208*, 35–42. [CrossRef]
11. Lim, S.; Chi, S.; Lee, J.; Lee, H.-J.; Choi, H. Analyzing psychological consitions of field-workers in the construction industry. *Int. J. Occup. Environ. Health* **2017**, *23*, 261–281. [CrossRef]
12. Tabernero, C.; Luque, B.; Cuadrado, E. A Multilevel study of alcohol consumption in young adults: Self-efficacy, peers' motivations and protective strategies. *Int. J. Environ. Res. Public Health* **2019**, *16*, 2827. [CrossRef]
13. Flannery, J.; Ajayi, S.O.; Oyegoke, A.S. Alcohol and substance misuse in the construction industry. *Int. J. Occup. Saf. Ergon.* **2019**, 1–16. [CrossRef] [PubMed]
14. Ken, P.; Vinita, D.; Roche, A.; Allan, T. Young construction workers: Substance use, mental health, and workplance psychosocial factors. *Adv. Dual. Diagn.* **2017**, *10*, 155–165. [CrossRef]
15. Wang, S.-C.; Chen, Y.-C.; Chen, S.-J.; Lee, C.-H.; Cheng, C.-M. Alcohol addiction, gut microbiota, and alcoholism treatment: A review. *Int. J. Mol. Sci.* **2020**, *21*, 6413. [CrossRef]
16. Tiwary, G.; Gangopadhyay, P.; Biswas, S.; Nayak, K.; Chatterjee, M.; Chakraborty, D.; Mukherjee, S. Socio-economic status of workers of building construction industry. *Indian J. Occup. Environ. Med.* **2012**, *16*, 66–71. [CrossRef] [PubMed]
17. Łyszczarz, B. Production losses associated with alcohol-attributable mortality in the European Union. *Int. J. Environ. Res. Public Health* **2019**, *16*, 3536. [CrossRef]
18. Demirel, B.; Ergin, M.; Özhasenekler, A.; Teke, C.; Tanrıverdi, F.; Kurtoğlu Çelik, G.; Şener, A.; Pamukçu Günaydın, G.; Gökhan, Ş. The influence of alcohol related accidents on health care cost. *Disaster Emerg. Med. J.* **2018**, *3*, 71–74. [CrossRef]
19. Biggs, H.; Williamson, A. Reducing the risk of alcohol and other drugs in construction: An Autralian assessment. In *New Developments in Structural Engineering and Construction*; Research Publishing Services: Singapore, 2013; pp. 1399–1404. [CrossRef]
20. Banwell, C.; Dance, P.; Quinn, C.; Davies, R.; Hall, D. Alcohol, other drug use, and gambling among Australian Capital Territory (ACT) workser in the building and related industries. *Drugs* **2006**, *13*, 167–178. [CrossRef]
21. Coggon, D.; Harris, E.; Brown, T.; Rice, S.; Palmer, K. Occupation and mortality related to alcohol, drugs and sexual habits. *Occup. Med.* **2010**, *60*, 348–353. [CrossRef]
22. Rebelo, M.; Silveira, F.; Czarnocka, E.; Czarnocki, K. Construction safety on scaffolding; Building Information Modeling (BIM) and Safety Management—A systematic review. *U Porto J. Eng.* **2019**, *2*, 46–60. [CrossRef]
23. Hoła, B.; Szóstak, M. Modeling of the accidentality phenomenon in the construction industry. *Appl. Sci.* **2019**, *9*, 1878. [CrossRef]
24. Statistical Office of the European Union. Available online: https://ec.europa.eu/eurostat/web/health/data/database (accessed on 10 August 2020).
25. Central Statistical Office. Health Status of Population in Poland in 2014. Warsaw. 2016. Available online: https://stat.gov.pl/en/topics/health/health/health-status-of-population-in-poland-in-2014,4,1.html (accessed on 23 April 2020).
26. Brenner, H.; Arndt, V.; Rothenbacher, D.; Schuberth, S.; Fraisse, E.; Fliedner, T.M. The association between alcohol consumption and all-cause mortality in a cohort of male employees in the German construction industry. *Int. J. Epidemiol.* **1997**, *26*, 85–91. [CrossRef] [PubMed]
27. Biggs, H.C.; Williamson, A.R. Safety impacts of alcohol and other drugs in construction: Development of an industry policy and cultural change management program. In Proceedings of the 28th Annual Conference (ARCOM 2012), Edinburg, UK, 3–5 September 2012; pp. 445–454.
28. du Plessis, K.; Corney, T.; Burnside, L. Harmful drinking and experiences of alcohol-related violence in Australian male construction industry apprentices. *Am. J. Mens. Health* **2013**, *7*, 423–426. [CrossRef] [PubMed]

29. Peacock, A.; Leung, J.; Larney, S.; Colledge, S.; Hickman, M.; Rehm, J.; Giovino, G.; West, R.; Hall, W.; Griffths, P.; et al. Global statistics on alcohol, tobacco and illicit drug use: 2017 status report. *Addiction* **2018**, *113*, 1905–1926. [CrossRef] [PubMed]
30. Roche, A.M.; Lee, N.K.; Battams, S.; Fischer, J.A.; Cameron, J.; McEntee, A. Alcohol use among workers in male-dominated industries: A systematic review of risk factors. *Saf. Sci.* **2015**, *78*, 124–141. [CrossRef]
31. Australian Bureau of Statistics National Survey of Mental Health and Wellbeing: Summary of Results; Sydney. 2008. Available online: www.abs.gov.au/AUSSTATS/abs@.nsf/Lookup/4326.0Main+Features32007?OpenDocument (accessed on 23 April 2020).
32. Central Statistical Office. Employment in National Economy in 2018. Warsaw. 2019. Available online: https://stat.gov.pl/en/topics/labour-market/working-employed-wages-and-salaries-cost-of-labour/employment-in-national-economy-in-2018,7,15.html (accessed on 23 April 2020).
33. Elliott, K.; Shelley, K. Effects of drugs and alcohol on behavior, job performance, and workplace safety. *J. Employ. Couns.* **2006**, *43*, 130–134. [CrossRef]
34. Karriker-Jaffe, K.; Tam, C.; Cook, W.; Greenfield, T.; Roberts, S. Gender equality, drinking cultures and second-hand harms from alcohol in the 50 US States. *Int. J. Environ. Res. Public Health* **2019**, *16*, 4619. [CrossRef]
35. Martin, J.L.; Gadegbeku, B.; Wu, D.; Viallon, V.; Laumon, B. Cannabis, alcohol and fatal road accidents. *PLoS ONE* **2017**, *12*. [CrossRef]
36. Marchand, A.; Parent-Lamarche, A.; Blanc, M.-È. Work and high-risk alcohol consumption in the Canadian workforce. *Int. J. Environ. Res. Public Health* **2011**, *8*, 2692–2705. [CrossRef]
37. Zhang, J.; Wu, L. Cigarette smoking and alcohol consumption among chinese older adults: Do living arrangements matter? *Int. J. Environ. Res. Public Health* **2015**, *12*, 2411–2436. [CrossRef] [PubMed]
38. Rodríguez-Rabassa, M.; López, P.; Sánchez, R.; Hernández, C.; Rodríguez, C.; Rodríguez-santiago, R.E.; Orengo, J.C.; Green, V.; Yamamura, Y.; Rivera-amill, V. Inflammatory biomarkers, microbiome, depression, and executive dysfunction in alcohol users. *Int. J. Environ. Res. Public Health* **2020**, *17*, 689. [CrossRef] [PubMed]
39. Huang, J.-H.; Li, R.-H.; Huang, S.-L.; Sia, H.-K.; Chen, Y.-L.; Tang, F.-C. Lifestyle factors and metabolic syndrome among workers: The role of interactions between smoking and alcohol to nutrition and exercise. *Int. J. Environ. Res. Public Health* **2015**, *12*, 15967–15978. [CrossRef] [PubMed]
40. van de Loo, A.; Mackus, M.; van Schrojenstein Lantman, M.; Kraneveld, A.; Brookhuis, K.; Garssen, J.; Scholey, A.; Verster, J. Susceptibility to alcohol hangovers: The association with self-reported immune status. *Int. J. Environ. Res. Public Health* **2018**, *15*, 1286. [CrossRef]
41. Howland, J.; Rohsenow, D.J.; Cote, J.; Gomez, B.; Mangione, T.W.; Laramie, A.K. Effects of low-dose alcohol exposure on simulated merchant ship piloting by maritime cadets. *Accid. Anal. Prev.* **2001**, *33*, 257–265. [CrossRef]
42. Marques, P.H.; Jesus, V.; Olea, S.A.; Vairinhos, V.; Jacinto, C. The effect of alcohol and drug testing at the workplace on individual's occupational accident risk. *Saf. Sci.* **2014**, *68*, 108–120. [CrossRef]
43. Yi, W.; Chan, A. Health profile of construction workers in Hong Kong. *Int. J. Environ. Res. Public Health* **2016**, *13*, 1232. [CrossRef]
44. Liu, R.; Chen, L.; Zhang, F.; Zhu, R.; Lin, X.; Meng, X.; Li, H.; Lei, X.; Zhao, Y. Trends in alcohol intake and the association between socio-demographic factors and volume of alcohol intake amongst adult male drinkers in China. *Int. J. Environ. Res. Public Health* **2019**, *16*, 573. [CrossRef]
45. Kilian, C.; Manthey, J.; Moskalewicz, J.; Sieroslawski, J.; Rehm, J. How attitudes toward alcohol policies differ across european countries: Evidence from the Standardized European Alcohol Survey (SEAS). *Int. J. Environ. Res. Public Health* **2019**, *16*, 4461. [CrossRef]
46. Berry, J.G.; Pidd, K.; Roche, A.M.; Harrison, J.E. Prevalence and patterns of alcohol use in the Australian workforce: Findings from the 2001 National Drug Strategy Household Survey. *Addiction* **2007**, *102*, 1399–1410. [CrossRef]
47. The National Program for Prevention of Alcohol Problems for the Years 2011–2015. 2010. Available online: http://www.parpa.pl/images/image/NP%202011_2015_11_011.pdf (accessed on 23 April 2020).
48. Ronksley, P.E.; Brien, S.E.; Turner, B.J.; Mukamal, K.J.; Ghali, W.A. Association of alcohol consumption with selected cardiovascular disease outcomes: A systematic review and meta-analysis. *BMJ* **2011**, *342*, 479. [CrossRef] [PubMed]

49. ICAP. International Recommendations on Alcohol Consumption. 2003. Available online: http://alkoholizm.eu/pub/zalecenia_ICAP.pdf (accessed on 23 April 2020). (In Polish).
50. Biggs, H.C.; Williamson, A.R.; Davey, T.M. The role of education and awareness in workplace alcohol and drug use in the Australian construction industry: Proposed program of research and preliminary results. *Vulnerable Groups Incl.* **2012**, *3*, 17284. [CrossRef]
51. Koob, G.; Arends, M.; Moal, M. Alcohol. In *Drugs, Addiction, and the Brain*; Academic Press: Cambridge, MA, USA, 2014; Chapter 6; pp. 173–219. [CrossRef]
52. The Penal Code of 06 June 1997 (Law Gazette, No. 88, it. 553, as amended). In Polish. 1997. Available online: http://prawo.sejm.gov.pl/isap.nsf/DocDetails.xsp?id=WDU19970880553 (accessed on 23 April 2020).
53. Promille Calculator. In Polish. Available online: www.medme.pl/kalkulatory/alkohol (accessed on 24 April 2020). (In Polish).
54. Caderbaum, A. Alcohol metabolism. *Clin. Liver. Dis.* **2012**, *16*, 667–685. [CrossRef] [PubMed]
55. Edenberg, H. The Genetics of alcohol metabolism: Role of alcohol dehydrogenase and aldehyde dehydrogenase variants. *Alcohol Res. Health* **2007**, *30*, 5–13.
56. Iriste, S.; Katane, I. Expertise as a research method in education. *Rural Environ. Educ. Person.* **2018**, *11*, 74–80. [CrossRef]
57. Krzysztof, C.; Bojar, H.; Silveira, F.; Rebelo, M.; Elżbieta, C. Health behaviours in scaffold use risk assessment model—SURAM. *Annals Agric. Environ. Med.* **2019**, *26*, 138–142. [CrossRef]
58. World Mecidal Association, WMA Declaration of Helsinki—Ethical Pricniples for Medical Research Involving Human Subjects, 1964 as Amended. Available online: www.wma.net/policies-post/wma-declaration-of-helsinki-ethical-principles-for-medical-research-involving-human-subjects/ (accessed on 10 August 2020).
59. The Supreme Medical Court. The Polish Code of Medical Ethics. 2003. In Polish. Available online: http://nil.org.pl/dokumenty/kodeks-etyki-lekarskiej (accessed on 10 August 2020).
60. Pannucci, C.; Wilkins, E. Identyfying and avoiding bias in research. *Plast. Reconstr. Surg.* **2011**, *126*, 619–625. [CrossRef]
61. Althubaiti, A. Information bias in jealth research: Definition, pitfalls, and adjustment methods. *J. Multidiscip. Healthc.* **2016**, *9*, 211–217. [CrossRef]
62. Davis, R.; Couper, M.; Janz, N.; Caldwell, C.; Resnicow, K. Interviewer effects in public health surveys. *Health Educ. Res.* **2010**, *25*, 14–26. [CrossRef]
63. Mishra, P.; Pandey, C.; Singh, U.; Gupta, A.; Sahu, C.; Keshri, A. Desciptive statistics and mornality tests for statistical data. *Ann. Card. Anaesth.* **2019**, *22*, 67–72. [CrossRef]
64. Sijtsma, K.; Emons, W. Nonparametric statistical methods. In *International Encyclopedia of Education*, 3rd ed.; Elsevier: Oxford, UK, 2010; pp. 347–353. [CrossRef]
65. Kelley, K. Confidence intervals for standardized effect sizes: Theory, application, and implementation. *J. Stat. Soft.* **2007**, *20*. [CrossRef]
66. Peng, C.-Y.; Lee, K.; Ingersoll, G. An introduction to logistic regression analysis and reporting. *Int. J. Educ. Res.* **2010**, *96*, 3–14. [CrossRef]
67. López-Martínez, A.; Exteve-Zarazaga, R.; Ramirez-Maestre, C. Perceived social support and coping responses are independent variables explaining pain adjustment among chronic pain patients. *J. Pain* **2008**, *9*, 373–379. [CrossRef] [PubMed]
68. Wang, H.; Peng, J.; Wang, B.; Lu, X.; Zheng, J.; Wand, K.; Tu, X.; Feng, C. Inconsistency between univariate and multiple logistic regressions. *Shanghai Arch. Psychiatr.* **2017**, *25*, 124–128. [CrossRef]
69. Mayo, D. *Statistical Inference as Severe Testing*; Cambridge University Press: Cambridge, UK, 2018; ISBN 9781107286184. [CrossRef]
70. Hoła, A.; Sawicki, M.; Szóstak, M. Methodology of classifying the causes of occupational accidents involving construction scaffolding using Pareto-Lorenz analysis. *Appl. Sci.* **2018**, *8*, 48. [CrossRef]
71. Yang, K.; Ahn, C.; Vuran, M.; Kim, H. Collective sensing of workers' gait patterns to identify fall hazards in construction. *Automat. Constr.* **2017**, *82*, 166–178. [CrossRef]
72. Albert, A.; Hallowell, M.; Kleiner, B.; Chen, A. Enhancing construction hazard recognition with high-fidelity augmented virtuality. *J. Constr. Eng. Manag.* **2014**, *140*, 4014024. [CrossRef]
73. Szóstak, M. Analysis of occupational accidents in the construction industry with regards to selected time parameters. *Open Eng.* **2019**, *9*, 312–320. [CrossRef]

74. Sanchez, F.; Pelaez, G.; Alis, J. Occupational safety and health in construction: A review of applications and trends. *Ind. Health* **2017**, *55*, 210–218. [CrossRef]
75. Act. of 26 October 1982 on Upbringing in Sobriety and Counteracting Alcoholism (Law Gazette, No. 35, it. 230 as amended). 1982. Available online: http://prawo.sejm.gov.pl/isap.nsf/DocDetails.xsp?id=WDU19820350230 (accessed on 23 April 2020).

applied sciences

MDPI

Article

Analysis of Defects in Residential Buildings Reported during the Warranty Period

Edyta Plebankiewicz * and Jarosław Malara

Cracow University of Technology, Warszawska 24, 31-155 Kraków, Poland; jmalara@l7.pk.edu.pl
* Correspondence: eplebank@L7.pk.edu.pl; Tel.: +48-12-628-2330

Received: 23 July 2020; Accepted: 29 August 2020; Published: 3 September 2020

Abstract: The aim of the article is to present the results of preliminary research of the defects in residential buildings occurring during the warranty period. Due to the small amount of data, the research results cannot be generalized but allow for the formulation of research hypotheses that will be verified in future studies. The data collected included reports of defects in three multifamily residential buildings constructed by the developer in one of the big cities in Poland, which were then examined. For the examination of defects, statistical analysis was used, which revealed that more than half of the reports contained reasonable defects. The results of the preliminary research also indicate that, on the one hand, owners are very active in making warranty claims in the first three months from the date of commissioning, and, on the other hand, with time, the percentage of reasonable defects increases. In terms of the significance of defects, the largest percentage was significant defects. The results showed little activity on the part of property managers in the initial phase of the operation of the buildings, which is the opposite of that of apartment owners. Reports of faults in windows and door joinery, moisture, scratches on walls, and in the area of balconies and terraces are characterized by a relatively low number of cases reported in the first half of the year after the building is commissioned and a gradual increase in the subsequent warranty period. On the other hand, reports related to electrical installation defects are most frequent in the initial period of the warranty, but, with time, their number decreases.

Keywords: residential buildings; warranty; acceptance of the work

---

1. Introduction

By concluding a contract of works, the contractor undertakes to hand over the facility provided for in the contract and the contracting authority to pay a specified remuneration. The key stage in mutual relations is the acceptance of works. However, the obligations of the parties do not end with the acceptance. Under legal regulations, as well as the usual relevant contractual provisions, the contractor takes responsibility for the defectiveness and noncompliance of the performed works with the contract. In the Polish legal environment, the rules of liability are determined by guarantee and warranty. These two types of liability of the contractor operate independently of each other, showing both similarities and significant differences. The main difference is that granting the guarantee is a voluntary declaration by the contractor to assume responsibility for defects in the works performed, which is regulated in the contract. On the other hand, the warranty is legally binding and is regulated in detail in the provisions of the Civil Code [1].

According to Article 556 of the Civil Code, the warranty is an inalienable right of the buyer, while the seller is liable to the buyer if the item sold has a physical or legal defect. A physical defect is defined as one of four possible situations concerning the sold item:

1. There are no features that such a thing should have in view of the purpose in the contract, as indicated or resulting from the circumstances or purpose.

2. There are no features that the seller has provided the buyer with, including the presentation of a sample or pattern.
3. Not fit for the purpose of which the buyer informed the seller at the conclusion of the contract, and the seller did not object to such a purpose.
4. The item is issued to the buyer in an incomplete state.

However, when identifying potential defects, it is important to keep in mind the limited time for reporting them. In the case of a property defect, it is a period of five years from the date of handing it over to the buyer, which, with regard to construction works, takes place at the time of acceptance of the works [2,3]. It should be noted that the detailed specifications of the contractor's liability are described in the national legislation of individual countries. While the paper refers to Polish regulations, the laws of other countries also include the concept of warranty, differing in only some formal and procedural details.

In the context of the warranty, the provisions of the Civil Code do not describe the fault but the defect of the item sold. In this paper, the notions of fault and defect are treated as equivalent due to the way the issue is described in Polish publications, where these terms are also interchangeable.

Defects are a common phenomenon in the construction industry in all countries. Contractors, as well as investors, should pay special attention to them as they can have a significant impact on the costs and required resources of the project. In the literature, there are many publications related to the frequency of defects in residential buildings conducted in various countries [4–26]. However, the vast majority of the publications are related to defects reported during acceptance. The research of the defects in buildings occurring during the warranty period may become an important contribution to the analysis of defects. The aim of the article is to present the results of preliminary research of the defects in residential buildings occurring during the warranty period. Due to the small amount of data, the research results cannot be generalized but allow for the formulation of research hypotheses that will be verified in future studies.

This paper is organized as follows. Section 2 contains the literature review. Section 3 introduces formal and legal conditions relating to defects reported during the warranty period. Section 4 describes the analysis method and results. Finally, Section 5 presents conclusions.

2. Literature Review

The acceptance of construction works is regulated differently in the legal systems of each country. For example, as follows from [5], in the German legal system, the subject of acceptance of construction works is regulated in more detail than in the Polish legal system. However, a certain fundamental similarity can be seen in all systems, which is the similarity of the effects of acceptance of construction works, including, above all, the payment of the contractor's remuneration and the calculation of the limitation period for claims concerning the liability for defects.

Defects are a common phenomenon in the construction industry. In the literature, there are many publications related to the frequency of defects in residential buildings. In the years 2009–2012, Ojo and Ijatuyi [6] conducted research into defects on the example of the Sunshine Gardens housing estate in Akure, Nigeria. The most common defects found in the roof structure and covering included the use of improper quality materials, improper wood treatment, poor quality of workmanship, and inaccurate supervision of construction workers. The walls of the analyzed buildings were made in the wrong way, low-quality materials were used, and short window and door lintels were applied. Moreover, the floors were made of low-quality materials. Rotimi et al. [7] presented the example of housing construction in New Zealand to determine the level of detection of defects by independent building inspectors at the time of handing over 216 new residential buildings in the years 2008–2011 and examined the number and types of defects found. The most common defects included uneven painting, nail marks, poor quality of room and floor finishes, incorrectly fixed handles in doors and windows, cracks in buildings, and incorrect installation of toilets. In [8], the results of a research project analyzing the location and type of 3647 faults located in 68 residential buildings in Spain are presented. The research showed that the most frequent defects were inappropriate materials or components, their bad location, surface defects, including uneven surfaces, scratches, cavities, faults of machines, and components affecting their functionality, such as doors rubbing against the floor or nonfunctioning air conditioning. Shirkavand et al. [9] presented the numbers of defects detected during acceptance in Norwegian construction projects, mainly in Trondheim. Seven buildings were analyzed: a kindergarten, four nursing homes, a school, and apartments. The most frequent defects were damage to surfaces and installation networks. Ismail et al. [10] performed a study to investigate the most common defects in 72 new terrace houses in Malaysia. The most common defects were ones on the corners of walls, uneven joints, lack of angles and planes of walls, unevenly painted walls, cavities in wall tiles, doors and windows not closing, unfinished works (such as the unfinished installation of railings), and dampness. The results of studies concerning the number and type of defects are also presented in [3,11–14].

A number of studies have also been devoted to the causes of defects. Ahzahar et al. [15] investigated the factors contributing to construction failures and defects in Malaysia. According to the authors, defects and faults in buildings are affected by, among others, certain building materials, errors during construction, corruption, lack of supervision, and design errors. Mesa Fernández et al. [16] analyzed various factors in quality control in residential construction projects in Spain. The authors proposed changes in control, for example, focusing not only on the management process but also on product quality and greater control of material deliveries. In [17], the defects reported by the representatives of the cooperative are presented and analyzed, focusing on the relationship between the characteristics of the building, the size of the developer/provider company, and the type of defect. According to the authors, the size of the developer company and the location of the building have a significant impact on construction defects. In [18], the main factors influencing the occurrence of defects at the design stage of residential buildings in the Gaza Strip are identified and ranked. For the purpose of analysis, a survey was conducted, which indicated three main design errors: ignoring or incorrectly performing the ground survey, lack of qualified supervision over the drawings, and conflicts between architectural and structural drawings.

On the basis of the literature study, it can be concluded that the main causes of construction defects are poor quality of materials, poor quality of workmanship, and inaccurate control of construction works. During acceptance, defects most often appear on walls: the greatest detection of defects on surfaces may result from the fact that these defects can be easily found visually, and no specialist equipment or great effort is required to detect them.

The purpose of the research on defects in housing construction is aimed at not only quantifying the defects and identifying their causes but also at estimating the costs to be incurred for the necessary repairs. Based on the research conducted by Mills et al. [19] in Australia from 1982 to 1997, it can be concluded that the cost of repairing the defects constitutes around 4% of the contract value. On the other hand, Josephson [20] concluded that the cost of defects corresponds to 4.4% of the construction costs of buildings, and the time to repair them amounts to 7% of the total working time. Kucharska-Stasiak and Mielczarek [21] studied the quality of housing construction on the example of Widzew EF housing estate in Łódź from 1977 to 1980. The authors calculated the costs of correction works on the basis of the collected data and estimated the losses incurred by the contractor due to the poor quality of workmanship.

Building defects are the main reason for exceeding the budget of a construction project, hence the search for a model approach to defect management. Park et al. [22] are working on a construction defect management system involving augmented reality (AR) and modeling information about the BIM (Building Information Modeling). In [23], a model for forecasting defects of multistory reinforced concrete buildings is presented using neural networks (NN–PSO classifier), while in [24], the causes of defects and faults are analyzed using trees and risk measures. In [25], a model based on LDA (Loss Distribution Approach) is presented for the assessment of the system of responsibility for repairing defects and faults in the period of responsibility for defects in residential buildings using the LDA loss distribution method. In turn, Oswald and Abel [4] discussed the procedure for the assessment of defects found in buildings. The authors created and described a block diagram for the evaluation of optical defects and presented a method of utility analysis consisting of considering several alternatives in terms of criteria and selection of the best solution and the Aurnhammer's method for determining the reduction of value when defects are found.

3. Formal and Legal Conditions

The buildings analyzed in the article were implemented by the developer. The contract between the developer and the buyer of an apartment, in accordance with Polish law, is governed by the Act on Protection of Buyers of Apartments or Single-Family Houses [4–27]. According to the Act, the transfer of ownership to the purchaser is preceded by acceptance in the presence of the customer or his proxy (notarial power of attorney is required). Before taking over the residential unit, the developer is obliged to notify the buyer of this fact. Based on the legal status, as well as court decisions, the authors have prepared a graph depicting the moment of commencement of the warranty period in the case of developer investments (Figure 1).

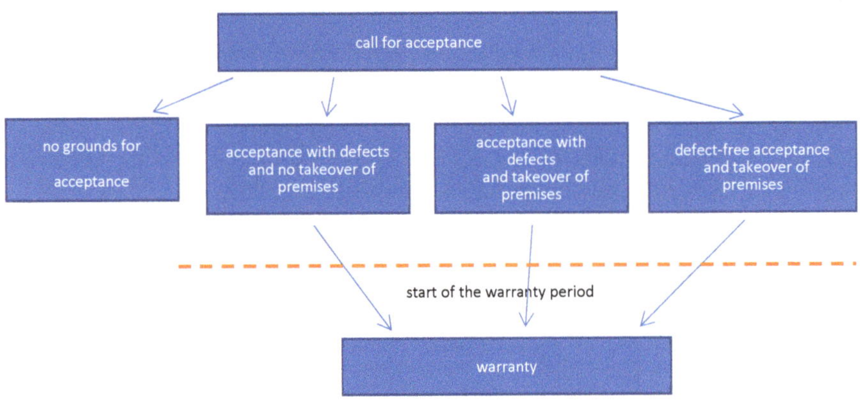

Figure 1. Diagram showing the start of the warranty after acceptance.

The buildings analyzed in the paper were executed by a developer. According to the law, the transfer of ownership to the buyer is preceded by acceptance in the presence of the customer. During the acceptance meeting, the client or the company supporting the acceptance inspects the apartment, after which an acceptance protocol is to be prepared, to which the buyer or their proxy may report any defects or faults found during the acceptance meeting. The customer, as the receiving party, has the right to decide which defects will be included in the acceptance protocol. He or she has the right to report all objections to the apartment, even the least important ones, such as dirt on the handles.

After the possible identification of minor defects and preparing the acceptance report, the apartment is handed over to the client. Before handing over the apartment, minor assembly work is usually done. In case of detecting significant defects and faults, the apartment is not handed over to the client, and the developer is obliged to repair the defects. At the time of acceptance, the customer agrees to the condition of the apartment. This means that if the customer has not noticed any defects during acceptance, or he or she has not decided to put a defect noticed in the acceptance report, the customer has no right to demand its repair. If the defect was, for example, hidden or a result from the operation of the building, the buyer of the apartment can take advantage of a 5-year warranty for construction works because, during this time, the developer is obliged to maintain the so-called "acceptance status" of the flat.

To a large extent, the results presented in this paper form a continuation of the research on defects occurring at the acceptance stage [3,27]. The research results presented here involve the case of a couple of buildings, one of which had a statistical analysis of defects performed at the acceptance stage.

4. Methods

4.1. Data Collection

The results presented in this paper are based on data that were collected through analysis of the reports on the state of defects made by the inspector during the warranty period. The area of the conducted research included a total of 3 residential buildings. The buildings are part of various projects but were produced by the same developer and contractor. Two of them possess 16 aboveground stories, which are divided into 3 staircases with 172 flats, each with areas ranging from 29 to 117 m^2. These buildings were accepted in 2017 and 2018, which, in further analysis, coincides with the beginning of the research. The third building is a 6-story one, with two staircases and 73 flats. The time of the building's acceptance was August 2018. Details of the facilities analyzed are presented in Table 1.

Table 1. Details of the residential buildings.

Building	The Number of Stories	The Number of Staircases	The Number of Flats	Total Area of Flats (m^2)	The Time of Acceptance
A	16	3	172	above 8000	2017/2018
B	16	3	172	above 8000	2017/2018
C	6	2	73	3000	August 2018
Total			432	19,000	

All buildings were finished to the same degree. The differences were in the type of finishing materials used in the common areas (that is, in staircases and the outdoor area). The development standard included plastering inside the premises, walls prepainted with white paint, screed, window and balcony joinery, entrance doors, embedded internal and external window sills, distributed ventilation system with diffusers for mechanical ventilation, distributed central heating installation together with installed radiators, water and sewage system without fittings, electrical installation together with plug and lighting sockets and switchboard, distributed teletechnical installation with a collective box, sockets and intercom, balustrades, and tiles on external floors. A precise definition of the elements of the shell unit standard is important in terms of warranty claims.

The collected results of the research present an assessment of the validity of defect reports made by customers. The data are the result of actions taken by the developer, the supervision inspector, and the contractor. Each defect included in the statistics underwent verification of the original position of the developer. The role of the supervision inspector is to make an independent assessment and qualification of the report of potential defects by customers and managers of individual properties. The results of the research presented in this article are, therefore, based on expert knowledge.

The collected data on warranty notifications cover the period from building acceptance in January 2018 and August 2018 to March 2020. A total of 560 notifications on the occurrence of defects during the warranty have been identified in the course of the conducted research. The collected data were analyzed statistically using GNU PSPP Statistical Analysis Software. First, the authors determined the percentage of valid and unfounded claims and, next, the percentage of claims under warranty cumulatively. Subsequently, the statistics of the three-stage qualification of defects were established—of low, medium, and high significance—as well as the relationship of defects to the place where they occur. The successive analyses concerned the type of defects. Pearson's parametric correlation and the determination factor R2 were computed to assess the relationship between the defects reported in different buildings and to establish the strength of the interdependence between different groups of defects.

4.2. Data Analysis

A total of 560 notifications on the occurrence of defects during the warranty were identified. The research concerned 353 defects qualified as valid, which accounted for 63.04% of all reports. Additionally, 207 defects were considered unfounded, which accounted for 36.96% of the reports. Table 2 presents the summary statistics of the reported defects. Buildings A and B are larger both in terms of the number of floors and the number of flats. Moreover, they were accepted earlier than Building C, which significantly affected the number of reports identified in the study.

Table 2. The statistics of defects.

Building	Number of Defects	Valid		Unfounded	
		No.	%	No.	%
A	250	165	66.00	85	34.00
B	247	147	59.51	100	40.49
C	63	41	65.08	22	34.92
Total	560	353	63.04	207	36.96

What draws attention to the data presented is the similarity in the percentage of valid defect reports in the individual buildings. The comparability of the results obtained is the result of the relatively large size of the research sample, which eliminates the divergence of results.

Taking into account the different warranty periods on individual buildings, in order to unify the presentation of statistical characteristics of the testing ground, the occurrence of both the reports of defects as well as their justification was calculated per one month of warranty per one commercial unit. The results of the calculations are presented in Table 3.

Table 3. Defect statistics per unit.

Building	Number of Defects/Month/One Flat	Number of Valid Defects/Month/One Flat	Number of Unfounded Defects/Month/One Flat
A	0.0543	0.0359	0.0185
B	0.0537	0.0320	0.0217
C	0.0486	0.0316	0.0170
Average	0.0531	0.0335	0.0196

The authors also analyzed the validity of the defects reported within 4 months from the date of acceptance, which is related to the typical time of finishing the premises by the owners (Table 4), and after 12 months, namely, the period when a significant number of premises are already finished and used (Table 5).

Table 4. Defect statistics after 3 months from acceptance.

Building	Number of Defects	Valid		Unfounded	
		No.	%	No.	%
A	61	34	55.74	27	44.26
B	43	24	55.81	19	44.19
C	12	9	75.00	3	25.00
Total	116	67	57.76	49	42.24

Table 5. Defect statistics after 12 months from acceptance.

Building	Number of Defects	Valid		Unfounded	
		No.	%	No.	%
A	153	96	62.75	57	37.25
B	135	79	58.51	56	41.48
C	46	27	58.70	19	41.30
Total	334	202	60.48	132	39.52

The analysis performed helped us to formulate two important statements. One is that the owners are very active in making warranty claims in the first 3 months from the date of acceptance. The analysis revealed that 116 reports took place in the first months of use, which accounts for 20.71% of all reports included in the study within only 11% of the study's timeframe. The other observation is the increasing percentage of reported defects, which for the first 3 months was 57.76%, for 12 months 60.48%, and at the end of the study (after half the warranty period) 63.04%. The validity of the defects is presented cumulatively in Figure 2. The deviation of the result for Building C for the first three months of the warranty was due to its smaller size in relation to the other two, which was also associated with a small number of reports, namely, 12 of them.

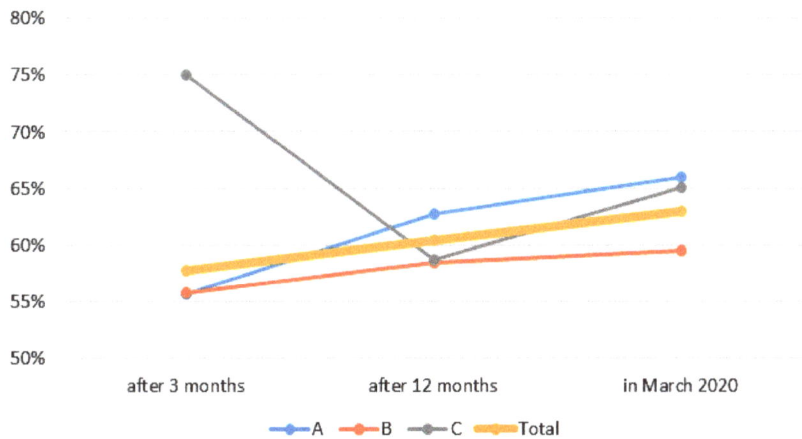

Figure 2. Percentage of valid claims under warranty, presented cumulatively.

4.3. The Classification of Defects

Based on the analysis of the literature, as well as on the results of the research conducted during the warranty period, the authors propose a three-stage qualification of defects: low, medium, and high significance. The division into these groups corresponds to many publications [3]. In this paper, the following definitions are adopted:

- Defects of high significance—defects causing the appearance of a fault that prevents proper operation of the premises, which may cause a threat to the health or life of people using the property. Removal of very important defects usually requires the elimination of the causes and their effects. Repair works are associated with large financial and material outlays and are time-consuming. Examples of very important defects include flooding of the property, cracking of a structural element, short-circuit of the electrical system, causing fire, or leakage of the roof sheathing.
- Significant defects—faults causing limitations in the proper operation of the premises. This formulation identifies a number of defects that reduce the possibility of unrestricted use.

However, they do not force the cessation of use because they do not constitute a direct threat to the health and life of the users. Due to the large dispersion of the types of these defects, it is not possible to clearly determine the degree of technical and technological complexity of their removal. Examples include the following: intercom failure, decaying tiles on the balcony, unsealing of window and balcony joinery, or slow main door locks.

- Defects of low significance—a minor defect that does not hinder the operation of the premises. This term is used by the authors to describe a defect that has a visual or cosmetic effect, which does not influence the functional properties of the premises. The removal of these defects does not generate the necessity of using any advanced technology, while the time of their removal may vary. They can be illustrated by the following examples: wall scratches, paint chips, spontaneous scratching of the glass, or loosening the silicone next to balcony tiles.

Taking into account the indicated classification, this paper also analyzes the significance of the defects occurring. The results of the significance tests are presented in Table 6.

Table 6. Statistics of defect significance.

Building	Defects of High Significance	Significant Defects	Defects of Low Significance
A	7.88%	55.15%	36.97%
B	19.05%	51.70%	29.25%
C	4.88%	70.73%	24.39%
Average	12.18%	55.53%	32.29%

As can be seen, the largest percentage, both globally and for individual buildings, was significant defects. It should be noted here that the class of these defects includes issues related to the regulation of woodwork, not the operation of the intercom system or improper connection of the electrical system. The defects of high significance, on the other hand, constitute a state of emergency, therefore in their case, a large discrepancy in frequency can be observed. For Building B, as much as 19.05% were highly significant defects, while Building C had only 4.88% of such reports. The results of the preliminary research also revealed that about 1/3 of the valid reports concerned minor defects, mainly of a cosmetic nature.

4.4. Relationship of Reported Defects and the Location

The next steps of the preliminary research were the investigation of the relationship between the reported defects during the warranty and their location. Two cases were considered:

- the defect has been reported by the manager, which is the same as its location in common parts,
- the fault has been reported by the owner of the flat and does not concern elements other than those covered by the boundaries of the flat (additional spaces such as storage rooms or lockers are excluded).

The authors, after analyzing the data from Tables 7 and 8, point out some interesting observations. The first one is the low activity of property managers in the initial phase of the building operation—an average of 9.20% of defects reported during the first 3 months—which stands in contrast to the activities of apartment owners—an average of 22.05% of the reports during the first 3 months from the date of building acceptance. Another observation is the practical equalization of the percentage of reports after 12 months: 57.47–58.95%. On this basis, it can be hypothesized that for the developer in the construction industry, the key period for stabilization of the occurrence of defects is the first 12 months.

Table 7. Defects reported by the manager concerning the common parts.

Building	Number of Defects	After 3 mth		After 12 mth	
		No.	%	No.	%
A	40	2	5.00	24	60.00
B	25	1	4.00	10	40.00
C	22	5	22.73	16	72.73
Total	87	8	9.20	50	57.47

Table 8. Defects reported by the owners of the flat concerning the residential part of the building.

Building	Number of Defects	After 3 mth		After 12 mth	
		No.	%	No.	%
A	204	54	26.47	120	58.82
B	213	40	18.78	121	56.81
C	41	7	17.07	29	70.73
Total	458	101	22.05	270	58.95

4.5. Statistical Analysis of Defect Types

Taking into account reports appearing during the warranty period, the defects were divided into several groups. The first three groups included reports concerning problems with electrical, plumbing, and central heating (c.h.) installations in apartments. The next groups are reports concerning window and door joinery. Quite often, there were defects consisting of dampness, mainly of walls, from different causes. Separately, reports of wall scratches and other problems related to elements of apartment finishing (such as floors) were grouped. The last group is various defects occurring in the area of balconies and terraces. For each of the buildings, the number of defects was determined in the range of 1–3, 4–6, 7–12, and 13–26 months from the date of acceptance. The list is presented in Table 9.

Table 9. The number of defects in groups.

Building	Time (mth)	Electr. Install.	Plumb. Install.	c.h. Install.	Joinery	Dampness	Scratches	Terraces	Other
A	1–3	26	2	9	12	3	4	5	0
	4–6	3	0	0	8	0	7	0	1
	7–12	5	16	5	21	9	3	12	2
	13–26	11	7	3	25	8	21	19	3
Total in A		45	25	17	66	20	35	36	6
B	1–3	20	2	3	9	6	0	1	2
	4–6	9	6	1	6	5	4	12	1
	7–12	11	4	4	8	5	5	10	1
	13–26	9	8	7	33	18	15	14	8
Total in B		49	20	15	56	34	24	37	12
C	1–3	4	2	1	4	0	0	0	1
	4–6	0	1	1	2	0	2	0	1
	7–12	2	2	3	4	8	7	1	0
	13–26	5	1	1	6	0	3	0	1
Total in C		11	6	6	16	8	12	1	3
Total		105	51	38	138	62	71	74	21

The most numerous group is woodwork defects, and the least numerous group is central heating system faults. These dependencies are the same in all analyzed buildings.

In order to confirm that the number of defects appearing in individual groups is not accidental, the correlation between the numbers of defects in Building A and Building B in the first three months after acceptance was checked (Table 10). Building C was not taken into account due to insufficient data. Then, a correlation was established between all defects reported in all three buildings over the entire observation period (Table 11).

Table 10. The number of defects in the groups in the first three months.

Time [m-th]	Electr. Install.	Plumb. Install.	c.h. Install.	Joinery	Dampness	Scratches	Terraces
1–3	26	2	9	12	3	4	5
4–6	20	2	3	9	6	0	1

Pearson correlation r = 0.93; significance level $p = 0.003$.

Table 11. The number of defects in the groups in all buildings.

Building	Electr. Install.	Plumb. Install.	c.h. Install.	Joinery	Dampness	Scratches	Terraces
A	45	25	17	66	20	35	36
B	49	20	15	56	34	24	37
C	11	6	6	16	8	12	1

Correlations: between A and B, r = 0.86, $p = 0.013$; between A and C, r = 0.66, $p = 0.106$; between B and C, r = 0.51, $p = 0.24$.

The analyses performed allow us to confirm the hypothesis about the lack of randomness of the collected data. Correlation for data from Table 10 allows us to confirm a very strong correlation, while the significance level allows us to reject the hypothesis of zero correlation. For the data in Table 11, the indicators show a very strong correlation between the data for Buildings A and B and a strong correlation for Building C.

For the data concerning all defects (Table 12), the correlation coefficients and the coefficients of determination for individual defect groups were checked. This was to find the relationship between the analyzed groups.

Table 12. The number of defects in the group in individual time intervals.

Time (mth)	Electr. Install.	Plumb. Install.	c.h. Install.	Joinery	Dampness	Scratches	Terraces
1–3	50	6	13	25	9	4	6
4–6	12	7	2	16	5	13	12
7–12	18	22	12	33	22	15	23
13–26	25	16	11	64	26	39	33

The strongest correlations were found between "joinery" and "scratches" (r = 0.91; $R^2 = 0.82$), "joinery" and "dampness" (r = 0.88; $R^2 = 0.78$), "joinery" and "dampness" (r = 0.88; $R^2 = 0.78$), "scratches" and "terraces" (r = 0.93; $R^2 = 0.86$), "scratches" and "dampness" (r = 0.76; $R^2 = 0.58$), and "dampness" and "terraces" (r = 0.92; $R^2 = 0.84$).

Figures 3–5 show the graphs of the number of defects in individual time intervals, respectively, for defects that appear in a similar distribution over time.

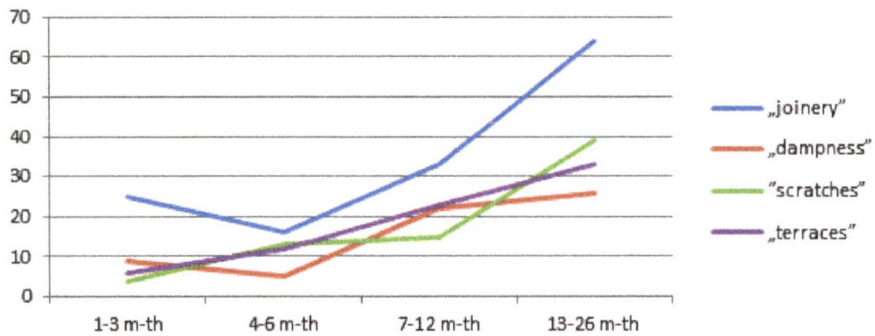

Figure 3. The graphs of the distribution of defects: "Joinery", "Dampness", "Scratches", and "Terraces".

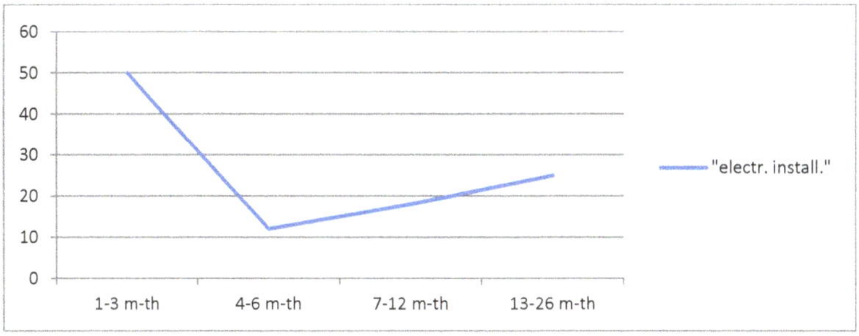

Figure 4. The graphs of the distribution of defects: "Electr. Install.".

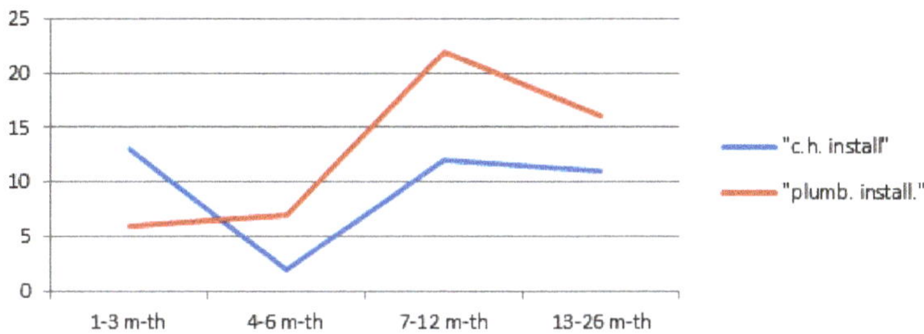

Figure 5. The graphs of the distribution of defects: "c.h. Install." and "Plumb. Install.".

Figure 3 illustrates the groups of defects with the highest correlation. These defects are characterized by a relatively small number of cases reported in the first half of the year after the building acceptance and a gradual increase in the subsequent warranty period. The characteristics of defects associated with the electrical installation are definitely different, where by far the largest number of defects are reported in the initial warranty period and their number decreases with time. The pattern of faults in other installations, for which the level of occurrence is similar throughout the observed warranty period, is slightly different.

5. Conclusions

A key stage in the mutual relations between the developer and the buyer of an apartment, which, however, does not put an end to the mutual commitment, is the acceptance of works. In the Polish legal environment, the rules of liability are defined by the guarantee and the warranty. The warranty is valid by law and is regulated in detail in the Civil Code.

The paper analyzes warranty notifications in three multifamily residential buildings in the period from their acceptance: January 2018 and August 2018 to March 2020. In total, 560 reports of defects during the warranty were identified. The analyses showed 353 defects qualified as valid and 207 considered as unfounded. Due to the small amount of data, the research results cannot be generalized but allow for the formulation of research hypotheses that will be verified in future studies. The results of the preliminary research reveal that owners are very active in making warranty claims in the first three months from the date of acceptance, and, on the other hand, the percentage of validity of the claims increases with time. In terms of the significance of defects, the largest percentage concerned significant defects. Approximately one-third of reasonable claims involved minor defects, mainly of a cosmetic nature. The results of the preliminary research revealed little activity of property managers in the initial phase of the buildings' operation, which is the opposite of that of apartment owners. Reports of defects in window and door joinery, moisture, wall scratches, and those appearing in the area of balconies and terraces are characterized by a relatively low number of cases reported in the first half of the year after the building is commissioned and a gradual increase in the subsequent warranty period. On the other hand, faults related to electrical installation are most often reported in the initial period of the warranty, and their number decreases with time.

Understanding defects is a vital prerequisite to their prevention and elimination. So far, the research has focused mainly on defects appearing at the stage of the acceptance of works. The analyses of the defects occurring during the warranty period could fill the research gap, which may significantly affect the development of defect management procedures and the creation of a knowledge map concerning the frequency of defects in particular places of the building and building elements. There are specific costs associated with repairing the defects. Knowledge about those occurring in residential buildings can, therefore, be used for better planning of the investment budget.

The limitation of the research is the small test sample, which only allows for the formulation of research hypotheses concerning the dependence of the appearance of defects during the warranty period. Initially, the research may have had some bias, being of the same builder, a very close period of time, and buildings with similar characteristics. The authors will conduct further research. A larger amount of data will allow us to confirm the conclusions of the present paper.

Author Contributions: The individual contribution and responsibilities of the authors were as follows: E.P.: literature review, writing—original draft preparation, writing—review and editing; J.M.: conceptualization, resources, methodology, data curation. All authors have read and agreed to the published version of the manuscript.

Funding: This research received no external funding.

Conflicts of Interest: The authors declare no conflict of interest.

References

1. Act of 23 April 1964—Civil Code (Dz.U.2019.0.1145). Available online: https://isap.sejm.gov.pl/isap.nsf/download.xsp/WDU20190001145/T/D20191145L.pdf (accessed on 1 September 2020).
2. Judgment of the Court of Appeal in Katowice of 21 April 2009, Reference Number Act ACa 88/2009. Available online: http://arbitraz-sidir.pl/orzecznictwo/sady-powszechne/sygnatura-akt-v-aca-8809/ (accessed on 1 September 2020).
3. Plebankiewicz, E.; Zima, K.; Malara, J.; Biel, S. A procedure of repairing housing defects in development investments. In Proceedings of the MATEC Web of Conferences, Krynica, Poland, 15 February 2019; Volume 262.
4. Oswald, R.; Abel, R. Wady i Usterki w Budynkach. In *Defects and Faults in Buildings*; Instalator Polski: Warszawa, Poland, 2000. (In Polish)
5. Plebankiewicz, E.; Mosio, T. Odbiór robót budowlanych zgodnie z polskim i niemieckim systemem prawnym [Construction works acceptance according to Polish and German legal systems]. *Mater. Bud.* **2016**, *6*, 159–161. (In Polish)
6. Ojo, A.M.; Ijatuyi, O.O. Defective Construction in Residential Buildings: A Study of Sunshine Gardens, Akure Nigeria. *Int. J. Civ. Eng.* **2014**, *1*, 16–30.
7. Rotimi, F.E.; Tookey, J.; Rotimi, J.O. Evaluating defect reporting in new residential buildings in New Zealand. *Buildings* **2015**, *5*, 39–55. [CrossRef]
8. Forcada, N.; MacArulla, M.; Gangolells, M.; Casals, M. Assessment of construction defects in residential buildings in Spain. *Build. Res. Inf.* **2014**, *42*, 629–640. [CrossRef]
9. Shirkavand, I.; Lohne, J.; Lædre, O. Defects at Handover in Norwegian Construction Projects. *Procedia Soc. Behav. Sci.* **2016**, *226*, 3–11. [CrossRef]
10. Ismail, I.; Ani, A.I.C.; Razak, M.Z.A.; Tawil, N.M.; Johar, S. Common building defects in new terrace houses. *J. Teknol.* **2015**, *75*, 83–88. [CrossRef]
11. Stangierski, K.; Mielczarek, W. Jakość nowowznoszonego budownictwa mieszkaniowego w świetle badań w RSM" Polesie" w Łodzi. *Acta Univ. Lodziensis. Folia Oeconomica* **1984**, *38*, 41–58. (In Polish)
12. Gurmu, A.; Paton-Cole, V. A Review of defects in low-rise residential buildings in the Australian State of Victoria. In Proceedings of the ICEC-PAQS 2018: Grassroots to Concrete Jungle, Sydney, Australia, 18–20 November 2018.
13. Forcada, N.; Macarulla, M.; Love, P.E.D. Assessment of residential defects at post-handover. *J. Constr. Eng. Manag.* **2013**, *139*, 372–378. [CrossRef]

14. Kasi, Z.K.; Mahar, W.A.; Khan, J.F. Structural defects in residential buildings: A study of Quetta, Pakistan. In Proceedings of the 1st International Conference on Advances in Engineering & Technology (ICAET 2018), Quetta, Pakistan, 2–3 April 2018; pp. 1–16.
15. Ahzahar, N.; Karim, N.A.; Hassan, S.H.; Eman, J. A study of contribution factors to building failures and defects in construction industry. *Procedia Eng.* **2011**, *20*, 249–255. [CrossRef]
16. Fernandez, J.M.M.; González, C.P.; Cabal, V.Á.; Balsera, J.V. Analysis of the quality control planning in residential construction projects in Spain. *Rev. Constr.* **2016**, *15*, 106–114.
17. Zalejska Jonsson, A.; Hungria Gunnelin, R. Defects in newly constructed residential buildings: Owners' perspective. *Int. J. Build. Pathol. Adapt.* **2019**, *37*, 163–185. [CrossRef]
18. Tayeh, B.A.; Maqsoom, A.; Issa Abu Aisheh, Y.; Almanassra, M.; Salahuddin, H.; Irshad Qureshi, M. Factors affecting defects occurrence in the construction stage of residential buildings in Gaza Strip. *Appl. Sci.* **2020**, *2*, 167.
19. Mills, A.; Love, P.E.D.; Williams, P. Defect costs in residential construction. *J. Constr. Eng. Manag.* **2009**, *135*, 12–16. [CrossRef]
20. Josephson, P.E. *Defects and Defect Costs in Construction*; Chalmers University of Technology: Gothenburg, Sweden, 1998.
21. Kucharska-Stasiak, E.; Mielczarek, W. Jakość wykonawstwa budownictwa mieszkaniowego na przykładzie osiedla Widzew EF w Łodzi. *Acta Univ. Lodziensis. Folia Oeconomica* **1983**, *26*, 53–62. (In Polish)
22. Park, C.S.; Lee, D.Y.; Kwon, O.S.; Wang, X. A framework for proactive construction defect management using BIM, augmented reality and ontology-based data collection template. *Autom. Constr.* **2013**, *33*, 61–71. [CrossRef]
23. Chatterjee, S.; Sarkar, S.; Hore, S.; Dey, N.; Ashour, A.S.; Balas, V.E. Particle swarm optimization trained neural network for structural failure prediction of multistoried RC buildings. *Neural Comput. Appl.* **2017**, *28*, 2005–2016. [CrossRef]
24. Aljassmi, H.; Han, S. Analysis of causes of construction defects using fault trees and risk importance measures. *J. Constr. Eng. Manag.* **2013**, *139*, 870–880. [CrossRef]
25. Lee, S.; Lee, J.; Ahn, Y. LDA-Based Model for assessing the defect liability system in residential buildings' maintenance phase. *J. Perform. Constr. Facil.* **2020**, *34*, 04020007. [CrossRef]
26. Plebankiewicz, E.; Zima, K.; Malara, J.; Biel, S. Analiza statystyczna usterek w budynkach mieszkalnych. [Statistical analysis of defects in residential buildings]. *Mater. Bud.* **2018**, *1*, 45–48. (In Polish)
27. Act of 16 September 2011 on the Protection of the Rights of Buyers of a Dwelling or a Single-Family House. Available online: http://isap.sejm.gov.pl/isap.nsf/DocDetails.xsp?id=WDU20112321377 (accessed on 1 September 2020).

applied sciences

MDPI

Article

Structural Analysis of Factors Influencing the Costs of Facade System Implementation

Agnieszka Leśniak * and Monika Górka

Faculty of Civil Engineering, Cracow University of Technology, 31-155 Krakow, Poland; mgorka@l7.pk.edu.pl
* Correspondence: alesniak@l7.pk.edu.pl

Received: 6 August 2020; Accepted: 28 August 2020; Published: 31 August 2020

Abstract: External facades of buildings and other structures shape the image of every building, creating the architecture of cities. Traditional concrete forms, as a symbol of durability and stability, have been replaced by lightweight enclosures—for example, in the form of aluminium–glass facades and ventilated facades. In this paper, the authors attempt to verify the strength of influence and relations between the identified factors shaping the costs of facade system implementation using structural analysis. On the basis of the collected quantitative and qualitative data obtained as a result of research on design documentation and cost estimates of implemented public buildings, as well as on the basis of interviews conducted among experts, factors which have a real impact on the costs of facade systems in the form of aluminium and glass facades and ventilated facades were identified. The indicated factors were analysed and classified using the method of structural analysis, namely the MICMAC method (refers to the French acronym for Cross-Impact Matrix Multiplication Applied to Classification). Particular influences and relations between factors were examined. Finally, six groups of factors influencing the costs of facade systems were identified, including regulatory factors that do not have a very strong impact on the level of costs, but which show a strong correlation with other factors; determinants that have a very strong impact on the costs; and a group of external factors that show the smallest influence on the estimation of façade cost.

Keywords: facade systems; cost management; structural analysis; MICMAC method

1. Introduction

External facades of buildings and other structures shape the image of every building, creating the architecture of cities. The dynamic development of technology, the discovery of new innovative building materials [1], as well as the ever-increasing technical, production, and execution capabilities have resulted in exterior walls with complex shapes and forms. Traditional concrete forms as a symbol of durability and stability have been replaced by lightweight enclosures—for example, in the form of aluminium–glass facades and ventilated facades [2]. Aluminium and glass facades are built from aluminium sections, and the spaces between the aluminium construction is filled with glass. Glass that is used in construction should meet the requirements of thermal protection, fire protection, burglary protection, protection against noise, and safety of use [3]. A solid building made in the form of aluminium–glass facades is an indispensable element of architecture for a modern city [4]. Also, ventilated facade systems have gained much attention. It is a cladding system, which incorporates the insulation layer, positioned usually on the external surface of the wall, the air cavity and the outer skin [5]. The air gap allows air to enter and flow through the facade. The external part of the ventilated façade is a panel suspended, glued, or screwed to the substructure. Ventilated facades allow for shaping external claddings from various materials, structures, textures, or colours. The panels can be made of many materials, including wood, fiber concrete, aluminium and ceramics. Due to their high aesthetics, ventilated facades are increasingly often used as external parts of newly built buildings, but they are also perfect for buildings undergoing renovations [2,6]. This makes the ventilated facade an effective form of finishing the building—it adds prestige and distinguishes it from the environment. Descriptions and photos or characteristics of the discussed facade systems can be found in a number of publications [4,7,8].

The modern trend in the design of exterior facades as lightweight structures is primarily reflected in examples of public buildings, and constitutes the mainstay of urban architecture. The construction of skyscrapers and high-rise buildings are symbols of all larger metropolises.

The external facades of such buildings are no longer just the outer cladding itself, but a combination of many interdisciplinary functions. They combine stability and reliability with aesthetics, usability [9], energy efficiency, and ecology [10]. All these aspects affect the quality of the assembly process, time of execution, and costs incurred [3]. These three elements determine the success of each construction project. Many studies presented in the literature concern the quality of construction projects [11], their costs [12,13], and time schedule [14], as well as the maintenance of the existing buildings [15] and the opportunity to improve the energy efficiency of buildings [16].

Reliable cost estimation is one of the most important aspects of construction projects, both from the investor's and the contractor's point of view [17]. Estimating the costs of facade systems is time-consuming and complicated [3]. The calculation process is influenced by architectural, construction, and system data, as well as parameters related to production and the type of assembly. It is necessary to have an individual approach to each investment, starting from the analysis of design documentation and ending with the location of the construction project. The cost amount is influenced not only by the prices of individual facade elements, but additionally by the costs of their execution in the production plant and assembly on the construction site. In the case of facade system implementation, one should take into account indirect costs, widely understood, which are closely related to the specific location of the investment, the availability of land around the building, or the option of setting scaffolding or others factors [18].

Many works present attempts to identify factors and analyse them in terms of research issues and problems in the construction industry. They concern, for example, factors shaping the costs of works [19], risk factors [20], factors affecting occupational safety [21], or determining energy consumption [22]. The study of these issues, their analysis, as well as the application of innovative solutions, mathematical methods, or artificial intelligence, enables the construction of models supporting decision-making in construction management [23,24].

In this paper, the authors attempted to verify the strength of influence and relations between the 15 factors shaping the costs of facade systems using structural analysis. The factors that have a real impact on the costs of aluminium and glass facades and ventilated facades were identified as results of research conducted in Poland. Quantitative and qualitative data were obtained as a result of research on design and cost documentation implemented between 2013–2019 for 80 public buildings. The details of data collection are described later in the paper. Finally, the factors were analysed and classified using a method of structural analysis, namely the MICMAC method (referring to the French acronym for Cross-Impact Matrix Multiplication Applied to Classification, developed by Michel Godet in 1971 [25]). This is one of the methods for organizing and analyzing a set of variables, which enables the study of relationships and mutual relationships between them.

2. Short Characteristic of Selected Facade Systems

Aluminium–glass facades are one of the solutions dedicated to public, office, commercial, and service buildings [4]. Such buildings have become the objects of designers' architectural ideas, attracting designers with their form and original shape. The facade of the building should give the character of the object, as well as ensure comfort in its use. The facades are used to shape external walls, thanks to which curvilinear facades can be designed and made; these are characterized by lightness of form. The core of the facade is made of aluminium sections, while the filling is glass. The most frequently used systems are mullion–transom, semistructural, and structural systems [18]. Types of aluminium–glass facades are illustrated in Figure 1.

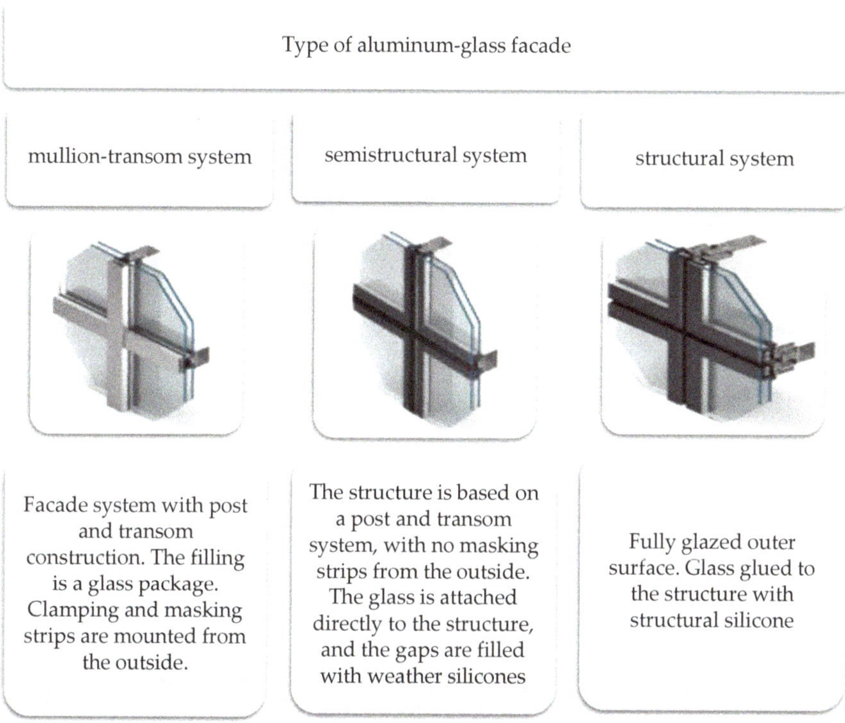

Figure 1. Type of aluminium–glass facades. Own study based on [26].

The first type of structure consists of mullion–transom, and the inner space is filled with glass. On the outside of the mullion–transom system are termination bars and masking strips. A semistructural façade is smooth. Glass is fixed directly to the construction, and the gap is filled with a special weather silicone. The structural system is characterized by the possibility of obtaining an external glass faced without visible elements. This effect can be achieved by gluing the glass directly to the aluminium structure.

The other facade system frequently used for public buildings in combination with aluminium and glass facades is a ventilated facade [18]. This solution is based on leaving a ventilation gap between the external cladding and the thermal insulation layer. A ventilated facade is defined as a set of elements for enclosing external walls, consisting of the following [2]:

- external panels (for instance, cement, stone, wood-based, plastic, metal, or laminate) attached to the grate (suspended, glued, or screwed to the grate);
- grate (made of metal or wood) attached—suspended, glued, or screwed—to the external wall;
- fastening elements;
- insulating material.

Due to the variety of materials used for the external panel and the substructure used (grate), it is possible to present the classification of ventilated facades, as shown in Figure 2.

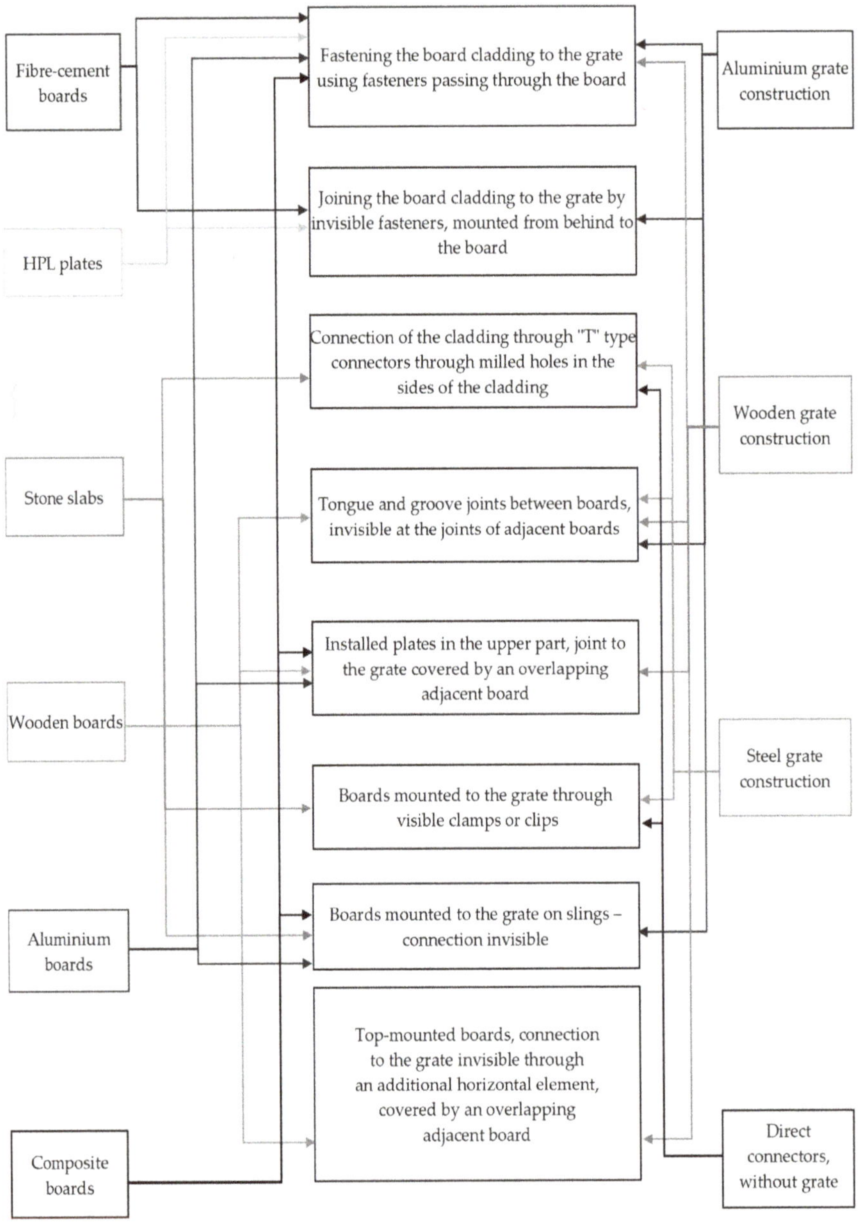

Figure 2. Classification of ventilated facades, according to the material of the external panel and the method of installation. Own study based on [8].

Ventilated facades should meet all the requirements for technical assessment, and above all, should be tested for fire safety, namely, reaction to fire, resistance to fire, and ability to continuously smoulder [27]. Ventilated fire facade elements should be tested, and their fire class should be equal to class "A". Aluminium and steel grates, which are substructures of ventilated facades, are classified as non-combustible in class A1, and they are most often used for fire-rated ventilated facades. On the other hand, the external lining itself, whether in the form of composite panels or HPL (High Pressure Laminate) panels, should have a fire-resistant core, which gives the fire-resistant characteristics to the element in question.

3. The Structural Analysis: Methodology

Structural analysis, namely the MICMAC method (refers to the French acronym Matrice d'Impacts Croisés Multiplication Appliquée á un Classement/Cross-Impact Matrix Multiplication Applied to Classification), allows us to examine particular influences and relations between specific variables or factors. The MICMAC method examines direct influences, but also analyses indirect relationships that are not always noticed by experts and analysts [28]. The input element of the analysis is the evaluation and description of direct impacts made by the experts, and then indirect impacts that may occur between factors are additionally examined. As a result of the application of this method, it is possible to separate those factors which are the most decisive and crucial for the examined area from the set of variables. The method also makes it possible to prioritize and organize the variables that seemingly do not influence each other, but thanks to cross analysis, it is possible to show their mutual interactions [29,30]. The individual steps of the structural analysis are depicted in Figure 3.

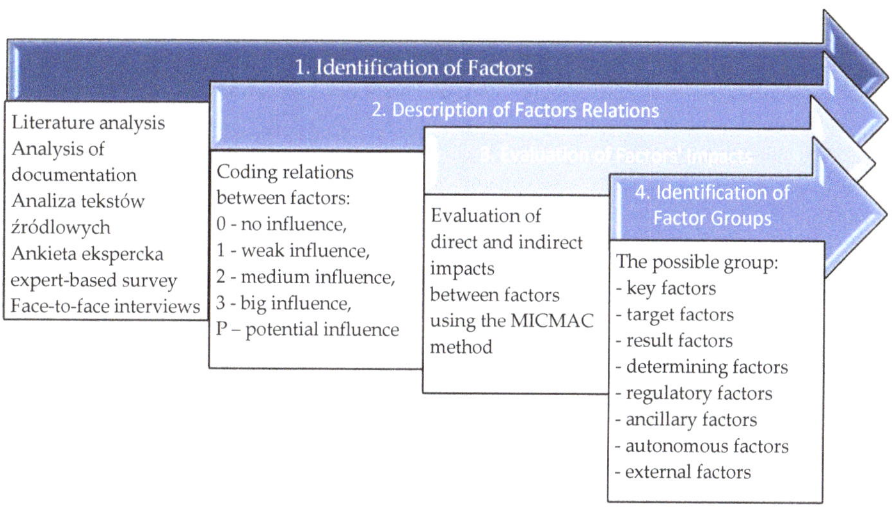

Figure 3. Structural analysis stages. Own study based on [28].

3.1. *The First Stage: Identfication of Factors*

The basis for structural analysis is the identification of the variables/factors that influence and shape a given research area. This is the first and most time-consuming stage. The collected research materials, documentation, literature studies, expert surveys, and face-to-face interviews allow us to distinguish a group of factors that have a decisive influence on the analysed problem.

3.2. *The Second Stage: Description of Factor Relationships*

The second stage of the analysis involves a description of mutual relationships between individual factors by coding the relationships. Mutual relations are defined as follows [28]: 0 = no influence, 1 = weak influence, 2 = medium influence (significant but not decisive), 3 = big influence (decisive), and P = potential influence. This step is usually performed by experts.

3.3. The Third Stage: Examination of Factors' Impacts

The next step examines the direct and indirect impacts that may occur between factors. To evaluate and describe the direct and indirect factor relationships, the authors used the MICMAC free online software developed by French Computer Innovation Institute 3IE and LIPSOR Prospective (foresight) Strategic and Organisational Research Laboratory [25]. The software allows for a quick calculation. Its operation is based on the algebraic principle of Boolle's logic, which is usually used to build scenarios at the initial stage of describing future trends.

On the basis of the experts' assessment, a direct impact matrix and a graph are built, the vertices of which represent the factors. In order to calculate the strength of each factor's influence on another factor, the number of paths (relationships) between them and their length is then determined as the strength of the relationship. Indirect relationships between the factors are obtained by successive exponentiation of the direct influence matrix. The total strength of influence is the sum of all elements of the matrix row of direct influence for a given factor, while the total strength of dependence is the sum of elements in the matrix column.

As a result, two matrices are built out of direct and indirect interactions, as well as two graphs of the direct and indirect impact strength of the variables.

3.4. The Fourth Stage: Identification of Factor Groups

A comparison of the results of different classifications of variables as direct (dependent), indirect, or potential impacts enables an in-depth analysis of the subject under consideration.

Additionally, the analysis allows us to distinguish the following in the structure of the research area [31]:

- Key factors, which are characterized by a very high impact on other factors;
- Target factors, which involve variables that change themselves to a large extent under the influence of factors other than those that affect them;
- Result factors, which have a low impact on the structure of the research area, but are very dependent on other factors;
- Determining factors (mainspring and barrier), which describe, characterize, and control the structure of the research topic. These have a real impact on the whole system, either driving it or being a barrier to the whole operation;
- Regulatory factors and ancillary factors, which do not affect the structure and do not have any strong dependencies on other factors, but help to define the strategic objective;
- Autonomous and external factors, which include secondary factors that have very little influence on the system and its forecasts.

The result of the MICMAC method is ordered variables, where one can distinguish such factors that have the greatest real impact on the examined system [32]. The system of influence–dependence factors within MICMAC method is shown in Figure 4.

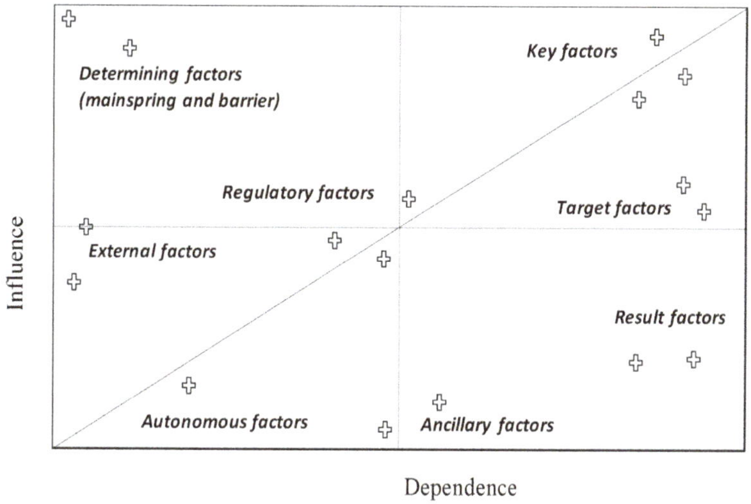

Figure 4. The system of influence–dependence factors within MICMAC (French acronym Matrice d'Impacts Croisés Multiplication Appliquée á un Classement, also known as Cross-Impact Matrix Multiplication Applied to Classification) method. Own study based on [29,30].

4. Results and Discussion

The methods used and the results obtained through the successive stages of the research procedure are presented in Figure 5. A detailed description is provided in the subchapters below.

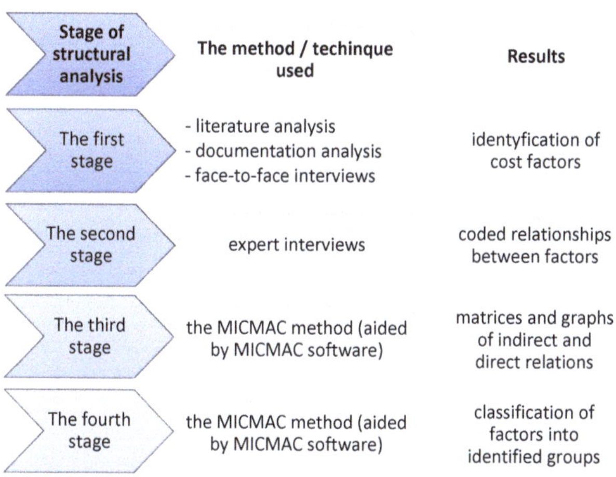

Figure 5. Methods and results of the conducted analysis.

4.1. *Identified Cost Factors*

The cost of the construction of elevations of public facilities using facade systems is influenced by many factors. The authors identified cost factors with the analysis of project and cost documentation and as-built settlements for cases, as well as direct interviews with contractors and investors. The factor identification studies presented in this paper constitute an extension of the preliminary studies, the results of which have been previously presented [18]. Finally, the completed study included

80 cases of public buildings constructed in the years 2013–2019 in Poland. The characteristics of the analysed buildings due to their different parameters are presented in Table 1.

Table 1. Characteristics of the analysed buildings.

Public Buildings	
Building function	Service and commercial Office
Location	City centre Outside city centre Out-of-town
Type of facade	Aluminium–glass facade Ventilated facade
Type of aluminium–glass facade	Mullion–transom Semistructural Structural
Type of glass	Fire glass Single glass Insulated glass
Type of ventilated facade cladding	Composite board Aluminium sheet Board fiber cement

The conducted research identified 15 factors. Fourteen of them were proposed by the authors, while the remaining one, the availability of subcontractors, was proposed by the contractors participating in direct interviews. The factors were then assigned to five groups. The number and type of groups were intuitively proposed by the authors, based on the studies of literature on factors in the construction industry and on experience from engineering practice. The proposed division of factors is illustrated in Figure 6. The factors were additionally given letter symbols to facilitate further analysis.

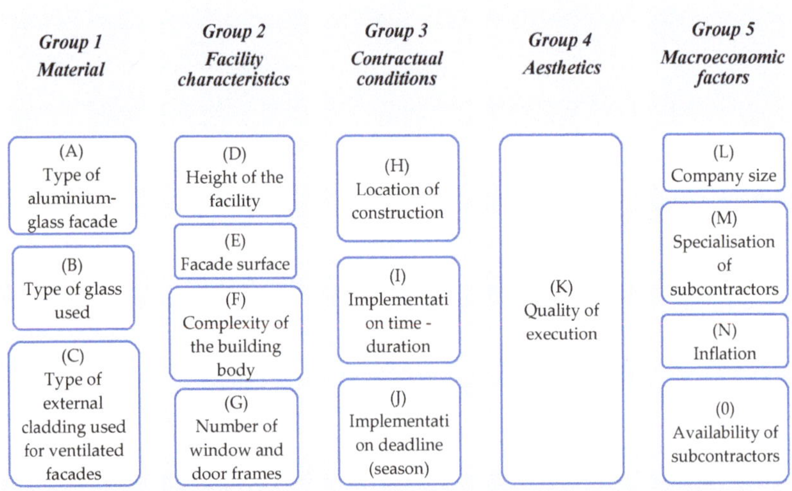

Figure 6. Proposed division of the identified factors.

Group 1—materials—contains the following factors: type of aluminium–glass façade, type of glass used, and type of external cladding used for ventilated facades. These are the main parameters directly influencing implementation costs.

Group 2—facility characteristics—includes the height of the facility, facade surface, complexity of the building body, and number of window and door frames. The factors of this group describe the characteristics of the building body. Not only does the size of the area have an impact on costs, but its height is also important. High buildings generate additional costs due to the employment of building scaffolding and a crane, the method of transporting facade elements inside the facility, and the need to use specialised equipment dedicated to the assembly of glass panels. Complication of the building body, such as inclined surfaces in combination with straight surfaces, will generate additional costs related to difficult installation. The factor "the number of window and door frames" concerns the surfaces of windows and doors (in m^2) and their condition. High functionality of the windows and doors requires the use of appropriate accessories, such as anti-panic hardware, actuators, automatic sliding door machines, or access control accessories.

Group 3 is contractual conditions. These factors have an indirect impact on the cost of the facade. The location of construction site (in or outside the city centre) has a significant impact on the costs of transporting and unloading large facade elements, e.g., aluminium sections or glass panes. Implementation time–duration is a factor that generates both labour costs (shorter implementation times will increase them) and indirect costs (such as construction site organisation). Implementation deadline (season), together with adverse weather conditions, can cause delays in facade installation.

Group 4—aesthetics—includes only one factor: quality of execution. It is expressed by the proper selection of materials, technology, and above all, by the correct execution of construction works. Its assessment can be obtained by performing a number of tests, including checking the facade planes and testing the tightness of a given building.

Group 5—macroeconomic factors—includes company size, specialisation of subcontractors, inflation, and availability of subcontractors. The factors listed in this group have an indirect impact on the cost of facade systems. Large companies have greater opportunities to access specialised machinery and equipment, which affects the time of assembly and prefabrication, and thus the cost of implementation. In addition, companies making facade systems use subcontractors, who support both the assembly and prefabrication process. Therefore, well-specialized subcontractors with large machine parks, experience, and highly qualified staff are valued companies, which results in their limited availability on the market. All these parameters describing subcontracting companies influence the price of their service, which additionally influences the final cost of the facade systems. Moreover, inflation and the related change in means of production—that is, the prices of materials and services—may determine the change of costs of facade execution in the form of lightweight casing.

4.2. Conduct and Results of Factor Analysis

The expert research was performed with 10 Polish companies dealing with and specialising in the implementation of external facades. In 2019, 15 experts were invited (including works managers, production managers, and bid managers) to describe the relationships between the identified factors. According to the principle of factor interdependence, one of the factors (X) may have a very strong influence on the other factor (Y), while the Y factor itself does not have to influence the X factor or the influence may be very weak. The evaluation involved a five-stage scale, where 0 = no impact, 1 = weak impact, 2 = medium impact (significant but not decisive), 3 = large impact (decisive), and P = potential impact.

On the basis of the results obtained, matrix A (direct influence) was constructed.

$$A = \begin{pmatrix}
0 & 2 & 0 & 0 & 0 & 0 & P & 0 & 2 & 0 & 2 & 0 & 0 & 0 & 0 \\
3 & 0 & 0 & 0 & 0 & 0 & 0 & 0 & 1 & 0 & 3 & 0 & 0 & 0 & 0 \\
0 & 0 & 0 & 0 & 0 & 0 & 0 & 0 & 2 & 0 & 3 & 0 & 0 & 0 & 0 \\
2 & 2 & 2 & 0 & 3 & 0 & 1 & 0 & 3 & P & 0 & 0 & 0 & 0 & 0 \\
1 & 1 & 1 & 0 & 0 & 0 & 0 & 0 & 3 & 0 & 0 & 0 & 0 & 0 & 0 \\
2 & 3 & 2 & 0 & 0 & 0 & P & 0 & 3 & 0 & 3 & 0 & 0 & 0 & 0 \\
1 & 0 & 0 & 0 & 0 & 0 & 0 & 0 & 2 & P & 0 & 0 & 0 & 0 & 0 \\
1 & 2 & P & 0 & 0 & 0 & 0 & 0 & 1 & P & 0 & 0 & 0 & 0 & 0 \\
1 & 2 & 2 & 0 & 0 & 0 & 0 & 0 & 0 & 0 & 2 & 0 & 0 & 0 & 1 \\
0 & 1 & 2 & 0 & 0 & 0 & 0 & 0 & 2 & 0 & 0 & 0 & 0 & 0 & 1 \\
0 & 0 & 0 & 0 & 0 & 0 & 0 & 0 & 2 & 0 & 0 & 0 & 0 & 0 & 0 \\
2 & 1 & 2 & 0 & 0 & 0 & 0 & 0 & 2 & 0 & P & 0 & 3 & 0 & 0 \\
3 & 2 & 2 & 0 & 0 & 0 & 0 & 0 & 3 & P & 3 & 0 & 0 & 0 & P \\
P & P & P & 0 & 0 & 0 & 0 & 0 & P & 0 & 0 & P & 0 & 0 & 2 \\
0 & 0 & 0 & 0 & 0 & 0 & 0 & 0 & 3 & 1 & 0 & 0 & 0 & 0 & 0
\end{pmatrix} \quad (1)$$

Using the example of the first row of the A matrix, the following interpretation of the influence of factor A on other factors can be made:

- Factor A affects on average factors B, I, and K, which is why the matrix was assigned a grade 2 according to a five-degree scale;
- Factor A does not affect factors C, D, E, F, H, J, L, M, N, and O—therefore, a degree of 0 is assigned in the matrix;
- Factor A potentially affects factor G; therefore, a grade P is assigned in the matrix.

Table 2 presents the results of the structural analysis as the total strength of influences and direct relationships for the analysed factors. The calculated values of the influence strength and dependencies are the dominants of the influence force of the factors on each other.

Table 2. Power of influence and dependence on identified factors.

No.	Group of Factors/Factor	Symbol	Total Strength of Influence	Total Strength of Dependencies
	Material			
1	Type of aluminium-glass facade	A	6	16
2	Type of glass used	B	7	16
3	Type of external cladding used for ventilated facades	C	5	13
	Facility characteristics			
4	Height of the facility	D	13	0
5	Facade surface	E	6	3
6	Complexity of the building body	F	13	0
7	Number of window and door frames.	G	3	1
	Contractual conditions			
8	Location of construction	H	4	0
9	Implementation time–duration	I	8	29
10	Implementation deadline (season)	J	6	1
	Aesthetics			
11	Quality of execution	K	2	16
	Macroeconomic factors			
12	Company size	L	10	0
13	Specialisation of subcontractors	M	13	3
14	Inflation	N	2	0
15	Availability of subcontractors	O	4	4

The following variables reveal the greatest strength of direct influence on other factors: height of the facility (D), complexity of the building body (F), and specialisation of subcontractors (M). On the other hand, the variable showing the greatest dependence on other factors and the subsequent impact on the costs of facade systems is implementation time–duration (I), while the following revealed less dependence on other factors, but also at a high level: type of aluminium–glass façade (A), type of glass used (B), and quality of execution (K).

The direct impact strength of the individual variables is illustrated in Figure 7.

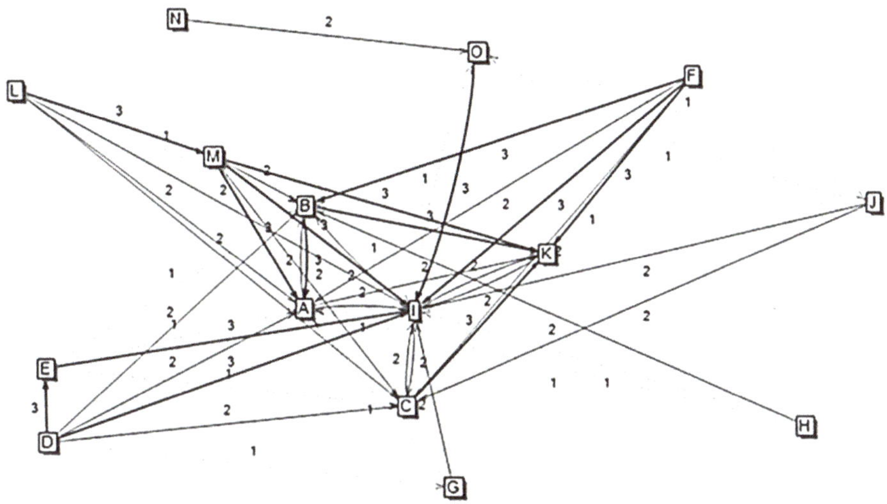

Figure 7. Direct impact graph.

The next stage of structural analysis of the identified variables was the analysis of impacts and indirect relationships between the identified variables. The MICMAC method allows us to analyse the spread of interactions through connections and feedback loops coming in and out of particular factors, which in turn extracts hidden relationships between variables, which are often not directly visible for experts or analysts. Using the MICMAC software [25], matrix B—indirect influence—was calculated.

$$B = \begin{pmatrix} 18{,}826 & 20{,}432 & 13{,}684 & 0 & 0 & 0 & 0 & 0 & 36{,}270 & 1410 & 40{,}876 & 0 & 0 & 0 & 6842 \\ 18{,}920 & 22{,}354 & 15{,}308 & 0 & 0 & 0 & 0 & 0 & 36{,}759 & 1233 & 41{,}018 & 0 & 0 & 0 & 7654 \\ 12{,}838 & 14{,}612 & 9988 & 0 & 0 & 0 & 0 & 0 & 24{,}630 & 882 & 27{,}748 & 0 & 0 & 0 & 4994 \\ 45{,}916 & 51{,}890 & 35{,}030 & 0 & 0 & 0 & 0 & 0 & 89{,}116 & 3257 & 99{,}833 & 0 & 0 & 0 & 17{,}515 \\ 22{,}675 & 23{,}988 & 16{,}470 & 0 & 0 & 0 & 0 & 0 & 41{,}989 & 1639 & 48{,}262 & 0 & 0 & 0 & 8235 \\ 40{,}015 & 44{,}136 & 29{,}694 & 0 & 0 & 0 & 0 & 0 & 77{,}269 & 2923 & 86{,}842 & 0 & 0 & 0 & 14{,}847 \\ 12{,}224 & 12{,}992 & 9100 & 0 & 0 & 0 & 0 & 0 & 22{,}096 & 838 & 25{,}694 & 0 & 0 & 0 & 4550 \\ 15{,}208 & 16{,}246 & 10{,}892 & 0 & 0 & 0 & 0 & 0 & 29{,}051 & 1145 & 32{,}878 & 0 & 0 & 0 & 5446 \\ 21{,}123 & 24{,}523 & 16{,}092 & 0 & 0 & 0 & 0 & 0 & 43{,}042 & 1570 & 47{,}020 & 0 & 0 & 0 & 8046 \\ 19{,}970 & 21{,}729 & 14{,}614 & 0 & 0 & 0 & 0 & 0 & 38{,}444 & 1484 & 43{,}296 & 0 & 0 & 0 & 7307 \\ 8740 & 8924 & 6280 & 0 & 0 & 0 & 0 & 0 & 15{,}474 & 618 & 18{,}232 & 0 & 0 & 0 & 3140 \\ 46{,}432 & 51{,}596 & 34{,}924 & 0 & 0 & 0 & 0 & 0 & 89{,}172 & 3318 & 10{,}047 & 0 & 0 & 0 & 17{,}462 \\ 39{,}822 & 44{,}346 & 30{,}048 & 0 & 0 & 0 & 0 & 0 & 76{,}545 & 2835 & 86{,}154 & 0 & 0 & 0 & 15{,}024 \\ 5490 & 7190 & 4680 & 0 & 0 & 0 & 0 & 0 & 12{,}008 & 380 & 12{,}620 & 0 & 0 & 0 & 2340 \\ 16{,}789 & 17{,}688 & 12{,}388 & 0 & 0 & 0 & 0 & 0 & 30{,}275 & 1170 & 35{,}303 & 0 & 0 & 0 & 6194 \end{pmatrix} \quad (2)$$

The interactions between variables are shown in Figure 8.

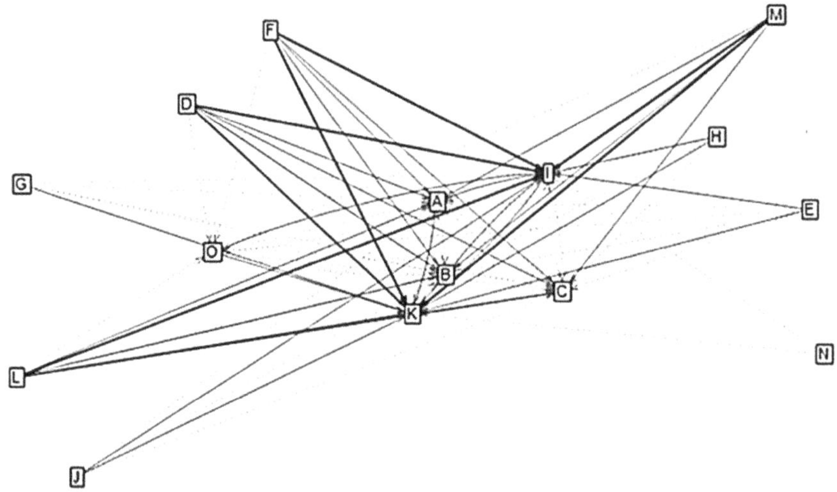

Figure 8. Indirect impact graph.

The analysis of the graph of indirect impacts of variables reveals that the following factors have the strongest influence on the factor quality of execution (K): height of the facility (D) and company size (L). Lower-strength impact on the factor quality of execution (K) is exerted by the following factors: complexity of the building body (F) and specialisation of subcontractors (M). The factor implementation time–duration (I) is strongly influenced by the factors company size (L), complexity of the building body (F), and specialisation of subcontractors (M).

The next stage of the research is to separate the structure of factors in the research area. Figure 9 presents the groups of factors that influence and describe the research problem of estimating the costs of facade systems.

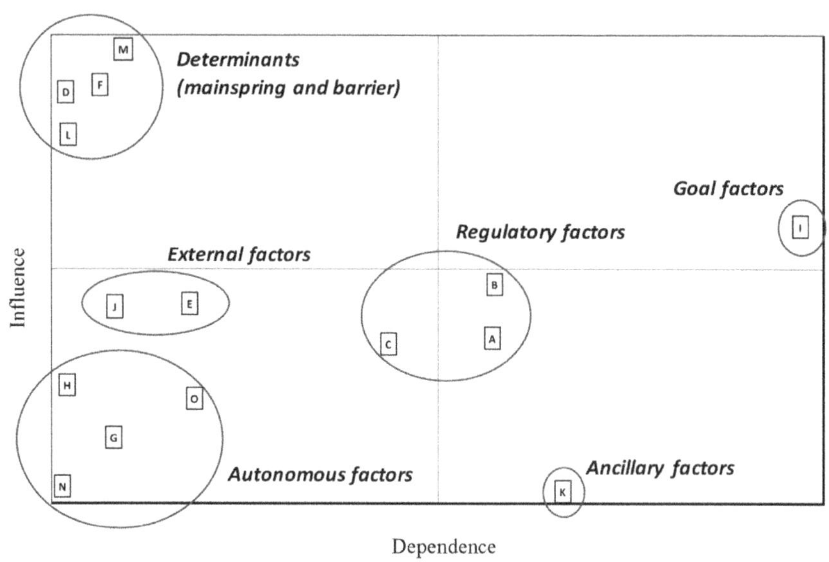

Figure 9. The system of influence–dependence factors on facade systems cost.

The analysis reveals that from among the identified factors shaping the costs of facade systems, six groups of defining factors can be distinguished. The first group of factors includes type of aluminium–glass facade (A), type of glass used (B), and type of external cladding (C). These are factors that do not have a very strong impact on the level of costs, but show a strong correlation with other factors, which in effect describes the strategic objective of the analysis, which is to estimate facade costs. The same role is played by a group of ancillary factors—quality of execution (K). The third group of factors involves determinants (mainspring and barriers), comprising the following variables: height of the facility (D), facade surface (E), complexity of the building body (F), and specialisation of subcontractors (M). These are factors that have a very strong impact on the costs of facade systems and characterise the facility and contractors, all of which have a real impact on the level of costs incurred. Another group of factors are the goal factors, where the factor implementation time (I) is the only one, and according to the analysis, changes as a result of the influence of other factors but does not influence them itself. The next group are external factors—company size (L) and implementation deadline (season) (J)—and autonomous determinants: number of window and door frames (G), location of construction (H), inflation (N), and availability of subcontractors (O). The factors belonging to these groups, according to the analysis, show the smallest influence on the estimation of facade costs.

5. Conclusions

The paper presents a structural analysis of mutual influences and relationships between factors identified on the basis of quantitative and qualitative studies of project documentation and cost estimates. This was done using the MICMAC free online software which, by means of cross-analysis of the interactions between variables, helped to determine the six groups of factors influencing the research area, being the estimation of costs of facade systems. The largest and strongest group of factors were determinants and regulatory factors. Determinants, also known as mainsprings and barriers, are factors that have a very strong influence on the cost estimation of the building facades under consideration, which include height of the facility, facade surface, complexity of the building body, and degree of specialisation of assembly companies. Another important group for the research problem are regulatory factors. They do not have a strong influence on the whole cost structure, but by interacting with each other and with other variables, they provide a basis for achieving the goal of estimating facade costs. The regulatory factors were the type of aluminium–glass facade, the type of glass used, and the external cladding for ventilated facades. During the analysis, a goal factor has also been identified, which is the implementation time. Limiting the assumptions for the isolated parameters may cause the implementation time to be shorter or longer. Such an effect makes the factor a goal in the time-and-cost relationship. The factors per group are presented in Table 3.

Table 3. Factors per group.

Regulatory Factors	Ancillary Factors	Determining Factors	Goal Factors	External Factors	Autonomous Factors
(A) Type of aluminium–glass façade (B) Type of glass used (C) Type of external cladding	(K) Quality of execution	(D) Height of the facility (E) Facade surface (F) Complexity of the building body (M) Specialisation of subcontractors	(I) Implementation time	(L) Company size (J) Implementation deadline (season)	(G) Number of window and door frames (H) Location of construction (N) Inflation (O) Availability of subcontractors

Additionally, the analysis of direct and indirect influence matrices and relationships between the variables revealed which interactions occur between individual variables. The direct relationships were a reflection of experts' opinions, which indicated that the strongest influence was shown by the following variables: height of the facility, complexity of the building body, and the level of specialisation of contractors. The analysis of the indirect influence matrix showed hidden relationships and interactions between the variables. The strongest influence on the quality of execution was exerted by the variable height of the facility and company (contractor) size.

Author Contributions: Conceptualization, A.L. and M.G.; formal analysis, A.L. and M.G.; investigation, A.L. and M.G.; methodology, M.G.; writing—original draft preparation, A.L. and M.G.; supervision, A.L. All authors have read and agreed to the published version of the manuscript.

Funding: This research received no external funding.

Conflicts of Interest: The authors declare no conflict of interest.

References

1. Saeedi, M. Study the Effects of Constructions New Techniques and Technologies on Time, Cost and Quality of Construction Projects from the Perspective of Construction Management. *J. Civ. Eng. Mater. Appl.* **2017**, *1*, 61–76. [CrossRef]
2. Colinart, T.; Bendouma, M.; Glouannec, P. Building renovation with prefabricated ventilated façade element: A case study. *Energy Build.* **2019**, *186*, 221–229. [CrossRef]
3. Leśniak, A.; Górka, M. Analysis of the cost structure of aluminum and glass facades. In *Advances and Trends in Engineering Sciences and Technologies III, Proceedings of the 3rd International Conference on Engineering Sciences and Technologies (ESaT 2018), Tatranské Matliare, Slovak Republic, 12–14 September 2018*; Al Ali, M., Platko, P., Eds.; CRC Press: Leiden, The Netherlands, 2019; p. 445.
4. Górka, M. Use of aluminium and glass facades in urban architecture. *Civ. Eng. Archit.* **2019**, *18*, 29–40. [CrossRef]
5. Balocco, C. A simple model to study ventilated facades energy performance. *Energy Build.* **2002**, *34*, 469–475. [CrossRef]
6. Schabowicz, K.; Sulik, P.; Zawiślak, Ł. Identification of the Destruction Model of Ventilated Facade under the Influence of Fire. *Materials* **2020**, *13*, 2387. [CrossRef]
7. Sandak, A.; Sandak, J.; Brzezicki, M.; Kutnar, A. State of the art in building façades. In *Bio-Based Building Skin*; Springer: Singapore, 2019; pp. 1–26.
8. Schabowicz, K.; Gorzelańczyk, T.; Mateusz Szymków, M. Modern ventilated facade systems—Types and requirements. *Izolacje* **2017**, *7*, 8.
9. Besir, A.B.; Cuce, E. Green roofs and facades: A comprehensive review. *Renew. Sustain. Energy Rev.* **2018**, *82*, 915–939. [CrossRef]
10. Ihara, T.; Gustavsen, A.; Jelle, B.P. Effect of facade components on Energy efficiency in office buildings. *Appl. Energy* **2015**, *158*, 422–432. [CrossRef]
11. Ikram, H.A.; Ahmad, M.; Qadeer, R.M.; Nawaz, M. Evaluation of key factors influencing process quality during construction projects in Pakistan. *Grey Syst. Theory Appl.* **2019**, *9*, 321–335. [CrossRef]
12. Juszczyk, M. Early Cost Estimates of Bridge Structures Aided by Artificial Neural Networks. In *VIII International Scientific Siberian Transport Forum*; Springer: Cham, Switzerland, 2019; Volume 1116. [CrossRef]
13. Wieczorek, D.; Plebankiewicz, E.; Zima, K. Model estimation of the whole life cost of a building with respect to risk factors. *Technol. Econ. Dev. Econ.* **2019**, *25*, 20–38. [CrossRef]
14. Krzemiński, M. Optimization of work schedules executed using the flow shop model, assuming multitasking performed by work crews. *Arch. Civ. Eng.* **2017**, *63*, 3–19. [CrossRef]
15. Nowogońska, B.; Korentz, J. Value of Technical Wear and Costs of Restoring Performance Characteristics to Residential Buildings. *Buildings* **2020**, *10*, 9. [CrossRef]
16. Sztubecka, M.; Skiba, M.; Mrówczyńska, M.; Bazan-Krzywoszańska, A. An Innovative Decision Support System to Improve the Energy Efficiency of Buildings in Urban Areas. *Remote Sens.* **2020**, *12*, 259. [CrossRef]
17. Juszczyk, M.; Zima, K.; Lelek, W. Forecasting of sports fields construction costs aided by ensembles of neural networks. *J. Civ. Eng. Manag.* **2019**, *25*, 715–729. [CrossRef]
18. Leśniak, A.; Górka, M.; Wieczorek, D. Identification of factors shaping tender prices for lightweight curtain walls. *Sci. Rev. Eng. Environ. Sci.* **2019**, *2*, 171–182. [CrossRef]
19. Leśniak, A.; Zima, K. Cost calculation of construction projects including sustainability factors using the Case Based Reasoning (CBR) method. *Sustainability* **2018**, *10*, 1608. [CrossRef]

20. Kowacka, M.; Skorupka, D.; Duchaczek, A.; Zagrodnik, P. Identification of geodetic risk factors occurring at the construction project preparation stage. *Open Eng.* **2019**, *9*, 14–17. [CrossRef]
21. Hoła, B.; Nowobilski, T. Analysis of the influence of socio-economic factors on occupational safety in the construction industry. *Sustainability* **2019**, *11*, 4469. [CrossRef]
22. Mrówczyńska, M.; Skiba, M.; Bazan-Krzywoszańska, A.; Sztubecka, M. Household standards and socio-economic aspects as a factor determining energy consumption in the city. *Appl. Energy* **2020**, *264*, 114680. [CrossRef]
23. Kapliński, O. Innovative solutions in construction industry. Review of 2016–2018 events and trends. *Eng. Struct. Technol.* **2018**, *10*, 27–33. [CrossRef]
24. Leśniak, A.; Kubek, D.; Plebankiewicz, E.; Zima, K.; Belniak, S. Fuzzy AHP application for supporting contractors' bidding decision. *Symmetry* **2018**, *10*, 642. [CrossRef]
25. Methods of Prospective. La Prospective. Available online: https://en.laprospective.fr/ (accessed on 10 April 2020).
26. Aluprof Aluminium Systems, Folders, Brochures, Cataloges. Available online: https://aluprof.eu/producenci/do-pobrania/foldery-broszury-katalogi (accessed on 15 April 2020).
27. ETAG 034. *Guideline for European Technical Approval of Kits of External Wall Claddings Part I: Ventilated Cladding and Associated Fixing*; EOTA: Brussel, Belgium, 2010.
28. Arcade, J.; Godet, M.; Meunier, F.; Roubelat, F. *Structural Analysis with the MICMAC Method & Factors Strategy with MACTOR Method*; Ac/UNU Millenium Project Futures Research Methodology: Paris, France, 1999.
29. Nazarko, J.; Ejdys, J.; Halicka, K.; Nazarko, Ł.; Kononiuk, A.; Olszewska, A. Structural Analysis as an Instrument for Identification of critical drivers of technology Development. *Procedia Eng.* **2017**, *182*, 474–481. [CrossRef]
30. Ranzi, G.; Gilbert, R. *Structural Analysis, Principles, Methods and Modelling*; CRC Press: Boca Raton, FL, USA, 2018.
31. Livesley, R.K. *Matrix Methods of Structural Analysis*, 2nd ed.; Pergamon International Library of Science, Technology, Engineering and Social Studies; Pergamon Press: Oxford, UK, 2013.
32. Benjumea-Arias, M.; Castaneda, L.; Valencia-Arias, A. Structural analysis of strategic variables through MICMAC use: Case Study. *Mediterr. J. Soc. Sci.* **2016**, *7*, 11. [CrossRef]

applied sciences

MDPI

Article

Influence of Maximum Aggregate Grain Size on the Strength Properties and Modulus of Elasticity of Concrete

Jacek Góra * and Małgorzata Szafraniec

Faculty of Civil Engineering and Architecture, Lublin University of Technology, Nadbystrzycka 40, 20-618 Lublin, Poland; m.szafraniec@pollub.pl

* Correspondence: j.gora@pollub.pl; Tel.: +48-81-538-44-46

Received: 21 May 2020; Accepted: 2 June 2020; Published: 5 June 2020

Abstract: Depending on the dimensions of concrete elements, aggregates of different grain sizes are used for the building structures. Taking this fact into account, the authors of the paper have undertaken in their work an issue concerning the analysis of the influence of maximum aggregate grain size on the strength properties and modulus of elasticity of concrete. This is also due to the fact that few published research results are available in this area. In this paper, the influence of the maximum grain size on the basic strength and deformation properties of concrete is discussed. The research concerns both concretes and gravel aggregates used for their construction with maximum grain sizes of 8 mm, 16 mm and 31.5 mm. The values of the compressive and splitting tensile strength, brittleness and modulus of elasticity of concretes with $w/c = 0.45$ were analysed. The analysis showed that the strength properties are proportional not only to the maximum size of aggregate grain, but also to the crushing strength of the aggregate. There were no analogous relations found with respect to the modulus of elasticity of the tested concretes. Tensile strength was particularly susceptible to the observed changes.

Keywords: the maximum aggregate grain size; modulus of elasticity; compressive and tensile strength of concrete; aggregate crushing value; gravel aggregate; interfacial transition zone

1. Introduction

Concrete is a multi–phase and heterogeneous composite whose behaviour varies according to the load applied [1]. It consists of the cement matrix, aggregates and mineral and chemical admixtures. Many factors are responsible for its durability and safety of concrete and reinforced concrete structures, including compression strength, tensile strength, modulus of elasticity and brittleness of concrete [2–4]. This feature is particularly important if concrete structures with minor defects in their structure are taken into account. If the construction has such defects in its structure, it may contribute to localised damage caused by loading with local discontinuities and sudden differences in the mechanical properties of the material. Within the micro–cracks or discontinuities of the crystal lattice, local stress concentrations may occur, which may contribute to the development of localised damage and, consequently, even to the failure of the structural component [1]. The most sensitive area is the transition zone of cement paste-coarse aggregates. The increased porosity of the area should also be emphasised [5] and the fact that sedimentation pores may form underneath the coarse aggregate grains. The formation of these pores is encouraged by the size of the coarse aggregate grains.

Damage to the structure of cement composites is a complex phenomenon, as it is very difficult to predict the stages of this process that may contribute to the destruction of the concrete element. The internal structure of the cement composite, which includes, among other things, aggregates, their distribution and diameter, is the basic factor that can cause an increase in damage [6]. In addition, about 75% of the concrete content is occupied by aggregates [7]. Most of which are coarse aggregates around 60–70% by volume.

The connection between the matrix and the coarse or fine aggregate in the concrete is the so-called interfacial transition zone (ITZ) [5]. According to many researchers, the mechanical properties and compressive strength of concrete are influenced by different sizes and types of aggregates [1,7,8]. Ferdous et al. [9] provided the existing models for compressive strength and elastic modulus. In paper [10], it is shown, among other things, that the tensile strength is influenced by the grain size of the aggregate used for concrete production. Piasta et al. [11] found that the type of coarse aggregate, as well as the aggregate shape, surface texture, porosity, ITZ size, and chemical bonding between the cement and the aggregate, affects the test subject deformation characteristics of ordinary concrete. The research has clearly shown that the worst deformation properties were obtained in case of concretes with granite aggregates. In the case of concretes with basalt, granite and pebble aggregates, a significant overestimation of normal modulus of elasticity values was found. Mahmoud et al. [12] determined the effect of the grain size of the aggregate used in concrete on the mass attenuation coefficient. It was specified that with the increase in grain size, the coefficient decreased by 4%. Neville [13] found that the larger the size of the aggregate, the less water is needed for the concrete mixture. As a result, the water-cement ratio is lowered and the strength of the concrete increases. However, in his work, he did not investigate what will happen if the same volume of cement paste is kept. Additionally, it is important to consider the water absorption capacity of the aggregates because highly absorptive aggregates can drastically affect water demand. Khotbehsara et al. [14] showed that the aggregate size has an impact on the mechanical properties.

The modulus of elasticity of concrete is directly related to structural deformations and may be defined on the basis of ASTM C469 standard [15]. These deformations can contribute to excessive deformations, which are a direct cause of cracks in concrete composites. The modulus of elasticity, which indicates the stiffness of the material and is associated with its strength, is one of the most important properties of concrete [6]. Determining the modulus of elasticity of concrete is not an easy task as the material is not completely flexible. Although its behaviour is flexible with low loads ranging from 30% to 40% of its final load capacity. Due to the non-linear behaviour of the stress-strain curve (σ-ε) of concrete, it is difficult to determine exactly the specific value of the static modulus of elasticity [16]. Since concrete is quasi-brittle, the stress-strain behaviour starts to be non–linear after 50–70% of the peak load.

The parameters affecting the modulus of elasticity of concrete depend on the characteristics of the aggregate matrix, cement paste, testing parameters and transition zone. According to [17], the weakest concrete components are hardened cement paste and the transition zone between cement paste and coarse aggregate, not the coarse aggregate itself. The study [18] examined the effect of four types of coarse aggregates—dolomitic and quartzitic limestone, steel slag and calcareous limestone—on the compressive strength and modulus of elasticity of high strength concrete. The influence of the type of coarse aggregate has more effect on the modulus of elasticity than on compression strength.

Based on the research, it was observed that the mineralogical composition of coarse aggregate strongly influences the modulus of elasticity of concrete. Indeed, the modulus of elasticity can be up to 30% different, depending on the type of aggregate and concrete composition. According to the data contained in the paper [19], the modulus of elasticity of concrete increases with its curing. The modulus of elasticity increases faster than the concrete's compressive strength, due to the higher density of the interfacial transition zone. The porosity of the matrix affects the individual strength of the cement paste, which results in a change in the modulus of elasticity [20].

The greatest difficulty in using theoretical models to determine the modulus of elasticity of concrete is that they require prior knowledge of the modulus of elasticity of cement paste and aggregates. Therefore, to solve this problem, normative empirical methods have been developed, which estimate the modulus of elasticity value based on the concrete's compressive strength [20].

To the best of our knowledge, in the literature, there are only a few examples of research evaluating the influence of maximum aggregate grain size on the strength properties and modulus of elasticity. Therefore, this study focuses on the assessment of the above correlations.

2. Materials and Methods

2.1. Materials

The concrete was prepared using CEM I 42.5 R Portland cement. The technical parameters of applied Portland cement are presented in Table 1. The cement used in the tests meets the requirements set out in the EN 197-1 [21] standard.

Table 1. The technical parameters of CEM I 42.5R Portland cement.

Parameters	Unit	Value
Specific surface	($cm^2 \cdot g^{-1}$)	4124
Initial setting time	(min)	184
Loss on ignition	(%)	3.33
Compressive strength		
after 2 days	(MPa)	30.1
after 28 days		60.2
Density	($g \cdot cm^{-3}$)	3.08
Volume stability	(mm)	1.00
SO_3 content	(%)	2.95
Cl content	(%)	0.089
Insoluble residue	(%)	0.57

Washed natural quartz sand (QS) (0 ÷ 2 mm) and gravel (2 ÷ 31.5 mm) was used to prepare concrete samples.

QS density was 2.65 $g \cdot cm^{-3}$, while the properties of gravel aggregate were included in Tables 2 and 3 show sieve analysis of the fine and coarse aggregates.

Table 2. Tests results of gravel properties.

Fraction	Specific Density	Bulk Density	Porosity	Aggregate Crushing Value	Content of Irregular Grains	Content of Mineral Dusts
(mm)	($kg \cdot dm^{-3}$)	($kg \cdot dm^{-3}$)	(%)	(ACV, %)	(Z_{ni}, %)	(Z_{pi}, %)
2 ÷ 8				7.9	0.7	0.22
8 ÷ 16	2.72	2.64	2.90	10.3	4.1	0.20
16 ÷ 31.5				12.9	3.9	0.13

Table 3. Sieve analysis of the fine and coarse aggregates, %.

Sieve Size mm	Sand	Gravel Aggregate		
		2 ÷ 8 mm	2 ÷ 16 mm	2 ÷ 31.5 mm
31.5	100	100	100	100
16	100	100	100	40
8	100	100	40	20
4	100	40	10	5
2	100	0	0	0
1	94	-	-	-
0.5	55	-	-	-
0.25	13	-	-	-
0.125	2	-	-	-

Three concrete mixtures were prepared with the composition presented in Table 4. The quantities of concrete mixture components were determined experimentally, with the assumption of a *w/c* value of

0.45 (minimum permissible maximum value regardless of the exposure class according to EN 206 [22]) and a slump class S1/S2, according to EN 206 [22]. The sand point of the crumble stack determined experimentally was 37.3% by weight. As shown in Table 4, the concrete made differed in the quantities and fractions of gravel aggregates used in their manufacture. Depending on the concrete used, the crossbow fractions of the samples were described as follows: fraction 2 ÷ 8 mm–GC2/8, fraction 2 ÷ 16 mm–GC2/16 and fraction 2 ÷ 31.5 mm GC2/31.5. A superplasticizer based on polycarboxylic ether combined with calcium lignosulfonate has also been added to all earnings.

Table 4. Composition of concrete mixtures.

Components	Unit	GC2/8	ACV [1]	GC2/16	ACV [1]	GC2/31.5	ACV [1]
			(%)		(%)		(%)
CEM I 42.5R cement	(kg·m^{-3})	364		364		364	
Quartz sand (0 ÷ 2 mm)	(kg·m^{-3})	708		708		708	
Gravel (2 ÷ 4 mm)	(kg·m^{-3})	476		119		60	
Gravel (4 ÷ 8 mm)	(kg·m^{-3})	714	7.9	357	9.5	178	
Gravel (8 ÷ 16 mm)	(kg·m^{-3})	-	-	714		238	11.6
Gravel (16 ÷ 31.5 mm)	(kg·m^{-3})	-	-	-	-	714	
Superplasticizier	(kg·m^{-3})	2.548		1.49		1.048	
Water	(kg·m^{-3})	165		165		165	

[1] ACV—aggregate crushing value for aggregates with grain diameters over 4 mm.

2.2. Methods

For each fraction of aggregates, the crushing strength, the content of irregular grains and mineral dust, as well as the bulk and specific density from which the total porosity was calculated were determined.

The crushing strength of aggregates was determined using the so-called aggregate crushing value (ACV) according to the PN-B-06714-40 standard [23], similarly as in the BS 812 standard [24]. An aggregate sample in a steel cylinder with an inside diameter of 150 mm was loaded with a force of 200 kN. The test is carried out on 4 ÷ 8 mm, 8 ÷ 16 mm and 16 ÷ 31.5 mm fractions. The aggregate crushing value is defined as the percentage of grains crushed into grains smaller than $\frac{1}{4}$ of the lower sieve size of a given fraction. The ACV can be used to classify the aggregate and assess its suitability for the proper concrete.

The content of irregular grains, i.e., flat and elongated grains (with the proportion of the smallest to the largest grain size exceeding 1:3) was determined according to the PN-B-06714-16 standard [25].

The mineral dusts content (percentage of the aggregate of grains smaller than 0.063 mm i.e., clays, clayey particles, etc.) was determined according to the PN-B-06714-13 standard [26].

The specific density was tested using the pycnometric method, after grinding the aggregate to dimensions smaller than 0.08 mm. The air contained between the grains of powdered material was removed by inserting a pycnometer in a vacuum chamber, and reducing the pressure to 2.33 kPa.

The apparent density was tested on the basis of the EN 1097-6 standard [27]. The total porosity was calculated from the value of apparent density and specific density.

The air content of the fresh concrete mix has been determined for each concrete, using the pressure method based on the EN 12350-7 standard [28]. The consistency was tested according to the EN 12350-2 [29] standard.

To determine the compressive strength and splitting tensile strength after 28 days of curing, 36 cubic samples with the edge length of 150 (6 for each concrete) [30,31]. Cylindrical specimens with a diameter of 150 mm and a height of 300 mm were used to determine the elastic modulus (18 specimens, 6 for each concrete) [32].

According to the EN 12390-3 [30] standard, the compressive strength test was carried out on cubic samples with an edge length of 150 mm. The values of elastic modulus were determined on the basis of the EN 12390-13 [32] standard (Figure 1). The splitting tensile strength was determined in accordance with the EN 12390-6 [31] standard on concrete blocks with an edge length of 150 mm.

Figure 1. The test stand for testing the static modulus of elasticity.

The concrete brittleness (K) can be calculated from the following formula:

$$K = \frac{f_{ct,ax}}{f_c}, \quad (1)$$

where: f_c—compressive strength (MPa), $f_{ct,ax}$—axial tensile strength (MPa).

The splitting tensile strength was determined in the presented tests. For the formula to find its application, the values of splitting tensile strength should be recalculated according to the below formula [33]:

$$f_{ct,ax} = 0.9 \cdot f_{ct,sp}, \quad (2)$$

where: $f_{ct,ax}$—axial tensile strength (MPa), $f_{ct,sp}$—splitting tensile strength (MPa).

3. Results and Discussion

Generally, when assessing the quality of the aggregate used in concrete, the requirements of the PN-B-06712 [34] standard can be referred to. It should be noted that the requirements contained in this standard have been developed by the Polish Committee for Standardization on the basis of several dozen years of testing experience and are still used by bridge or road construction. According to this standard, the basic property classifying the quality of aggregate is its crushing strength, which is measured by the so-called aggregate crushing value (ACV). On the basis of its value, classification is made into the so-called aggregate brand, on which, in turn, the limit requirements relating to the remaining properties of the aggregate, such as the content of irregular grains or content of mineral dusts, depend. Taking into account the average ACV values of the gravel aggregate used for the tested concretes (Table 4), it is classified to the highest brand 30 (ACV value < 12%), according to the PN-B-06712 [34] standard. The other results of the properties, which can directly affect both the strength properties of concrete and the adhesion of cement paste to aggregate grains, should be considered as very good, the content of irregular grains is significantly lower than the 20% limit and the content of mineral dusts is lower than the 1.5% limit (Table 2).

Table 5 shows the physical properties of gravel concrete, while, in Table 6, the mechanical properties are shown. The obtained results were compared to the values quoted on the basis of the EN 1992 standard [33] (Table 6).

When assessing the air content in the tested mixtures, slight differences were found, which did not exceed 0.3%, and it was considered that they do not affect the mechanical properties of the tested concrete. Similarly, for concrete mixture consistency, the differences did not exceed 1.0 cm (Table 5).

Table 5. Properties of gravel concrete.

Type of Concrete	Air Content (%)	Slump Test (cm)
GC2/8	2.40	5.0
GC2/16	2.60	4.0
GC2/31.5	2.70	4.0

The concrete strength classes determined from the results of the compressive strength tests are the same, and correspond to the same class C40/50 (Table 6). The difference between extreme f_{cm} values is 2.1 MPa, and no statistically significant differences in compression strength results were also confirmed by ANOVA test results ($p = 0.22 > p = 0.05$).

Different relations were found between splitting tensile strengths–the greater the maximum size of aggregate grain (D_{max}) the lower the value of $f_{ctm,sp}$ of concrete. The difference between the extreme values of $f_{ctm,sp}$ is 1.73 MPa and the GC2/31.5 concrete splitting tensile strength is 34.5% lower than the GC2/8 concrete strength. The significant differences in splitting tensile strength values were confirmed by ANOVA test results. The value of $p = 5.6 \times 10^{-9}$ is significantly lower than $p = 0.05$, which clearly and strongly confirms the statistically significant differences between the concrete tensile strength results. The least significant difference (LSD) test showed that each pair of averages differs significantly from each other (differences are statistically significant).

Table 6. Mechanical properties of the concretes.

Type of Concrete/Descriptive Statistics		Compressive Strength f_{cm}, (MPa)	Strength Class	Splitting Tensile Strength $f_{ctm,sp}$ ($f_{ctm,ax}$), (MPa)	EN 1992 [33]	Elastic Modulus E_{cm}, (GPa)	EN 1992 [33]	K (-)
GC2/8	Mean	54.0	C40/50	5.01 (4.51)	3.5	33.5	35	0.09
	CV	2.8		7.2		2.6		6.7
	SD	1.52		0.36		0.86		0.006
GC2/16	Mean	54.7	C40/50	3.69 (3.32)	3.5	33.1	35	0.06
	CV	4.8		3.5		1.8		5.0
	SD	2.61		0.13		0.58		0.003
GC2/31.5	Mean	56.1	C40/50	3.28 (2.95)	3.5	32.7	35	0.05
	CV	5.0		4.6		4.0		6.0
	SD	2.81		0.15		1.30		0.003

CV—coefficient of variation (%), SD—Standard deviation.

The obtained values of $f_{ctm,sp}$ were compared with the respective values of $f_{ctm,sp}$ according to the EN 1992 standard [33] corresponding to specific classes of concrete. It was found that the concretes GC2/16 and GC2/31.5 were characterized by lower values by 5% and 16% respectively, i.e., by one and two classes of concrete strength lower. From a practical point of view, this means that increasing the D_{max} of the aggregate can cause higher susceptibility to earlier scratches of the element.

The modulus of elasticity of concretes was practically not susceptible to changes caused by different sizes of coarse aggregate. The difference between the extreme values was 0.8 GPa, with the highest E_{cm} value for GC2/8 being only 1% higher than the lowest modulus of elasticity for the GC2/31.5 concrete. Therefore, the effect of the maximum grain size of the aggregate on the tested values of modulus of elasticity should be treated as insignificant. This was also confirmed by the ANOVA test, which shows that the differences between the values of concrete E_{cm} modules are statistically insignificant ($p = 0.43 > p = 0.05$).

By referring the results of the E_{cm} test to the values according to the EN 1992 standard [33], all the specified modules correspond to class C30/37, i.e., by two concrete strength classes below the specified

C40/50. However, it should be noted that the modulus of elasticity of concrete is not only dependent on its compressive strength. The influence of the type of coarse aggregate on the value of modulus of elasticity is very strongly visible, which is reflected not only in the EN 1992 standard [33], and also in the literature [35–38]. In particular, this effect may be difficult to assess in the case of gravel aggregates, which are polyminerals with a mineral composition often very different depending on the origin (location of the mine).

The brittleness, i.e., the ratio of the splitting tensile strength to the compressive strength f_{ctm}/f_{cm}, of the tested concretes is also varied. A material is considered brittle when the f_{ctm}/f_{cm} ratio is less than 1/8 (0.125) [39]. The lower the *K*-value, the more brittle the material is. The K value calculated on the basis of the EN 1992 standard [33] is 0.07 for all concretes. In the case of two types of aggregates, it was found that the brittleness of the concrete GC2/16 and GC2/31.5 is higher than the value calculated according to standard values (for a given concrete strength class). Only concrete GC2/8, with a brittleness index equal to 0.09, met the requirements calculated on the basis of the EN 1992 standard [33]—it turned out to be less brittle than other concretes (Table 6). In the case of concrete, the matrix, as well as aggregates, are brittle materials. The higher brittleness of concretes GC2/16 and GC2/31.5 may be caused by concrete defects, such as air voids, cracks, discontinuities in the crystal network that may have formed between aggregates of different sizes. Another reason may be that the aggregate used in their manufacture has lower mechanical resistance (Figure 2) than the aggregate used in GC2/8 concrete. The aggregate with the 2 ÷ 8mm fraction has a lower ACV coefficient by about 23% and 63%, compared to the 8 ÷ 16mm and 16 ÷ 31.5mm aggregates, respectively.

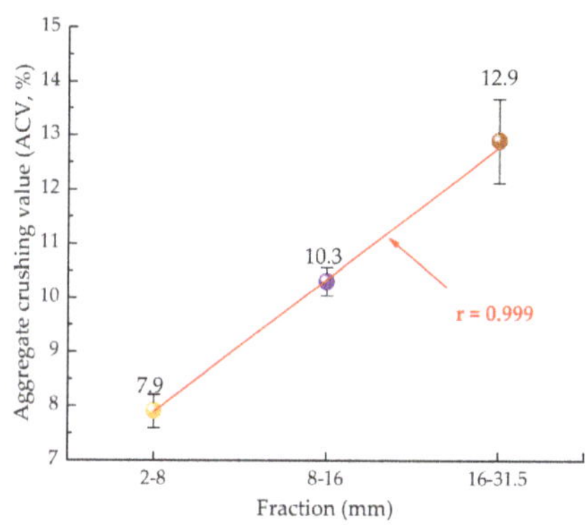

Figure 2. Correlation between aggregate crushing value and aggregate fraction.

In order to explain the differences found in the properties of concrete, the properties of gravel aggregate examined were analysed. The ACV was considered to be the most variable, and its values depend on the thickness of coarse aggregate grains, and there is a very strong correlation between them (Figure 2), with statistically significant differences ($p = 5.6 \times 10^{-5} < p = 0.05$). It should be noted that the increase in the ACV value of the aggregate corresponds to its lower mechanical resistance (lower crushing strength resistance).

For this reason, the potential for relationships between the properties of the concrete and the corresponding ACV values of the aggregate and how correlated they are was analysed.

The strongest correlation was found between the splitting tensile strength $f_{ctm,sp}$ of concretes and ACV values of aggregates (Figure 3). The correlation coefficient $r = 0.91$ indicates a very strong correlation. The correlation shows that the increase in D_{max} of the aggregate and, at the same time, the decrease in the mechanical resistance of the aggregate (increase in ACV) corresponds to the decrease in the splitting tensile strength.

There is a strong correlation between the compressive strength and ACV values ($r = 0.84$) (Figure 4). The increase in D_{max} of the aggregate has corresponded to a slight increase in the compressive strength of the concrete. However, it is surprising that in this case, despite the fact that the mechanical resistance of the aggregate significantly decreases, the compressive strength of the concrete increases slightly.

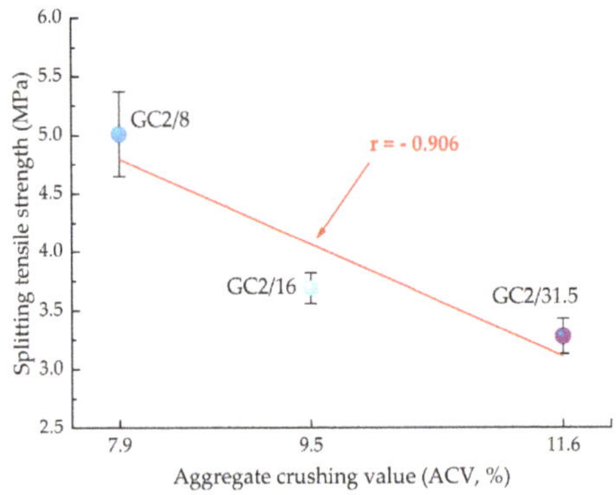

Figure 3. Correlation between aggregate crushing value and splitting tensile strength of concrete.

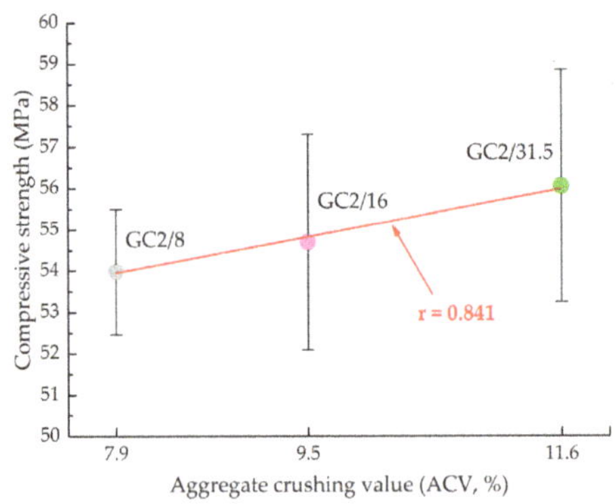

Figure 4. Correlation between aggregate crushing value and compressive strength of concrete.

A very weak correlation was found between aggregate crushing value and elastic modulus of concretes (Figure 5). These parameters do not differ significantly statistically.

Analysing the influence of D_{max} of aggregate on the values of strength properties of the tested concretes, an opposite relation to the commonly accepted one was found, a slight increase in compression strength (within the same concrete strength class) corresponds to a clear decrease in splitting tensile strength.

The relations described above indicate that the influence of D_{max} of the aggregate on the values of strength properties is greater and more significant for splitting tensile strength. It is widely recognised that the most significant impact on the mechanical properties of concrete is exerted by the interfacial transition zone (ITZ), where the first micro-cracks form under load. As the load increases, micro-cracks propagate, the increasing scratches start to merge and expand until the concrete is destroyed.

In the splitting test, aggregate grains are very often split (cracked), and also in the case of gravel, especially the weaker ones. Therefore, apart from the area of ITZ with increased porosity [5], which plays the most important role in the destruction, the share of aggregate grains with lower mechanical resistance (higher ACV) will also be significant.

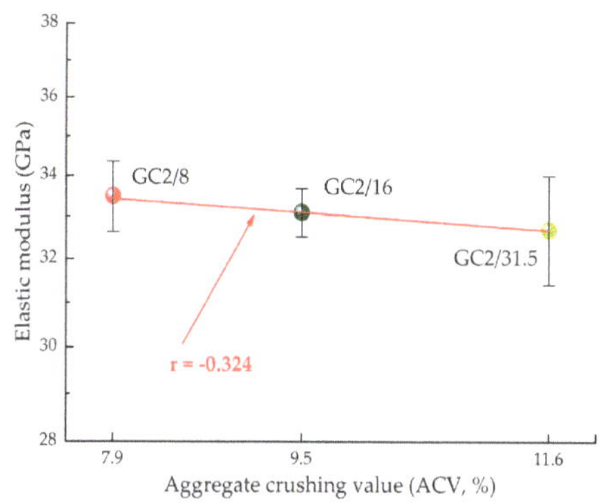

Figure 5. Correlation between aggregate crushing value and elastic modulus E of concrete.

Elsharief et al. demonstrated that the reduction of D_{max} of the aggregate causes the formation of ITZ, with lower porosity around the grains [40].

This explains the results obtained in the tests. The growth of D_{max} of the aggregate from 8 mm to 31.5 mm resulted in the formation of a more porous ITZ around the aggregate grains in GC2/31.5 and GC2/16 in comparison to ITZ in GC2/8. Simultaneously, with the growth of D_{max}, the mechanical resistance of the aggregate also deteriorated, and the lowest ACV value was determined for the aggregate in the GC2/31.5 concrete. The overlapping of these two factors caused that, despite the same class of all three concretes and similar values of their compressive strength, there was a clear decrease in the splitting tensile strength as the maximum aggregate grain size increased.

Analogous relations to those obtained in this study were obtained by Akçaoğlu et al. [41], but instead of natural aggregate they used steel balls with diameters of 9, 12, 19, 25 and 32 mm, i.e., with very similar D_{max} to the gravel used in the concretes studied. Low strength concrete (LSC, 25 MPa) and high strength concrete (HSC, 47 MPa) were tested. They also found a significant decrease in tensile strength (to about 15%) as the aggregate size increased, while the compression strength increased only slightly. The observed decrease in tensile strength along with the increase in grain size in both LSC and HSC was justified by the decrease in bond strength. This is due to the increased volume of the aggregate in relation to the total volume of the composite, which makes the significant difference between the elastic modules of the two phases more pronounced, thus creating an increased stress concentration and more micro-cracks near the aggregate. The negative influence of the smooth texture of the aggregate surface on the binding force due to the increased aggregate surface was also emphasized. The smooth surface texture and high modulus of elasticity of the aggregate result in a greater reduction of tensile strength in composites with a lower w/c ratio (w/c = 0.42). According to the authors [41], the interfacial bond was considered decisive in the reduction of tensile strength and played a minor role in the influence on the values of the compressive strength. It was found that the tensile strength decreases as the size of the aggregate increases. The rate of tensile strength reduction with increasing size of a single aggregate becomes higher in HSC [41].

In turn, the results of this study and Akçaoğlu et al. [41] differ significantly from the results of the research on mortars and concretes presented by Reinhardt [10]. The tests carried out in the study by Reinhardt [10] showed that when aggregates from 2 mm to 8 mm were used in concrete production, the tensile strength value increased from 2.35 MPa to 2.7 MPa, respectively. This shows a 15% increase in this value. However, when aggregates with a larger diameter of 8 mm to 32 mm were used, the tensile strength of the concrete remained constant at 2.86 MPa. It was found that there is a systematic difference between the strength of mortar (aggregate size up to 4 mm) and concrete (from 8 mm upwards). As noted, whether this conclusion is valid for all types of concrete and aggregate sizes should be confirmed by experiments and numerical simulations, which have not been conducted so far [10].

Figure 6a shows a mathematical/experimental model of the splitting tensile strength of concretes which depends on two other concrete characteristics: x_1—compressive strength and x_2—aggregate crushing value. The interaction of compressive strength and aggregate crushing value, as described in the three-dimensional graph, indicates that within the tested ranges a higher compressive strength with a lower aggregate crushing value would significantly increase splitting tensile strength. The statistical analysis carried out in the STATISTICA 12 program (StatSoft, Inc., Tulsa, USA) showed that the whole presented regression model is statistically significant. The dependent variable-splitting tensile strength-is explained in 80% using the above model, although the correlation matrix before the analysis indicated that one of the independent variables is statistically insignificant-compressive strength.

As can be seen in Figure 6b, very similar relationships occur between splitting tensile strength, elastic modulus and aggregate crushing value. The dependent variable-splitting tensile strength is expressed by two independent variables: x_1—elastic modulus and x_2—aggregate crushing value. As in the previous case, as E_{cm} increases and ACV value decreases, splitting tensile strength increases simultaneously. This model also describes 80% of the dependent variable, although initially, the predictor variable E_{cm} seems to be statistically insignificant.

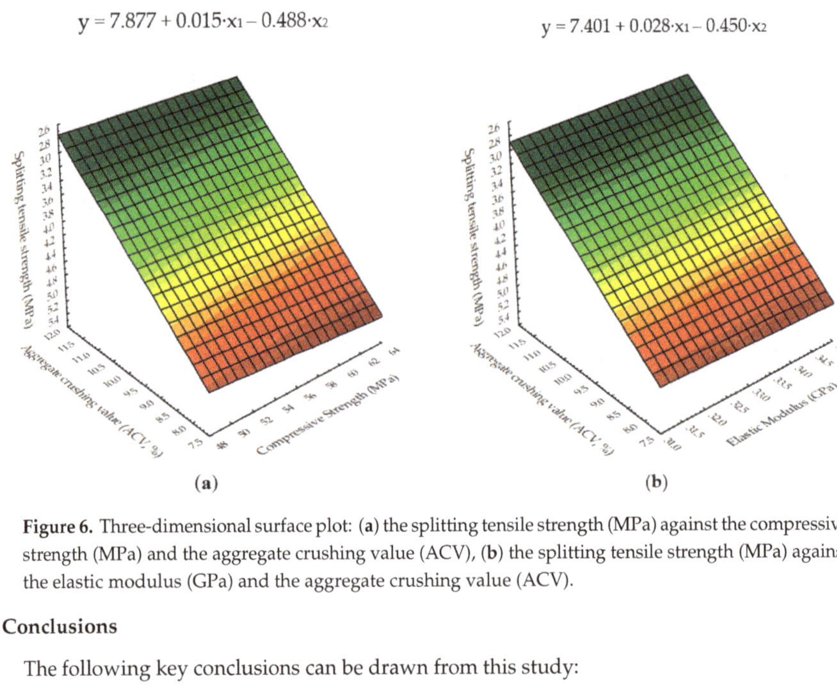

(a) (b)

Figure 6. Three-dimensional surface plot: (**a**) the splitting tensile strength (MPa) against the compressive strength (MPa) and the aggregate crushing value (ACV), (**b**) the splitting tensile strength (MPa) against the elastic modulus (GPa) and the aggregate crushing value (ACV).

4. Conclusions

The following key conclusions can be drawn from this study:

1. The influence of D_{max} of the aggregate on the values of strength properties is greater and more significant for splitting tensile strength. Analysing the obtained results, it was found that, with the increase of D_{max}, the splitting tensile strength of $f_{ctm,sp}$ deteriorated-for concrete GC2/31.5 is 34.5% lower than the strength of concrete GC2/8. An increase in aggregate D_{max} with a simultaneous decrease in the mechanical resistance of the aggregate (increase in ACV) significantly reduces the tensile strength.
2. Comparing the obtained results of $f_{ctm,sp}$ with standard values, it was found that GC2/16 and GC2/31.5 concretes are characterized by 5% and 16% lower values respectively, i.e., one and two classes of concrete strength lower. This may indicate that an increase in D_{max} may make the element more susceptible to earlier scratching.

3. The influence of the maximum size of the aggregate grain on the modulus of elasticity was negligible. The changes in the size of the coarse aggregate did not affect the modulus of elasticity of concrete-the E_{cm} for GC2/8 was only 1% higher than the GC2/31.5 concrete module, which was the lowest of all.
4. With the growth of D_{max} of the aggregate, the brittleness of the tested concretes increases.
5. The opposite to commonly accepted relationships have been observed—despite a decrease in the mechanical resistance of the aggregate, the compressive strength of the concrete has increased slightly, and a slight increase in the compressive strength corresponds to a clear decrease in the splitting tensile strength.

To summarise, the issue of the influence of the maximum size of aggregate grain on the properties of concrete requires further extensive research. As can be seen, the few test results presented are not consistent, and thick aggregates of different fractions are used for structural concretes. Undoubtedly, it is also necessary to analyse what effect on the changes caused by the maximum grain size will be exerted by the use of crushed aggregate, instead of pebbles, with irregular grains and rough texture. Mechanical adhesion of the grout to the aggregate will be improved in this case, but how this will affect the properties of the concrete needs to be explained by means of experiments.

Author Contributions: Conceptualization, J.G.; methodology, J.G.; validation, J.G.; formal analysis, All; investigation, J.G.; data curation, J.G.; writing—original draft preparation, All; writing—review and editing, All; visualization, M.S.; translation, M.S.; supervision, J.G.; project administration, All; funding acquisition, J.G. All authors have read and agreed to the published version of the manuscript.

Funding: This work was financially supported by the Ministry of Science and Higher Education, within the statutory research number FN14/ILT/2019.

Conflicts of Interest: The authors declare no conflict of interest.

References

1. Topçu, İ.B.; Uğurlu, A. Elasticity Theory of Concrete and Prediction of Static E-Modulus for Dam Concrete Using Composite. *J. Eng. Stud. Res.* **2007**, *18*, 4055–4067.
2. van Mier, J.G.M. *Fracture Processes of Concrete: Assessment of Material Parameters for Fracture Models*; CRC Press: Boca Raton, FL, USA, 2017.
3. Oucif, C.; Mauludin, L.M. Continuum damage-healing and super healing mechanics in brittle materials: A state-of-the-art reviews. *Appl. Sci.* **2018**, *8*, 2350. [CrossRef]
4. Li, J.; Huang, Q.; Ren, X. Dynamic initiation and propagation of multiple cracks in brittle materials. *Materials* **2013**, *6*, 3241–3253. [CrossRef] [PubMed]
5. Scrivener, K.L.; Lyon, A.K.C.; Laugesen, F.P.; Crumbie, A.K.; Laugesen, P.; Lyon, A.K.C.; Laugesen, F.P. The interfacial transition zone (ITZ) between cement paste and aggregate in concrete. *Interface Sci.* **2004**, *12*, 411–421. [CrossRef]
6. Zhou, C.; Li, K.; Ma, F. Numerical and statistical analysis of elastic modulus of concrete as a three-phase heterogeneous composite. *Comput. Struct.* **2014**, *139*, 33–42. [CrossRef]
7. Sengul, O.; Sengul, C.; Keskin, G.; Akkaya, Y.; Tasdemir, C.; Tasdemir, M.A. Fracture and microstructural studies on normal and high strength concretes with different types of aggregates. In Proceedings of the 8th International Conference on Fracture Mechanics of Concrete and Concrete Structures, FraMCoS 2013, Toledo, Spain, 10–14 March 2013; pp. 714–725.
8. Saouma, V.E.; Broz, J.J.; Brühwiler, E.; Boggs, H.L. Effect of Aggregate and Specimen Size on Fracture Properties of Dam Concrete. *J. Mater. Civ. Eng.* **1991**, *3*, 204–218. [CrossRef]
9. Ferdous, W.; Manalo, A.; Wong, H.S.; Abousnina, R.; AlAjarmeh, O.S.; Zhuge, Y.; Schubel, P. Optimal design for epoxy polymer concrete based on mechanical properties and durability aspects. *Constr. Build. Mater.* **2020**, *232*, 117229. [CrossRef]
10. Reinhardt, H.W. Factors affecting the tensile properties of concrete. In *Understanding the Tensile Properties of Concrete*; Woodhead Publishing Limited: Cambridge, UK, 2013; pp. 19–51. [CrossRef]
11. Piasta, W.; Budzyński, W.; Góra, J. The effect of coarse aggregate on the deformation properties of normal concretes. *Przegląd Bud.* **2012**, *83*, 35–38. (In Polish)

12. Mahmoud, K.A.; Tashlykov, O.L.; El Wakil, A.F.; El Aassy, I.E. Aggregates grain size and press rate dependence of the shielding parameters for some concretes. *Prog. Nucl. Energy* **2020**, *118*, 103092. [CrossRef]
13. Neville, A. *Properties of Concrete*, 5th ed.; Pearson Education Limited: Harlow, UK, 2012.
14. Khotbehsara, M.M.; Manalo, A.; Aravinthan, T.; Reddy, K.R.; Ferdous, W.; Wong, H.; Nazari, A. Effect of elevated in-service temperature on the mechanical properties and microstructure of particulate-filled epoxy polymers. *Polym. Degrad. Stab.* **2019**, *170*, 108994. [CrossRef]
15. ASTM. *ASTM C469/C469M-14, Standard Test Method for Static Modulus of Elasticity and Poisson's Ratio of Concrete in Compression*; ASTM International: West Conshohocken, PA, USA, 2014.
16. Diógenes, H.J.F.; Cossolino, L.C.; Pereira, A.H.A.; El Debs, M.K.; El Debs, A.L.H.C. Determinação do módulo de elasticidade do concreto a partir da resposta acústica. *Rev. IBRACON Estruturas Mater.* **2011**, *4*, 803–813. [CrossRef]
17. de Larrard, F.; Belloc, A. Are Small Aggregates Really Better for Making High-Strength Concrete? *Cem. Concr. Aggregates* **1992**, *14*, 62. [CrossRef]
18. Beshr, H.; Almusallam, A.A.; Maslehuddin, M. Effect of coarse aggregate quality on the mechanical properties of high strength concrete. *Constr. Build. Mater.* **2003**, *17*, 97–103. [CrossRef]
19. Mehta, P.K.; Monteiro, P.J.M. *Concrete Microstructure, Properties and Materials*, 4th ed.; Mc Grwa Hill Education: Michigan, MI, USA, 2014.
20. Monteiro, P.J.M.; Helene, P.R.L.; Kang, S.H. Designing concrete mixtures for strength, elastic modulus and fracture energy. *Mater. Struct.* **1993**, *26*, 443–452. [CrossRef]
21. European Committee for Standardization. *EN 197-1:2012. CementPart 1: Composition, Specifications and Conformity Criteria for Common Cements*; CEN: Brussels, Belgium, 2012.
22. European Committee for Standardization. *PN-EN 206+A1:2016-12. Concrete—Specification, Performance, Production and Conformity*; CEN: Brussels, Belgium, 2012.
23. Polish Committee for Standardization. *PN-B-06714-40:1978. Mineral Aggregates—Testing—Determination of Crushing Strength*; PKN: Warsaw, Poland, 1978. (In Polish)
24. British Committee for Standardization. *BS 812-110:1990. Testing Aggregates. Methods for Determination of Aggregate Crushing Value (ACV)*; BSI: London, UK, 1990.
25. Polish Committee for Standardization. *PN-B-06714-16:1978. Mineral Aggregates—Testing—Determination of Grain Shape (in Polish)*; PKN: Warsaw, Poland, 1978.
26. Polish Committee for Standardization. *PN-B-06714-13:1978. Mineral Aggregates—Testing—Determination of Mineral Dusts Content (in Polish)*; PKN: Warsaw, Poland, 1978.
27. European Committee for Standardization. *EN 1097-6:2013-11. Tests for Mechanical and Physical Properties of Aggregates—Part 6: Determination of Particle Density and Water Absorption*; CEN: Brussels, Belgium, 2013.
28. European Committee for Standardization. *EN 12350-7:2019-08. Testing Fresh Concrete—Part 7: Air content—Pressure Methods*; CEN: Brussels, Belgium, 2019.
29. European Committee for Standardization. *EN 12350-2:2019-07. Testing Fresh Concrete—Part 2: Slump Test*; PKN: Warsaw, Poland, 2019.
30. European Committee for Standardization. *EN 12390-3:2019-07. Testing Hardened Concrete-Part 3: Compressive Strength of Test Specimens*; CEN: Brussels, Belgium, 2019.
31. European Committee for Standardization. *EN 12390-6:2011. Testing Hardened Concrete—Part 6: Tensile Splitting Strength of Test Specimens*; CEN: Brussels, Belgium, 2011.
32. European Committee for Standardization. *EN 12390-13:2014-02. Testing Hardened Concrete—Part 13: Determination of Secant Modulus of Elasticity in Compression*; CEN: Brussels, Belgium, 2014.
33. European Committee for Standardization. *EN 1992-1-1:2008. Eurocode 2: Design of Concrete Structures—Part 1-1: General Rules and Rules for Buildings*; CEN: Brussels, Belgium, 2008.
34. Polish Committee for Standardization. *PN-B-06712:1986. Mineral Aggregates for Concrete (In Polish)*; PKN: Warsaw, Poland, 1986.
35. Piasta, W.; Góra, J.; Budzyński, W. Stress-strain relationships and modulus of elasticity of rocks and of ordinary and high performance concretes. *Constr. Build. Mater.* **2017**, *153*, 728–739. [CrossRef]
36. Beushausen, H.; Dittmer, T. The influence of aggregate type on the strength and elastic modulus of high strength concrete. *Constr. Build. Mater.* **2015**, *74*, 132–139. [CrossRef]
37. Baalbaki, W.; Benmokrane, B.; Chaallal, O.; Aietcin, P.C. Influence of coarse aggregate on elastic properties of high-performance concrete. *ACI Mater. J.* **1991**, *88*, 499–503. [CrossRef]

38. Aitcin, P.C.; Mehta, P.K. Effect of coarse-aggregate characteristics on mechanical properties of high-strength concrete. *ACI Mater. J.* **1990**, *87*, 103–107. [CrossRef]
39. Szymański, E. *Materiały Budowlane. Tom I*, 2nd ed.; Oficyna Wydawnicza WSEiZ w Warszawie: Warszawa, Poland, 2004. (In Polish)
40. Elsharief, A.; Cohen, M.D.; Olek, J. Influence of aggregate size, water cement ratio and age on the microstructure of the interfacial transition zone. *Cem. Concr. Res.* **2003**, *33*, 1837–1849. [CrossRef]
41. Akçaoğlu, T.; Tokyay, M.; Çelik, T. Effect of coarse aggregate size and matrix quality on ITZ and failure behavior of concrete under uniaxial compression. *Cem. Concr. Compos.* **2004**, *26*, 633–638. [CrossRef]

applied sciences

MDPI

Article

A Methodology for Determining the Rehabilitation Needs of Buildings

Beata Nowogońska

Architecture and Environmental Engineering, Faculty of Civil Engineering, University of Zielona Góra, Szafrana 1, 65-516 Zielona Góra, Poland; b.nowogonska@ib.uz.zgora.pl; Tel.: +48-68-3282-290

Received: 4 May 2020; Accepted: 1 June 2020; Published: 2 June 2020

Abstract: The appropriate rehabilitations planning of buildings should be based on the analysis of rehabilitation needs. This article proposes a methodology for Determining the Rehabilitation Needs of Buildings (DRNB). The DRNB method can be used for buildings made with traditional technology. The methodology provides the possibility to prioritize the analyzed objects and their elements as well as to determine the sequence of rehabilitation needs of any buildings and their elements. The method can be used for a single building or several buildings. The obtained results can be compared and order relations between them can be determined, which will allow the planning of repair works. In setting the priorities in the DRNB method, the implementation of the Analytical Hierarchy Process (AHP) was used. The article presents also the application the DRNB method and results of determination of rehabilitation needs for residential buildings that are located in Poland in Zielona Góra. Determining the rehabilitation needs of building components should be the first stage of planning repair works. The DRNB method helps to determine which elements in which buildings need necessary rehabilitation now, which elements of rehabilitation are important now, and which elements can be rehabilitated later—i.e., if the repair works are only useful and are not currently necessary.

Keywords: buildings; buildings elements; rehabilitation needs; rehabilitation planning; degree of technical condition; degradation of buildings

1. Introduction

Neglect of repairs is one of the main reasons for the decrease in the technical value of buildings [1–4]. The main task of building maintenance is rehabilitation planning [5–9]. In order to maintain existing buildings, it is necessary to solve problems related to forecasting rehabilitation needs [10–19]. Decision making is always difficult due to limited financial resources [20–23]. Buildings and their components are damaged to varying degrees. Building elements have different service lives. The estimation of the service life of the construction elements of a building and its materials [24–29] is an essential part of maintenance programs.

Research is needed to identify the most urgent rehabilitation needed. Making decisions connected with the choice of the type, scope and date of repairs on buildings is very problematic for managers. Algorithms supporting decision-making regarding repair works are necessary. Morelli and Lacasse proposed combines two methods: failure mode and effect analysis (FMEA) with the limit states (LS) to assess the durability of the given retrofit action [30]. Different models for maintenance management have been developed; e.g., the Building Envelope Life Cycle Asset Management BELCAM project by Vanier, Lacasse et al. that employs a stochastic decision-support system for roofing service life maintenance management [31,32]. Shen and Spedding presented [33] a model for priority setting in planned maintenance of large building stocks, and successful validation of the model in the UK and Hong Kong has been demonstrated. A different approach is found in work by Alshubbak, Pellicer,

Catala, Teixeira [34], where a model is developed that allows for the identification of the owner's needs in all phases of the building life cycle. Vanier, Tesfamariam, Sadiq, Lounis presented a number of prioritization techniques that can be used to compare and rank repair projects [35]. Bucoń and Sobotka proposed a decision model of the choice of the scope of repairs based on three assessments of a building [36]. Sherwin reviews overall models for maintenance management from the viewpoint of one who believes that improvements can be made by regarding maintenance as a contributor to profits rather than a necessary evil [37]. Jones and Sharp draw attention to the weakness inherent in the current theoretical model underpinning built asset maintenance and to propose a new performance-based model that aligns maintenance expenditure to corporate performance [38].

The correct building maintenance strategy should include a multi-annual maintenance action plan optimized for various criteria that match the owners' goals for existing restrictions. The developed model is used to compare the economics of various renovation plans in a selected scenario. The model of team Farahani is used to compare the economy of different maintenance and renovation plans in a chosen scenario in order to determine the optimal maintenance interval for a single and a combination of building components [39]. Bento Pereira, Calejo Rodrigues and Fernandes Rocha presented a post-occupancy evaluation (POE) method focused on building maintenance [40]. The three main purposes are: to obtain useful data for optimizing the buildings' maintenance plans; to search for correlations between the occupants' characteristics and their expectations toward the building; to study the occupants' willingness to pay for maintenance procedures.

In order to provide a specific time schedule for a sustainable building maintenance, [41] Daniotti and Lupica Spagnolo developed a specific method for Service Life Prediction based on the correlation between users' requirements and measured decays of building components' performance characteristics. Team Daniotti and Lupica Spagnolo outline benefits and challenges in adopting Building Information Modeling BIM based processes for the operation and maintenance of buildings [42]. Madureira, Flores-Colen, de Brito and Pereira presented a methodology to implement a maintenance plan for buildings' facades [43]. Later obsolescence of buildings can be reduced by taking into account the obsolescence criteria in the construction of new buildings [44].

All the above-mentioned studies are interesting and necessary in practice. This article presents a different approach. The proposed method applies to more buildings in use and takes into account the interdependence of rehabilitation works. Each building consists of components damaged in various degrees.

In the case of a large number of buildings, where each element is damaged to varying degrees, there are problems in planning rehabilitation work. There are many problems:

- Which building needs rehabilitation more and which should be next?
- Which element (e.g., windows, roofing) should be rehabilitated first?
- Is it possible to wait and rehabilitate several building components together?
- Which element of which building needs rehabilitation immediately?
- The proposed a methodology for Determining the Rehabilitation Needs of Buildings (DRNB) should be the answer to these questions.

2. Materials and Methods

The methodology for Determining the Rehabilitation Needs of Buildings (DRNB) consists of the following sequence of actions:

1. establishing criteria that take into account factors affecting the rehabilitation needs;
2. determination of importance of decision criteria;
3. selection of analyzed buildings;
4. determination the meters of the criteria;
5. determination of the mathematical equation determining the importance of rehabilitation needs;

6. determination of indicators determining the order of rehabilitation needs of the elements of the buildings analyzed.

The proposed methodology for Determining the Rehabilitation Needs of Buildings (DRNB) was developed in order to rescue damaged public residential buildings. However, it can also be used for private buildings.

The main aim of developing the DRNB method was to rescue damaged existing historic buildings and buildings located in representative parts of the city. For this reason, the choice of criteria was limited.

The following criteria were adopted: the degree of technical condition of a building element (Section 2.1); the type of element in the building structure (Section 2.2); the durability of the element (Section 2.3); the influence of the technical condition of the element on the damage to other elements (Section 2.4); the interdependence of the rehabilitation of the element related to the rehabilitation of another one (Section 2.5); the value of the building due to its location (Section 2.6); the value of the building heritage (Section 2.7).

It should be noted, however, that these criteria are selected as the main indicators for commissioning rehabilitation work. Many criteria have been omitted; for example, costs. It is assumed that historic buildings and those located in representative locations must be renovated. It has been assumed that the limitations of financial resources will only result in postponement of the rehabilitation works, but with the order obtained after applying the DRNB method.

The criteria were established on the basis of consultations with persons involved in the maintenance of residential buildings: building managers, university employees, appraisers, conservationists, designers and contractors of renovation works. The criteria adopted are presented in Table 1.

Table 1. Decision criteria.

Criteria Symbol c_i	Criteria Name
c_1	the degree of technical condition of the building element
c_2	type of element in the building structure
c_3	durability of element
c_4	impact of the technical condition of the element on damage to other elements
c_5	interdependence of the rehabilitation of an element related to the rehabilitation of another
c_6	the value of the building due to its location
c_7	the heritage value of the building

Criteria is determined by criteria meters. Matrix of data set D_j is determined by meters $m_{i,j}$:

$ti_{i,j}$ meters of the degree of technical condition of an i-th element in the j-th building;
$s_{i,j}$ meters of type of element in the building structure of an i-th element in a j-th building;
$d_{i,j}$ meters of the durability of an i-th element in a j-th building;
$o_{i,j}$ meters of the impact of the technical condition of an i-th element on damage to other elements in a j-th building;
$r_{i,j}$ meters of interdependence between of the repair of an i-th element related to the repair of other elements in a j-th building;
$l_{i,j}$ meters of the value of a j-th building due to its location;
$h_{i,j}$ meters of the heritage value of a j-th building;

where

i an ordinal number of an element in a building, $i = 1, 2, 3, \ldots, m$;
j an ordinal number of the building; $j = 1, 2, 3, \ldots, n$.

For the criteria to be comparable, the sum of the measures for individual elements in the building is 1.0 for all criteria.

2.1. Criterion of the Degree of Technical Condition of the Building Element

The criterion of the degree of technical condition of individual elements in buildings is based on the percentage values of wear of elements established during the evaluation of the technical condition of buildings.

The meters of the degree of technical condition wear $t_{i,j}$ is given by the formula:

$$t_{i,j} = \frac{t^{*}_{i,j}}{100\ m} \quad (1)$$

where:

- $t^{*}_{i,j}$ the degree of technical condition of an i-th element in the j-th building, resulting findings of experts (given in percent);
- i denotes an ordinal number of an element in a building, i = 1, 2, 3, … , m;
- j denotes an ordinal number of the building; j = 1, 2, 3, … , n;
- m total number of building elements analyzed.

2.2. Criterion of Type of Element in the Building Structure

The criterion type of element in the building structure assumes division of the building into structural elements, cladding, equipment and finishing. Function indicators have been assigned to each group of elements $s^{*}_{i,j}$:

for structural elements 1.0;
for shielding elements 0.75;
for equipment elements 0.50;
for finishing elements 0.25.

The meters of the structure of the building $s_{i,j}$ is determined by the equation:

$$s_{i,j} = \frac{s^{*}_{i,j}}{\sum_{i=1}^{m} s^{*}_{i,j}} \quad (2)$$

where:

- $s^{*}_{i,j}$ indicator of structure for i-th element in the j-th building.

Numerical values meters for criterion structure $s_{i,j}$ are given in Table 2.

Table 2. Numerical values meters for criterion structure $s_{i,j}$.

Number	Element Symbol e_i	Element Name	Indicator of Structure $s^{*}_{i,j}$	Meters of Structure $s_{i,j}$
1	e_1	walls	0.25	0.028
2	e_2	ceilings	0.75	0.083
3	e_3	stairs	0.75	0.083
4	e_4	roof construction	0.50	0.056
5	e_5	roof covering	0.50	0.056
6	e_6	gutters and drainpipes	0.50	0.056
7	e_7	floors	0.25	0.028
8	e_8	windows	0.25	0.028
9	e_9	doors	0.25	0.028
10	e_{10}	water installations	0.75	0.083
11	e_{11}	sewage installations	0.75	0.083
12	e_{12}	heating installations	0.50	0.056
13	e_{13}	internal plasters	0.50	0.056
14	e_{14}	external plasters	0.50	0.056
		total	9.00	1.00

2.3. Criterion of Durability

Durability criterion includes the diverse processes of technical wear and tear of the building elements due to their different durability periods. The durability meter $d_{i,j}$ is determined by the Equation (3).

$$d_{i,j} = \frac{d^*_{i,j}}{\sum_{i=1}^{m} d^*_{i,j}} \quad (3)$$

where:

$d^*_{i,j}$ indicator of durability periods of the i-th element in the j-th building;
m the number of all building elements analyzed.

The indicator of durability is determined by Equation (4).

$$d^*_{i,j} = \frac{\sum_{i=1}^{m} D_{i,j}}{D_{i,j}} \quad (4)$$

where:

$D_{i,j}$—average life of i-th element.

Numerical values meters for criterion of durability $d_{i,j}$ are given in Table 3.

Table 3. Numeric values of durability meters $d_{i,j}$.

Number	Element Name	Time of Durability $D_{i,j}$	Indicator of Durability $d^*_{i,j}$	Meters of Durability $d_{i,j}$
1	walls (brick)	140	5.804	0.023
2	ceilings (wooden)	70	11.607	0.046
3	stairs (wooden)	35	23.214	0.093
4	roof construction	80	10.156	0.041
5	roof covering	70	11.607	0.046
6	gutters and drainpipes	17.5	46.429	0.186
7	floors	50	16.250	0.065
8	windows	50	16.250	0.065
9	doors	90	9.028	0.036
10	water installations	37.5	21.667	0.087
11	sewage installations	37.5	21.667	0.087
12	heating installations	35	23.214	0.093
13	internal plasters	55	14.773	0.059
14	external plasters	45	18.056	0.072
	total	812.50	249.72	1.00

The shorter the element's lifetime, the sooner the element should be rehabilitation due to the progressing technical wear process. Therefore, the element's durability coefficient was determined to be inversely proportional to its durability.

2.4. Criterion of Impact of the Technical Condition of the Element on Damage to Other Elements

The effect of destruction of element on the condition of other elements was determined on the bases of the impact of the damaged element on the destruction of other elements in the building. The meter $o_{i,j}$ is determined by the Equation (5).

$$o_{i,j} = \frac{o^*_{i,j}}{\sum_{i=1}^{m} o^*_{i,j}} \quad (5)$$

where:

$o^*_{i,j}$ indicator of impact of wear condition of the i-th element in the j-th building on other elements;

m the number of all building elements analyzed.

Indicator of impact $o^*_{i,j}$ of wear condition of the i-th element in the j-th building on other elements is determined:

$$o^*_{i,j} = \sum_{iz=1}^{izm} g_{i,j} \quad (6)$$

where:

$g_{i,j}$ indicator of importance of the i-th element in the j-th object (determined on the basis of the importance of the elements given in the literature according to [45]);

iz the number of elements that are damaged due to the un-renovated i-element;

izm the number of all elements that are damaged due to the un-renovated i-element.

Numerical values of indicators of importance of elements $g_{i,j}$ are given in Table 4.

Table 4. Indicators of importance of elements $g_{i,j}$.

Number	Element Symbol e_i	Element Name	Importance Element According [45]	Indicator of Importance Element $g_{i,j}$
1	e_1	walls	3.0	0.294
2	e_2	ceilings	1.0	0.098
3	e_3	stairs	1.2	0.118
4	e_4	roof construction	1.0	0.098
5	e_5	roof covering	0.9	0.088
6	e_6	gutters and drainpipes	0.3	0.029
7	e_7	floors	0.2	0.020
8	e_8	windows	0.7	0.069
9	e_9	doors	0.5	0.049
10	e_{10}	water installations	0.3	0.029
11	e_{11}	sewage installations	0.3	0.029
12	e_{12}	heating installations	0.3	0.029
13	e_{13}	internal plasters	0.2	0.020
14	e_{14}	external plasters	0.3	0.029
		total	10.2	1.00

The meters of the impact of damage to the same elements in different buildings were assumed to be equal (e.g., the effect of the destruction of worn roofing on other elements is the same in all buildings) for all buildings regardless of design solutions, number of floors, type of heating, etc.

The meters of the impact of the damage of the other elements $o_{i,j}$ for individual building elements calculated according to formula (5) are included in Table 5.

Table 5. Meters of the impact of the damage $o_{i,j}$.

Number	Damaged Element	Elements Affected by Damage	Indicator of the Impact of the Damage $o_{i,j}$ * (according to formula 6)	Meters of the Impact of the Damage $o_{i,j}$ (according to formula 5)
1	walls	e_8, e_9, e_{13}, e_{14}	0.166	0.082
2	ceilings	$e_7, e_9, e_{10}, e_{11}, e_{12}, e_{13}$	0.176	0.087
3	stairs	e_7, e_{13}	0.039	0.019
4	roof construction	e_2, e_3, e_5	0.304	0.149
5	roof covering	$e_2, e_3, e_4, e_7, e_{13}$	0.353	0.173
6	gutters and drainpipes	e_{14}	0.029	0.014
7	floors	e_2	0.098	0.048
8	windows	e_7, e_{13}	0.039	0.019
9	doors	e_7, e_{13}	0.039	0.019
10	water installations	e_2, e_7, e_{13}	0.137	0.067
11	sewage installations	e_2, e_7, e_{13}	0.137	0.067
12	heating installations	e_2, e_7, e_{13}	0.137	0.067
13	internal plasters	e_7	0.020	0.010
14	external plasters	e_1, e_8	0.363	0.178
	total		2.038	1.00

Meters of the impact of the damage were calculated according to an example for roof covering:

$$o_5^* = g_2 + g_3 + g_4 + g_7 + g_{13} = 0.353 \quad (7)$$

$$o_5 = \frac{0.353}{2.038} = 0.173 \quad (8)$$

2.5. Criterion of Interdependence of the Rehabilitation of an Element Related to the Rehabilitation of Another

The meters of the interdependence of the rehabilitation of the i-th element $r_{i,j}$ is given by the formula:

$$r_{i,j} = \frac{r_{i,j}^*}{\sum_{i=1}^{m}(1 - r_{i,j}^*)} \quad (9)$$

where:

$r_{i,j}^*$—indicator of impact of interdependence of the rehabilitation of the i-th element by formula (10):

$$r_{i,j}^* = \sum_{ir=1}^{irm} g_{i,j} \quad (10)$$

where:

$g_{i,j}$ indicator of importance of the i-th element in the j-th object;
ir the number of elements that to be repaired before rehabilitation of the i-th element;
irm the number of all elements that need to be repaired before rehabilitation of the i-th element.

Numerical values meters for criterion of interdependence of the rehabilitation $r_{i,j}$ are given in Table 6.

Table 6. Meters of the of the interdependence of the rehabilitation $r_{i,j}$.

No.	Element Building of Rehabilitation	Interdependent Elements	Indicator of the Interdependence of the Rehabilitation $r_{i,j}$	1-$r_{i,j}$	Meters of Interdependence of the Rehabilitation $r_{i,j}$
1	walls	-	0.000	1.000	0.108
2	ceilings	-	0.000	1.000	0.108
3	stairs	e_1	0.294	0.706	0.076
4	roof construction	-	0.000	1.000	0.108
5	roof covering	e_4	0.098	0.902	0.097
6	gutters and drainpipes	e_1, e_4, e_5, e_{14}	0.510	0.490	0.053
7	floors	$e_1, e_2, e_3, e_8, e_9, e_{10}, e_{11}, e_{12}, e_{13}$	0.735	0.265	0.029
8	windows	e_1	0.294	0.706	0.076
9	doors	e_1, e_2	0.392	0.608	0.065
10	water installations	e_1, e_2	0.392	0.608	0.065
11	sewage installations	e_1, e_2	0.392	0.608	0.065
12	heating installations	e_1, e_2	0.392	0.608	0.065
13	internal plasters	$e_1, e_2, e_3, e_8, e_9, e_{10}, e_{11}, e_{12}$	0.716	0.284	0.031
14	external plasters	e_1, e_5, e_8, e_9	0.500	0.500	0.054
	total		4.715	9.285	1.00

Meters of the interdependence of the rehabilitation were calculated according to an example for roof covering:

$$r_5^* = g_4 = 0.098 \quad (11)$$

$$r_5 = \frac{1 - 0.098}{9.285} = 0.097 \quad (12)$$

2.6. Criterion of Locality

Some buildings need to be renovated first because of their location because they are located, for example, near the city center or on a transit route. In the method, the location meter for the entire building l_j can be equal to 0.00 or 1.00. For individual building elements, the location meter $l_{i,j}$ is given by the formula:

$$l_{i,j} = \frac{l_j}{m} \tag{13}$$

where:

l_j location indicator for the entire facility (i.e., the j-th building);
m the number of all building elements analyzed.

2.7. Criterion of the Heritage Value of the Building

Many of the buildings have a heritage value. The heritage value meter h_j for the whole object is assumed to be equal to 1.00 or 0.00 depending on whether the building has values related to the time of creation, type of use or history, legend, etc. For individual building elements the heritage value meter $h_{i,j}$ is given by the formula:

$$h_{i,j} = \frac{h_j}{m} \tag{14}$$

where:

h_j the heritage indicator for the entire facility (i.e., the j-th building);
m the number of all building elements analyzed.

2.8. Importance of Criteria

The importance of decision criteria was determined by using the Analytical Hierarchy Process AHP. The data was obtained on the basis of consultations with persons related to the rehabilitation of residential buildings: building managers, university research workers, appraisers, monument conservators, employees of design offices and executive rehabilitation companies. The results obtained are presented in Table 7.

Table 7. The value of assessing the validity of the criterion.

	c1	c2	c3	c4	c5	c6	c7	Total	Relative Importance of Criteria	Importance of Criteria
c_1	1.00	5.00	3.00	5.00	0.14	9.00	7.00	30.14	0.300	0.30
c_2	0.20	1.00	7.00	5.00	3.00	3.00	1.00	20.20	0.201	0.20
c_3	0.33	0.14	1.00	0.14	1.00	5.00	7.00	14.61	0.145	0.15
c_4	0.20	0.20	7.00	1.00	0.33	1.00	1.00	10.73	0.106	0.10
c_5	7.00	0.33	1.00	3.00	1.00	1.00	1.00	14.33	0.142	0.15
c_6	0.10	0.33	0.20	1.00	1.00	1.00	1.00	4.63	0.046	0.05
c_7	0.33	1.00	0.33	1.00	1.00	1.00	1.00	5.66	0.056	0.05

The results have been checked. Consistency ratio CR is less than 1.0, equal to 0.094.

2.9. Mathematical Model of Method for Determining the Rehabilitation Needs of Buildings DRBN

Decision criteria for rehabilitation needs $c_1, c_2, \ldots, c_7$ determined by meters of decision criteria of $t_{i,j}$, $s_{i,j}$, $d_{i,j}$, $o_{i,j}$, $r_{i,j}$, $l_{i,j}$, $h_{i,j}$ and the importance of these criteria $w_1, w_2, \ldots, w_7$ are the output for determining the matrix of indicators of the order of rehabilitation needs $k_{i,j}$.

The sequence of importance of rehabilitation needs of building elements can be determined by prioritizing the order indicators $k_{i,j}$. Order indicators for n elements in the j-th object can be obtained by solving the matrix equation.

$$K_j = B_j W_p \tag{15}$$

$$[k_{ip,j}]_{mx1} = [b_{ip,j}]_{mxu} \times [w_p]_{ux1} \tag{16}$$

where:

K_j, $[k_{ip,j}]_{mx1}$ matrix of indicators determining the need of rehabilitation of elements in the j-th building
B_j, $[b_{ip,j}]_{mxu}$ matrix of criteria measures for elements in the j-th building;
W_p, $[w_p]_{ux1}$ matrix of importance of criteria;

i —denotes an ordinal number of an element in a building, i = 1, 2, 3, ... , m;
j —denotes an ordinal number of the building; j = 1, 2, 3, ... , n;
p —denotes an ordinal number of the criteria; p = 1, 2, 3, ... , u.

The matrix of criteria measures B_j (for the j-th building) is a rectangular finite matrix with dimensions m × u.

$$B_j = \begin{bmatrix} t_{1,j} & s_{1,j} & d_{1,j} & o_{1,j} & r_{1,j} & l_{1,j} & h_{1,j} \\ t_{2,j} & s_{2,j} & d_{2,j} & o_{2,j} & r_{2,j} & l_{2,j} & h_{2,j} \\ t_{3,j} & s_{3,j} & d_{3,j} & o_{3,j} & r_{3,j} & l_{3,j} & h_{3,j} \\ \ldots & \ldots & \ldots & \ldots & \ldots & \ldots & \ldots \\ t_{m,j} & s_{m,j} & d_{m,j} & o_{m,j} & r_{m,j} & l_{m,j} & h_{m,j} \end{bmatrix} \tag{17}$$

The W_p matrix with u-importance criteria:

$$W_p = \begin{bmatrix} w_t \\ w_s \\ w_d \\ w_o \\ w_r \\ w_l \\ w_h \end{bmatrix} \tag{18}$$

The result of Equation (14) is the matrix K_j. Matrix K_j contains indicators determine the order of needs for rehabilitation of m-elements in the j-th building.

$$K_j = \begin{bmatrix} k_{1,j} \\ k_{2,j} \\ k_{3,j} \\ \ldots \\ k_{m,j} \end{bmatrix} \tag{19}$$

The task is to solve the matrix Equation (14). By multiplying the numerical meters of the criteria by the importance of these criteria, we obtain the numerical values assigned to each examined element in a given building that was included in the study. The numerical values resulting from research are indicators of the order of rehabilitation needs—$k_{i,j}$. The higher the index, the more it is necessary to renovate the i-th element in the j-th object. The indicator is not, however, any physical rehabilitation value of the elements, it is only used to rank the building elements due to the proposed order of rehabilitation needs.

The result for a large number of buildings is a set of matrixes corresponding to the number of buildings. Specific words of the matrix give the possibility of ordering the analyzed objects and their elements in any number of buildings, as well as determining the relation of the order of any two objects and their elements to each other.

The proposed method can be used in planning rehabilitation works. Analyzing the results of the method of using indicators, one can give the order of rehabilitation of individual elements in a damaged building. The method does not help in the determination of the date of the repair of any of the examined object or its elements, as the term depends on the funds that the buildings administrators may spend on the repair and also on the costs of the overhauls of particular building elements.

2.10. Scale Range of Indicators of Need Rehabilitation

In order to analyze the size of the rehabilitation needs of any examined building, i.e., determining whether the rehabilitation order index is high or low, two extreme theoretical (fictitious) models of buildings were adopted. The worst and best possible conditions were adopted and the largest and smallest result of the scale of rehabilitation indexes was obtained.

For the building where rehabilitation is the least necessary, the most favorable values for the building were adopted:

$t_i = 0.0$ (degree of wear of all elements in the building is 0%);
$l_j = 0.0$ (rehabilitation is not necessary due to the location);
$h_j = 0.0$ (the building without any cultural values).

The measures of the impact of damage on other components o_i, measure of structure s_i, measure of the durability t_i, measure of interdependence of the rehabilitation r_i and importance of criteria are permanent.

For this data, the K_{min} matrix was obtained, which contains the smallest possible indicator of rehabilitation needs that can be obtained by subsequent elements in the building:

$$K_{min} = \begin{bmatrix} 0.050 \\ 0.054 \\ 0.049 \\ 0.059 \\ 0.056 \\ 0.048 \\ 0.024 \\ 0.040 \\ 0.034 \\ 0.041 \\ 0.041 \\ 0.042 \\ 0.020 \\ 0.042 \end{bmatrix} \quad (20)$$

For the building where rehabilitation is the most necessary, the most unfavorable values for the building were adopted:

$t_i = 1.0$ (degree of wear of all elements in the building is 100%);
$l_j = 1.0$ (the building is located in the city center);
$h_j = 1.0$ (the building is in the register of monuments).

A matrix of rehabilitation needs was obtained for the building thus adopted:

$$K_{max} = \begin{bmatrix} 0.079 \\ 0.083 \\ 0.078 \\ 0.088 \\ 0.084 \\ 0.077 \\ 0.053 \\ 0.068 \\ 0.062 \\ 0.069 \\ 0.069 \\ 0.070 \\ 0.048 \\ 0.071 \end{bmatrix} \quad (21)$$

Individual matrix words are assigned to building elements. The largest possible indicator of the order of needs is for the roof structure.

After determining the extreme values that rehabilitation order indicators can achieve, you can relate the indicator of any element in any building to these two values. You can check the place of the numerical range determined according to the number denoting the order of a particular element in a given object.

3. Results and Discussion

In accordance with the proposed principles of the methodology for Determining the Rehabilitation Needs of Buildings (DRNB), an analysis of rehabilitation needs was carried out for 50 residential buildings in Zielona Góra (a city in Poland, population 140,000). All tested buildings are made in traditional technology. The buildings are characterized by similar material and construction solutions. The buildings are 3-storey, the walls are made of solid brick, with wooden ceilings and stairs, a wooden truss framing, roof covering with ceramic tiles. The buildings were built in the years 1850–1915 as town houses. Some of the buildings are in the register of the State Monument Protection Service. The manager of all the analyzed buildings is the Department of Public Utilities and Housing in Zielona Gora.

Periodic technical condition assessment [46] was carried out by experts for all buildings. During the assessment, percentage wear of the technical condition of individual building elements was determined (Table 8). The method DRBN uses these results. For the buildings studied, a cultural value index of 1.0 was adopted for buildings constructed before 1900 (for 68% of buildings), and 0.0 was adopted for the remaining buildings. A location indicator of 1.0 was used for buildings located in the city center (for 20% of facilities), and 0.0 was used for the remaining buildings.

Table 8. Average technical wear and tear for the components of the buildings tested.

No	Element Building	Average Value	Minimum	Maximum	Standard Deviation
1	walls	46.1	30	70	11.4
2	ceilings	46.3	30	80	12.7
3	stairs	45.2	25	90	14.4
4	roof construction	44.6	25	85	15.6
5	roof covering	48.0	0	75	16.9
6	gutters and drainpipes	53.7	30	80	14.2
7	floors	46.1	30	85	12.1
8	windows	61.8	30	90	13.2
9	doors	49.8	20	80	11.0
10	water installations	34.1	20	100	12.5
11	sewage installations	35.1	20	100	12.5
12	heating installations	37.6	20	100	14.6
13	internal plasters	39.1	20	75	9.9
14	external plasters	49.7	20	100	16.3

Appropriate calculations were carried out. The results are the numbers specifying the rehabilitation needs of the 14 elements in each of 50 buildings. The results are shown in Figure 1. The results obtained are indicators of rehabilitation needs. The indicators have been arranged from the highest to the lowest value. The order indicates the rehabilitation needs of the elements in the buildings under study. In this order, rehabilitation of all building group components should be carried out.

No	Element building	ORDINAL NUMBER OF THE BUILDING																								
		1	2	3	4	5	6	7	8	9	10	11	12	13	14	15	16	17	18	19	20	21	22	23	24	25
1	walls	0,068	0,066	0,061	0,060	0,066	0,066	0,065	0,066	0,065	0,067	0,068	0,062	0,062	0,061	0,061	0,062	0,066	0,060	0,067	0,067	0,072	0,063	0,060	0,060	0,062
2	ceilings	0,070	0,072	0,065	0,064	0,072	0,070	0,070	0,070	0,070	0,071	0,072	0,065	0,066	0,065	0,065	0,065	0,072	0,064	0,071	0,071	0,078	0,067	0,064	0,066	0,066
3	stairs	0,067	0,067	0,063	0,059	0,066	0,067	0,066	0,066	0,065	0,066	0,072	0,061	0,059	0,062	0,061	0,061	0,066	0,058	0,066	0,066	0,074	0,063	0,059	0,059	0,062
4	roof construction	0,077	0,078	0,070	0,069	0,076	0,077	0,077	0,074	0,078	0,074	0,078	0,072	0,070	0,070	0,070	0,070	0,077	0,068	0,074	0,074	0,084	0,070	0,073	0,069	0,072
5	roof covering	0,072	0,074	0,072	0,062	0,074	0,073	0,074	0,071	0,075	0,070	0,074	0,075	0,072	0,071	0,070	0,068	0,074	0,059	0,070	0,070	0,079	0,067	0,071	0,068	0,072
6	gutters and drain pipes	0,065	0,059	0,060	0,059	0,067	0,064	0,064	0,060	0,060	0,064	0,068	0,065	0,063	0,060	0,058	0,065	0,067	0,056	0,065	0,066	0,070	0,063	0,058	0,060	0,069
7	floors	0,041	0,041	0,035	0,035	0,041	0,040	0,040	0,038	0,041	0,039	0,041	0,035	0,037	0,037	0,041	0,037	0,042	0,034	0,039	0,040	0,048	0,037	0,039	0,037	0,037
8	windows	0,058	0,058	0,060	0,056	0,057	0,056	0,059	0,058	0,059	0,058	0,059	0,058	0,058	0,059	0,058	0,058	0,059	0,052	0,058	0,059	0,063	0,055	0,054	0,053	0,060
9	doors	0,050	0,050	0,048	0,046	0,048	0,048	0,048	0,048	0,050	0,048	0,049	0,047	0,046	0,050	0,050	0,050	0,049	0,045	0,052	0,053	0,054	0,049	0,045	0,044	0,050
10	water installations	0,052	0,053	0,051	0,051	0,055	0,053	0,053	0,053	0,052	0,052	0,055	0,052	0,051	0,053	0,052	0,051	0,054	0,050	0,052	0,054	0,056	0,051	0,051	0,051	0,053
11	sewage installations	0,053	0,053	0,051	0,051	0,053	0,053	0,053	0,053	0,052	0,052	0,057	0,052	0,051	0,053	0,052	0,052	0,054	0,050	0,052	0,054	0,056	0,051	0,054	0,052	0,053
12	heating installations	0,053	0,056	0,056	0,054	0,056	0,055	0,054	0,055	0,053	0,053	0,056	0,060	0,052	0,052	0,053	0,054	0,055	0,052	0,053	0,055	0,056	0,052	0,053	0,052	0,056
13	internal plasters	0,031	0,033	0,032	0,032	0,033	0,030	0,039	0,030	0,032	0,035	0,032	0,030	0,031	0,032	0,039	0,032	0,031	0,030	0,035	0,036	0,038	0,032	0,032	0,030	0,032
14	external plasters	0,059	0,054	0,061	0,057	0,054	0,054	0,061	0,060	0,054	0,060	0,061	0,063	0,062	0,061	0,053	0,058	0,064	0,060	0,062	0,060	0,061	0,057	0,054	0,050	0,059

No	Element building	ORDINAL NUMBER OF THE BUILDING																								
		26	27	28	29	30	31	32	33	34	35	36	37	38	39	40	41	42	43	44	45	46	47	48	49	50
1	walls	0,062	0,062	0,061	0,062	0,063	0,067	0,065	0,064	0,063	0,059	0,065	0,058	0,063	0,063	0,059	0,058	0,063	0,059	0,059	0,063	0,059	0,061	0,061	0,063	0,060
2	ceilings	0,066	0,065	0,065	0,069	0,067	0,071	0,069	0,068	0,067	0,062	0,069	0,062	0,067	0,067	0,063	0,062	0,063	0,063	0,062	0,068	0,060	0,065	0,065	0,067	0,064
3	stairs	0,059	0,061	0,061	0,062	0,063	0,066	0,064	0,064	0,063	0,056	0,066	0,057	0,063	0,063	0,056	0,057	0,056	0,060	0,057	0,061	0,055	0,060	0,060	0,063	0,059
4	roof construction	0,074	0,074	0,070	0,073	0,070	0,074	0,074	0,074	0,070	0,067	0,078	0,067	0,070	0,070	0,068	0,066	0,067	0,068	0,068	0,074	0,067	0,070	0,070	0,070	0,067
5	roof covering	0,071	0,068	0,069	0,071	0,067	0,070	0,070	0,067	0,067	0,067	0,069	0,063	0,067	0,067	0,068	0,066	0,066	0,066	0,072	0,072	0,068	0,067	0,063	0,067	0,063
6	gutters and drain pipes	0,059	0,058	0,058	0,060	0,063	0,065	0,065	0,063	0,063	0,061	0,061	0,058	0,062	0,060	0,057	0,055	0,056	0,059	0,061	0,063	0,057	0,061	0,059	0,061	0,058
7	floors	0,037	0,037	0,039	0,046	0,037	0,040	0,040	0,037	0,035	0,031	0,038	0,033	0,035	0,035	0,033	0,033	0,035	0,033	0,032	0,036	0,035	0,036	0,033	0,036	0,033
8	windows	0,059	0,058	0,056	0,056	0,055	0,058	0,058	0,055	0,055	0,056	0,056	0,052	0,054	0,054	0,053	0,048	0,055	0,048	0,055	0,058	0,046	0,054	0,052	0,054	0,052
9	doors	0,051	0,050	0,050	0,055	0,049	0,052	0,052	0,049	0,045	0,042	0,043	0,042	0,047	0,045	0,045	0,040	0,043	0,042	0,043	0,047	0,038	0,048	0,046	0,048	0,046
10	water installations	0,051	0,053	0,051	0,066	0,051	0,054	0,052	0,051	0,051	0,049	0,049	0,045	0,049	0,049	0,047	0,047	0,047	0,047	0,048	0,049	0,047	0,051	0,047	0,048	0,047
11	sewage installations	0,051	0,053	0,052	0,066	0,051	0,054	0,052	0,051	0,051	0,049	0,049	0,045	0,049	0,049	0,047	0,047	0,047	0,048	0,048	0,049	0,047	0,051	0,047	0,051	0,047
12	heating installations	0,053	0,053	0,053	0,067	0,052	0,055	0,053	0,052	0,052	0,051	0,049	0,046	0,049	0,049	0,048	0,048	0,057	0,050	0,057	0,051	0,048	0,052	0,048	0,052	0,048
13	internal plasters	0,030	0,032	0,031	0,040	0,032	0,036	0,036	0,032	0,031	0,029	0,030	0,029	0,031	0,031	0,026	0,026	0,026	0,026	0,026	0,027	0,024	0,032	0,029	0,032	0,029
14	external plasters	0,064	0,054	0,048	0,062	0,057	0,060	0,060	0,057	0,057	0,057	0,055	0,053	0,055	0,055	0,064	0,044	0,053	0,055	0,056	0,059	0,058	0,056	0,053	0,056	0,053

LEGEND

0,040 - indicators with values lower than 0.067 and higher than 0.036

0,068 - indicators with values lower than 0.071 and higher than 0.067

0,072 - indicators with values greater than 0.071

Figure 1. The results obtained for the buildings tested in Zielona Góra.

Assuming the division of building elements into order groups, it is possible to assess the size of rehabilitation needs. It is proposed to adopt 4 order groups of elements:

- elements for which rehabilitation is absolutely necessary, with sequence values greater than 0.068,
- elements for which rehabilitation is important, with values of order indicators from 0.051 to 0.068,
- elements for which rehabilitation is useful, with values of order indicators from 0.031 to 0.05,
- elements for which rehabilitation is not needed now, with sequence index values less than 0.031.

The first two groups include buildings elements which technical wear is greater than 50%. Additionally, rehabilitation is absolutely recommended for them due to the cultural and historical values of the buildings themselves or their location.

The "rehabilitation is useful" group contains elements with the degree of wear lower than 50%, but due to their location and cultural conditions, their (or their elements) rehabilitation must be performed earlier than the repair for the elements of the fourth group.

Assuming the division of the buildings into the four groups, it is possible to assess the size of the rehabilitation needs for the analyzed group of buildings.

Results are presented in Figure 2.

The analysis included 14 elements in 50 buildings, 700 in total. The results in Figure 1 indicate that as much as 92% of building components are eligible for rehabilitation. Rehabilitation is absolutely necessary for 12% of elements, rehabilitation is important for 52% elements, rehabilitation is useful for 28% and only for 8% of elements (usually internal plasters), rehabilitation is not needed now.

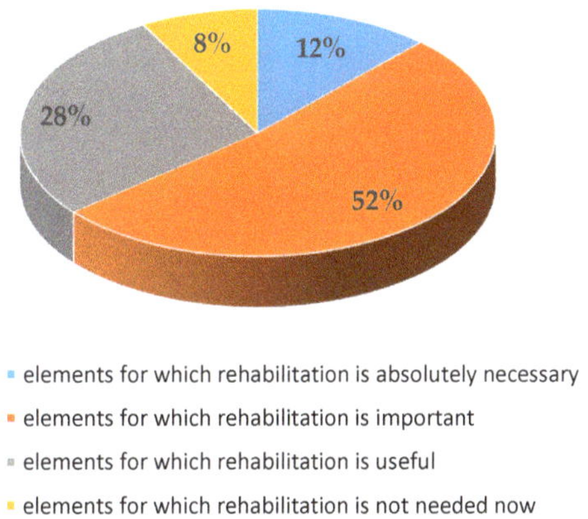

Figure 2. Rehabilitation needs of elements of the analyzed buildings in Zielona Góra.

The methods presented in the literature are helpful in planning rehabilitation works. They rely primarily on the assessment of the choice of the type of material in rehabilitation, assessment of costs over time or planning preventive rehabilitation of buildings. All methods are needed. The method for Determining the Rehabilitation Needs of Buildings (DRNB) presented in the article is based on the assessment of rehabilitation needs. The needs of rehabilitation may result from the poor technical condition of the building element, ending material durability, needs of rehabilitation due to the location or protection of cultural heritage.

The method DRBN may also be used in planning complex repairs for all quarters of a town. The result of such an applied method will be one matrix of sequence for all the buildings, which can be obtained after determining the weighted average wear for all the buildings.

4. Conclusions

The method DRNB presented in the article is based on the assessment of rehabilitation needs. The needs of repairs may result from the poor technical condition of the building element, ending material durability, needs of rehabilitation due to the location or protection of cultural heritage.

The most important advantages of the method are:

1. The proposed methodology for Determining the Rehabilitation Needs of Buildings DRNB for an arbitrary group of buildings allows us to arrange the examined objects in terms of the need for their rehabilitation and to determine the order between any two buildings.
2. This method can be used for rehabilitation work schedules for a single building and for a larger group of buildings, e.g., quarter of a town.
3. Financial resources are always too small. Choosing a building element or building from among many is always an important problem. The proposed method will help in making decisions.
4. Rehabilitation works carried out using the DRNB method will stop building degradation.

The method DRNB helps to determine which elements in which buildings need necessary rehabilitation now and for which elements rehabilitation is important now and which elements can be rehabilitated later—i.e., the repair works are only useful and are not currently needed. The results for buildings in Zielona Góra indicate that as much as 92% of building components are eligible for rehabilitation. Rehabilitation is absolutely necessary for 12% of elements, rehabilitation is important for 52% elements, rehabilitation is useful for 28% and only for 8% of elements (usually internal plasters), rehabilitation is not needed now.

The conclusions resulting from the conducted analysis are the basis for further research of methods studying the costs of rehabilitating buildings.

Funding: This research received no external funding.

Conflicts of Interest: The authors declare no conflict of interest.

References

1. Lacasse, M.A. Advances in Service Life Prediction—An Overview of Durability and Methods of Service Life Prediction for Non-Structural Building Components. In Proceedings of the Annual Australasian Corrosion Association Conference, Wellington Convention Centre, Wellington, New Zealand, 16–19 November 2008; pp. 1–13.
2. Silva, A.; de Brito, J.; Gaspar, P.L. *Methodologies for Service Life Prediction of Buildings*; Springer International Publishing: Cham, Switzerland, 2016; Volume VII, p. 432. [CrossRef]
3. Zavadskas, E.; Antuchevičienė, J.; Kapliński, O. Multi-criteria decision making in civil engineering: Part I—A state-of-the-art survey. *Eng. Struct. Technol.* **2016**, *7*, 103–113. [CrossRef]
4. Nowogońska, B. The Method of Predicting the Extent of Changes in the Performance Characteristics of Residential Buildings. *Arch. Civ. Eng.* **2019**, *65*, 81–89. [CrossRef]
5. Paulo, P.; Branco, F.; Brito, J.; Silva, A. Buildings Life—The use of genetic algorithms for maintenance plan optimization. *J. Clean. Prod.* **2016**, *121*, 84–98. [CrossRef]
6. Matulionis, R.C.; Freitag, J.C. *Preventive Maintenance of Buildings*; John Wiley & Sons, Inc.: Hoboken, NJ, USA, 1990.
7. Ortega, L.; Serrano, B.; Fran, J.M. Proposed method of estimating the service life of building envelopes. *Rev. De La Construcción* **2015**, *14*, 60–68. [CrossRef]
8. Christen, M.; Schroeder, J.; Wallbaum, H. Evaluation of strategic building maintenance and refurbishment budgeting method Schroeder. *Int. J. Strateg. Prop. Manag.* **2014**, *18*, 393–406. [CrossRef]
9. Leśniak, A.; Wieczorek, W.; Górka, M. Selection of the Variant of the Aluminium-Glass Facade Implementation Using the AHP Method. In *Advances in Intelligent Systems and Computing*; Springer International Publishing: Cham, Switzerland, 2020. [CrossRef]
10. Ibadov, N. The alternative net model with the fuzzy decision node for the construction projects planning. *Arch. Civ. Eng.* **2018**, *64*, 3–20. [CrossRef]
11. Drozd, W.; Kowalik, M. Comparison of technical condition of multi-family residential buildings of various ages. *Arch. Civ. Eng.* **2020**, *66*, 55–67. [CrossRef]
12. Nowogońska, B. Proposal for determing the scale of renovation needs of residential buildings. *Civ. Environ. Eng. Rep.* **2016**, *22*, 137–144. [CrossRef]
13. Prieto, A.J.; Vásquez, V.; Silva, A.; Horn, A.; Alejandre, F.J.; Macías-Bernal, J.M. Protection value and functional service life of heritage timber buildings. *Build. Res. Inf.* **2019**, *47*, 567–584. [CrossRef]

14. Drozd, W.; Leśniak, A. Ecological Wall Systems as an Element of Sustainable Development—Cost Issues. *Sustainability* **2018**, *10*, 2234. [CrossRef]
15. Nowogońska, B. Performance characteristics of buildings in the assessment of revitalization needs. *Civ. Environ. Eng. Rep.* **2019**, *29*, 119–127. [CrossRef]
16. Biolek, V.; Hanák, T. LCC Estimation Model: A Construction Material Perspective. *Buildings* **2019**, *9*, 182. [CrossRef]
17. Radziszewska-Zielina, E.; Śladowski, G. Proposal of the Use of a Fuzzy Stochastic Network for the Preliminary Evaluation of the Feasibility of the Process of the Adaptation of a Historical Building to a Particular Form of Use. In *IOP Conference Series: Materials Science and Engineering*; Materials Science and Engineering; IOP Publishing Ltd: Bristol, UK, 2017; Volume 245/7, p. 072029.
18. Colen, I.F.; de Brito, J.; Freitas, V. Discussion of Criteria for Prioritization of Predictive Maintenance of Building Façades: Survey of 30 Experts. *J. Perform. Constr. Facil.* **2010**, *24*. [CrossRef]
19. Paulo, P.; Branco, F.; Brito, J. Buildings Life: A building management system. *Struct. Infrastruct. Eng.* **2014**, *10*, 388–397. [CrossRef]
20. Moretti, N.; Re Cecconi, F. A Cross-Domain Decision Support System to Optimize Building Maintenance. *Buildings* **2019**, *9*, 161. [CrossRef]
21. Nowogońska, B.; Korentz, J. Value of Technical Wear and Costs of Restoring Performance Characteristics to Residential Buildings. *Buildings* **2020**, *10*, 9. [CrossRef]
22. Frangopol, D.M.; Lin, K.-Y.; Estes, A.C. Life-cycle cost design of deteriorating structures. *J. Struct. Eng. Asce* **1997**, *123*, 1390–1401. [CrossRef]
23. Wieczorek, D.; Plebankiewicz, E.; Zima, K. Model estimation of the whole life cost of a building with respect to risk factors. *Technol. Econ. Dev. Econ.* **2019**, *25*, 20–38. [CrossRef]
24. Rivera-Gómez, H.; Montaño-Arango, O.; Corona-Armenta, J.R.; Garnica-González, J.; Hernández-Gress, E.S.; Barragán-Vite, I. Production and Maintenance Planning for a Deteriorating System with Operation-Dependent Defectives. *Appl. Sci.* **2018**, *165*, 8. [CrossRef]
25. Hoła, A.; Sadowski, Ł. A method of the neural identification of the moisture content in brick walls of historic buildings on the basis of non-destructive tests. *Autom. Constr.* **2019**, *106*, 102850. [CrossRef]
26. Mrówczyńska, M.; Gibowski, S. Indicating Vertical Deviation of Historical Buildings Using Geodetic Methods—Case Study of Brick and Wood Tower in Nowe Miasteczko. *Civ. Environ. Eng. Rep.* **2016**, *22*, 127–136. [CrossRef]
27. Monczynski, B.; Ksit, B.; Szymczak-Graczyk, A. Assessment of The Effectiveness of Secondary Horizontal Insulation Against Rising Damp Performed by Chemical Injection. *IOP Conf. Ser. Mater. Sci. Eng.* **2019**, *471*, 52063. [CrossRef]
28. Radziszewska-Zielina, E.; Kania, E.; Śladowski, G. Problems of the Selection of Construction Technology for Structures of Urban Aglomerations. *Arch. Civil Eng.* **2018**, *64*, 55–71. [CrossRef]
29. Rudbeck, K. Methods for designing building envelope components prepared for repair and maintenance. In *Department of Buildings and Energy*; Technical University of Denmark: Lyngby-Taarbæk, 1999.
30. Morelli, M.; Lacasse, M.A. A systematic methodology for design of retrofit actions with longevity. *J. Build. Phys.* **2019**, *42*, 585–604. [CrossRef]
31. Lounis, Z.; Vanier, D.J.; Lacasse, M.A.; Kyle, B.R. *Decision-Support System for Service Life Asset Management: The BELCAM Project. Durability of Building Materials and Components*; National Research Council Canada: Ottwa, ON, Canada, 1999; Volume 4, pp. 2338–2347.
32. Vanier, D.J.; Lacasse, M.A. BELCAM project: Service life, durability, and asset management research. In Proceedings of the 7th Conference on Durability of Building Materials and Components, Stockholm, Sweden, 19–23 May 1996; Volume 2, pp. 848–856.
33. Shen, Q.; Spedding, A. Priority setting in planned maintenance—Practical issues in using the multi-attribute approach. *Build. Res. Inf.* **1998**, *26*, 169–180. [CrossRef]
34. Alshubbak, A.; Pellicer, E.; Catala, J.; Teixeira, J. A Model for identifying owner's needs in the building life cycle. *J. Civ. Eng. Manag.* **2015**, *21*, 1046–1060. [CrossRef]
35. Vanier, D.; Tesfamariam, S.; Sadiq, R.; Lounis, Z. Decision models to prioritize maintenance and renewal alternatives. In Proceedings of the Joint International Conference on Computing and Decision Making in Civil and Building Engineering, Montréal, QC, Canada, 14–16 June 2006; pp. 2594–2603.

36. Bucoń, R.; Sobotka, A. Decision-making model for choosing residential building repair variants. *J. Civ. Eng. Manag.* **2015**, *21*, 893–901. [CrossRef]
37. Sherwin, D. A review of overall models for maintenance management. *J. Qual. Maint. Eng.* **2000**, *6*, 138–164. [CrossRef]
38. Jones, K.; Sharp, M. A new performance-based process model for built asset maintenance. *Facilities* **2007**, *25*, 525–535. [CrossRef]
39. Farahani, A.; Wallbaum, H.; Olof Dalenbäck, J. Optimized maintenance and renovation scheduling in multifamily buildings—A systematic approach based on condition state and life cycle cost of building components. *Constr. Manag. Econ.* **2019**, *37*, 139–155. [CrossRef]
40. Bento Pereira, N.; Calejo Rodrigues, R.; Fernandes Rocha, P. Post-Occupancy Evaluation Data Support for Planning and Management of Building Maintenance Plans. *Buildings* **2016**, *6*, 45. [CrossRef]
41. Daniotti, B.; Lupica Spagnolo, S. Service Life Prediction Tools for Buildings' Design and Management. In Proceedings of the 11DBMC International Conference on Durability of Building Materials and Components, Istanbul, Turkey, 11–14 May 2008. [CrossRef]
42. Daniotti, B.; Pavan, A.; Lupica Spagnolo, S.; Caffi, V.; Pasini, D.; Mirarchi, C. Benefits and Challenges Using BIM for Operation and Maintenance. In *BIM-Based Collaborative Building Process Management*; Springer Tracts in Civil Engineering; Springer: Berlin/Heidelberg, Germany, 2020; pp. 103–142.
43. Madureira, S.; Flores-Colen, I.; de Brito, J.; Pereira, C. Maintenance planning of facades in current buildings. *Constr. Build. Mater.* **2017**, *147*, 790–802. [CrossRef]
44. Chen, C.; Juan, Y.; Hsu, Y. Developing a systematic approach to evaluate and predict building service life. *J. Civ. Eng. Manag.* **2017**, *23*, 890–901. [CrossRef]
45. Arendarski, J. *Durability and Reliability of Residential Buildings*; Arkady: Warszawa, Poland, 1978. (In Polish)
46. *Periodic Inspection Protocol 2018*; Department of Municipal and Housing Administration in Zielona Góra: Zielona Góra, 2018. (In Polish)

applied sciences

MDPI

Article

Creep Assessment of the Cement Matrix of Self-Compacting Concrete Modified with the Addition of Nanoparticles Using the Indentation Method

Paweł Niewiadomski * and Damian Stefaniuk

Faculty of Civil Engineering, Wroclaw University of Science and Technology, Wybrzeże Wyspiańskiego 27, 50-370 Wroclaw, Poland; damian.stefaniuk@pwr.edu.pl

* Correspondence: pawel.niewiadomski@pwr.edu.pl; Tel.: +48-508-859-290

Received: 2 March 2020; Accepted: 1 April 2020; Published: 3 April 2020

Abstract: In recent years, there has been an increased interest in the modification of cement composites with finer materials, including nanoparticles. Multi-scale studies are needed to fully assess the effect of nanoparticles and provide a complete overview of their impact on both the structure of an obtained material and its important mechanical parameters, such as creep. Therefore, the purpose of this paper is to fill the knowledge gap in the literature concerning the assessment of the creep of a cement matrix of self-compacting concrete modified with the addition of SiO_2, TiO_2, and Al_2O_3 nanoparticles using the indentation method. Depending on the type of used nanoparticles, we found an increase or decrease of the creep coefficient C_{IT} in comparison to the reference series. The obtained results were scrupulously analyzed in terms of statistics, which enabled the conclusion that the addition of nanoparticles does not significantly affect the creep of the cement matrix of self-compacting concrete. The methodology used in this paper allowed us to shorten the time needed to assess the creep phenomenon compared to traditional methods and fill the corresponding knowledge gap in the literature.

Keywords: self-compacting concrete; cement matrix; nanoparticles; indentation; creep

1. Introduction

In recent years, there has been a growing interest in the possibility of using nanometric materials as a potential additive for the production of cement composites [1]. This is mainly due to the unique properties of nanoparticles, such as their large specific surface area and chemical activity [2]. Until now, attempts have been made to determine the effect of some nanoparticles on the selected physical and mechanical properties of cement composites obtained with their use. Research included the influence of nanoparticles on the following concrete parameters: porosity [3], water absorption [4], compressive strength [5], bending strength [6], and the destruction process [7]. In the literature, however, in the case of concretes made with the addition of nanoparticles, there is no research concerning mechanical parameters related to rheology, such as creep.

Knowledge of the creep effect and of the possibility of its conscious control is important with regard to the durability and maintenance of objects made of concrete [8]. As a rule, when assessing the creep of concrete at the macro-scale, this material is treated as homogeneous, consisting mainly of aggregates. The standard methods that are used for such materials to determine concrete creep are time- and labor-consuming [9]. However, it has been proved that the effective mechanical parameters of concrete are significantly affected by the properties of the structure of the cement matrix observed at micro and nano levels [10], which should be assessed using appropriate techniques. One such technique is the indentation method.

Indentation is a method for measuring the mechanical properties of a material. This method is based on pressing a hard indenter in the form of, e.g., a diamond pyramid (Berkovich indenter) into a sample. Indentation research can be carried out at several levels of observation, from macro, to micro and to nano. The value of force applied to the surface of the sample being tested and the corresponding size of the obtained imprint, are directly related to the volume of the material subjected to the test. In the process of indentation, three stages can be distinguished: loading—the indenter is pressed into the material at a given speed until reaching the assumed force, which depends on the type of material; retaining the indenter in the material for a specified period of time; and unloading—usually at the same speed at which the indenter was pressed. The possibility of continuous monitoring of the indentation process enables the following material parameters to be determined: hardness μH, observed at micro-scale, and the indentation module μE [11]. It also allows rheological parameters such as creep [12] or relaxation [13] to be calculated.

When determining the creep of a cement matrix using the indentation method, which is a form of microstructural analysis, it is necessary to take into account several factors that affect the obtained results [14]. They include the preparation of the samples, the oxidation of the tested surface, the friction and adhesion of the indenter tip, the roughness of the tested surface, the surface of the imprint, and also the development of displacements. Particular attention should also be paid to the time of loading and unloading increase, the time of constant loading, and the number of loading and unloading cycles [14]. It is also worth noting that the assessment of creep using the method of indentation correlates well with the macroscopic assessment and also significantly accelerates the time of the study [15].

So far, many studies have concerned the subject of creep in ordinary, high-performance and self-compacting concrete (SCC). In a paper [16], it was proved that creep is greatly reduced by the use of ultrafine ground granulated blast-furnace slag (GGBS) and silica fume (SF). In addition, a reduction of creep by even more than 50% can be expected when the fly ash replacement level is increased from 35% to 60% [17]. Short-term tensile creep was investigated in concrete with the addition of steel fibers in a study [18], where it was demonstrated that incorporation of short steel fibers decreased the tensile creep coefficient and the specific creep in 14 days. The cement type has a significant influence on creep, as described in a study [19]. In turn, Persson [20] demonstrated that creep and shrinkage of self-compacting concrete did not differ significantly compared to those of ordinary concrete. Nevertheless, there is less research concerning the influence of nanoparticles on creep in self-compacting concrete.

Considering the above, the purpose of the paper was the filling of a knowledge gap in the literature, concerning the assessment of the creep of a self-compacting concrete cement matrix modified with the addition of nanoparticles, using the indentation method.

2. Materials and Methods

The following components were used to make the self-compacting concrete mixes for testing: Portland cement CEM I 52.5R with a density of 3.10 g/cm^3, which meets the requirements of PN-EN 197-1 [21] and has the chemical and phase composition shown in Table 1; an innovative third-generation superplasticizer (SP) Glenium Sky 600 produced by BASF, based on polycarboxylic ether (PCE) polymers, with a density of 1.08 g/cm^3, used in a quantity equal to 4.0% of the cement weight; potable tap water; a granite aggregate with an average density of 2.61 g/cm^3, supplied by Tampereen Kovakivi Oy, which had fractions of 10–5, 5–2, 2–1, 1.2–0.5, 0.6–0.1 mm, and one fraction with a grain size <0.1 mm acting as a fine filler. The fine aggregate, with a fraction of up to 2 mm, constituted 48% of the total aggregate. The particle size distribution curve of the granite aggregate that was used for testing is shown in Figure 1. For the designed concrete mixes, the water to cement W/C ratio was equal to 0.42.

Table 1. Chemical and phase composition of Portland cement CEM I 52.5R.

Component	[wt. %]	Phase	[wt.%]
CaO	65.0	C_3S	56.3
SiO_2	22.0	C_2S	20.8
Al_2O_3	5.4	C_3A	8.7
Fe_2O_3	3.3	C_4AF	10.0
MgO	3.2		
Cl-	≤0.08		
Loss of ignition	3.0		

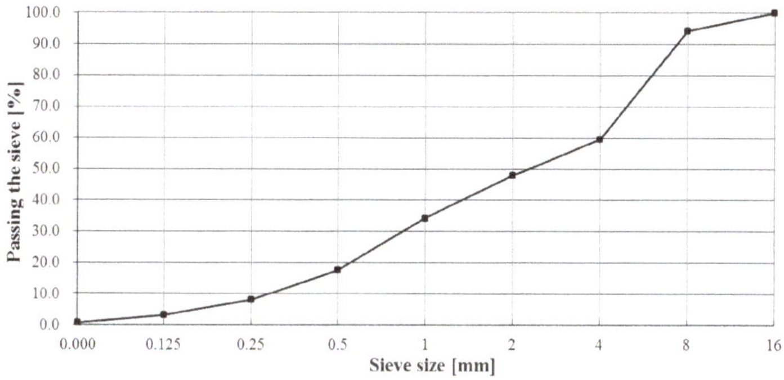

Figure 1. Grading curve of the aggregate.

The composition of the concrete mix was modified with the following nanoparticles in the form of dry powder in the amount of 4.0% of the cement weight: SiO_2 (Figure 2a) with particle size <20 nm, density of 2.4 g/mL at 25 °C, purity of 99.5%, and specific surface area equal to 450 m^2/g; TiO_2 (Figure 2b), with particle size <25 nm, density of 3.9 g/mL at 25 °C, purity of 99.7%, and specific surface area equal to 50 m^2/g; Al_2O_3 (Figure 2c), with particle size <50 nm and specific surface area of 40 m^2/g. The authors decided to use the highest content of nanoparticles (4.0%) used in previous studies [22] to see if a significant modification of SCC composition with the addition of nanoparticles has an impact on the properties of the hardened cement matrix. Generally speaking, using a lower content of nanoparticles (0.5% and 2.0%) resulted in slightly visible effects in previous tests [22], while using more than 4.0% of nanoparticles required a bigger amount of superplasticizer (>4.0%, which is not reasonable) to obtain the proper characteristics of a self-compacting concrete mix (slump flow > 550 mm). The nanoparticles used in the tests were provided by Sigma Aldrich.

Figure 2. SEM images of nanoparticles: (**a**) SiO_2; (**b**) TiO_2; (**c**) Al_2O_3.

Four self-compacting concrete mixes were designed and made with the above ingredients, one of which was prepared without the addition of nanoparticles, as a control. The compositions of all the designed mixes per 1 m^3 are presented in Table 2.

Table 2. Composition of the designed mixes.

Concrete Mix	Nanoparticles and their Amount [%]	Cement [kg/m^3]	Aggregate [kg/m^3]	Water [kg/m^3]	Nanoparticles [kg/m^3]	SP [kg/m^3]
S0	-	460.0	1640.0	193.2	-	18.4
S1	SiO_2—4.0%	441.6	1640.0	193.2	18.4	18.4
S2	TiO_2—4.0%	441.6	1640.0	193.2	18.4	18.4
S3	Al_2O_3—4.0%	441.6	1640.0	193.2	18.4	18.4

From each mix, with the designation and composition given in Table 2, a series of 12 cuboid specimens with dimensions of 40 × 40 × 160 mm were made. They were then matured in a climatic chamber at an air temperature of 20 °C (± 1 °C) and a relative humidity of 95% (± 5%). These series were marked identically as mixes S0, S1, S2, and S3. After one year, cylindrical samples of 25 mm in diameter and 20 mm in height were cut from the previously prepared cuboid specimens. The samples were cut from the centre of the specimens to avoid the boundary effect. In order to properly prepare the surface for the creep tests using the indentation method, the cylindrical samples were first immersed in epoxy resin in a vacuum machine that had an inside pressure of 0.07 bar, and their surfaces were then ground using 320 grit sandpaper and a 9, 3, and 1 µm graded diamond slurry until a smooth surface was obtained. The samples for the creep tests using the indentation method are shown in Figure 3.

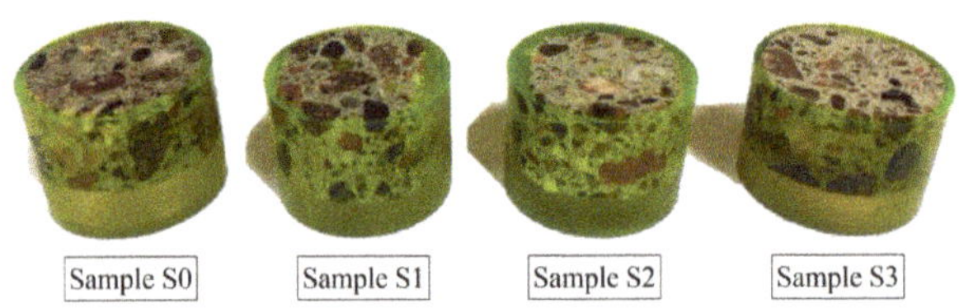

Figure 3. Samples prepared for creep tests based on the indentation method.

Due to the large diversity of the tested material, a large number of measurements was necessary for the results to be statistically well interpreted [23]. Therefore, each sample was subjected to a minimum of 80 creep measurements in a vast area of the cement matrix, while at the same time maintaining an imprint spacing of a minimum of 20 µm. For the measurements, a Poisson's ratio value of 0.3 was adopted for the tested concretes.

The TTX-NHT nanoindenter with a Berkovich indenter was used during the creep tests. According to the definition of creep, the test involved the introduction of an additional period of time during the standard process of indentation. During this time, the maximum load value was maintained, and the depth increase was measured, which is schematically shown in Figure 4.

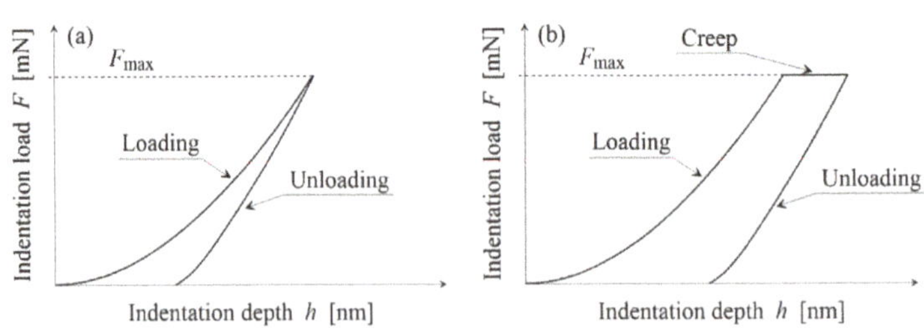

Figure 4. (**a**) Standard indentation curve and (**b**) indentation curve that includes creep.

During the test, the samples were loaded at a speed of 400 mN/min until the maximum load of 200 mN was obtained. Afterwards, at the creep stage, the maximum load value was maintained for a period of 300 s. Such a period of time enabled the deformation due to creep to be reduced to 5.0% [14]. The last stage involved unloading at a speed of 400 mN/min. The applied load range resulted in imprints up to 6 μm deep. The creep qualitative difference in the tested cement matrixes of the self-compacting concretes was measured using the creep coefficient C_{IT}, which is expressed as:

$$C_{IT} = \frac{h_2 - h_1}{h_1} \cdot 100\ [\%], \tag{1}$$

where h_1 is the depth reached by the indenterover time t_1 (immersion), and thus when obtaining the maximum load F_{max}, and h_2 is the indenter immersion between the creep phase and the beginning of the unloading phase t_2 (see Figure 5).

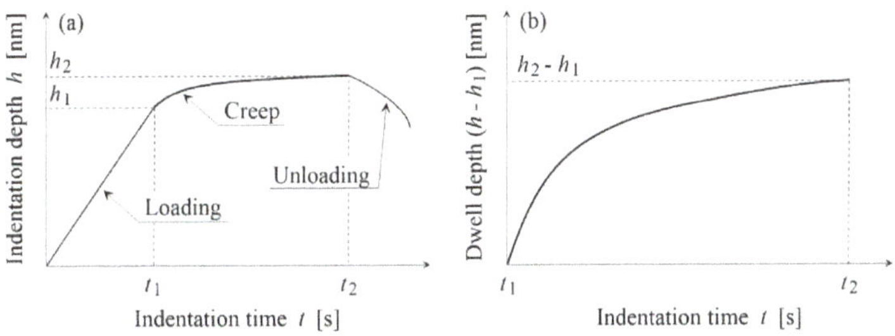

Figure 5. (**a**) Indentation curve including loading, creep, and unloading and (**b**) extracted creep phase only.

It should be noted that the lower the value of the creep coefficient C_{IT}, the greater the resistance of the cement matrix to long-term loading, and this is a positive phenomenon.

We analyzed not only the C_{IT} coefficient but also the course of the creep phase curves. This is because two different samples may have the same C_{IT} coefficient for a given creep time, but different rheological behaviour is nature (see Figure 6). This could occur if the time for the creep phase was not chosen correctly (in particular, if the creep time was too short).

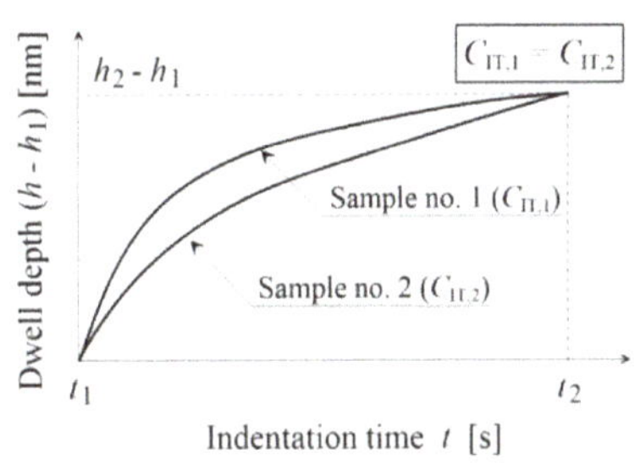

Figure 6. Indentation curve for the creep phase of two hypothetical samples with the same creep coefficient C_{IT}.

During indentation, continuous monitoring can reveal that some results are inconsistent, e.g., due to the occurrence of discontinuities in the matrix structure under the examined point [10]. Therefore, the curves that were inconsistent during the measurements were rejected and not further analyzed. As a result, approximately 5%–8% of the measurements—depending on the analysed sample—were rejected.

To better describe the tested series of self-compacting concretes, basic rheological, physical, and mechanical properties are shown in Tables 3–5 (a full description of the tests is available in a previous paper [23]). Selected characteristics included the time of reaching a slump with a diameter of 500 mm, the maximum slump diameter, the total air content, the compressive and bending tensile strengths after 28 days of probes maturing.

Table 3. Rheological properties of self-compacting concrete mixes (based on [22]).

Concrete Mix	Nanoparticles and their Amount [%]	Time of Reaching a Slump with a Diameter of 500 mm T_{500} [s]	Slump Diameter D [mm]
S0	-	2.9	760
S1	SiO_2—4.0%	-	370
S2	TiO_2—4.0%	2.6	735
S3	Al_2O_3—4.0%	5.5	510

Table 4. Physical properties of self-compacting concrete mixes (based on [22]).

Concrete Mix	Nanoparticles and their Amount [%]	Total Air Content A [%]
S0	-	1.29
S1	SiO_2—4.0%	2.39
S2	TiO_2—4.0%	1.37
S3	Al_2O_3—4.0%	1.75

Table 5. Mechanical properties of self-compacting concrete mixes (based on [22]).

Concrete Mix	Nanoparticles and their Amount [%]	Compressive Strength f_{cm} after 28 Days [MPa]	Bending Tensile Strength $f_{ctm,fl}$ after 28 Days [MPa]
S0	-	67.2	8.9
S1	SiO_2—4.0%	72.4	9.0
S2	TiO_2—4.0%	66.7	8.4
S3	Al_2O_3—4.0%	64.1	8.5

3. Results

Figure 7 shows examples of indentation prints for all the tested series of self-compacting concrete.

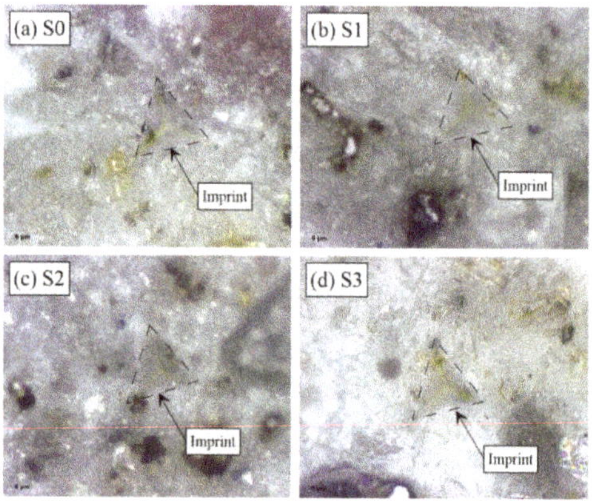

Figure 7. Exemples of indentation imprints for samples of the series (**a**) S0, (**b**) S1, (**c**) S2, and (**d**) S3.

To accurately track and visualize the test results, we obtained a set of indenter immersion curves, in particular for the creep phase, as well as the averaged outline of the above-mentioned curves and the standard deviation for the reference sample S0 (without the addition of nanoparticles), as shown in Figure 8.

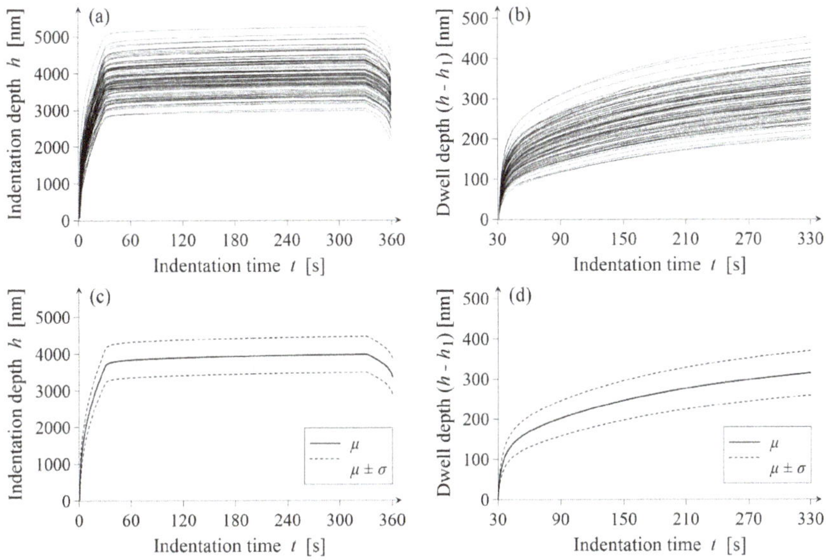

Figure 8. (**a**) Indenter immersion curves for all the indentation phases, (**b**) indenter immersion curves for the creep phase only, (**c**) mean (μ) indenter immersion for all the indentation phases (σ means one standard deviation), and (**d**) average immersion for the creep phase only. Reference sample S0.

In turn, Figures 9–11 show a comparison of the average indenter immersion during all the indentation phases as well as in the creep phase only for series S0 and S1, S0 and S2, and S0 and S3, respectively.

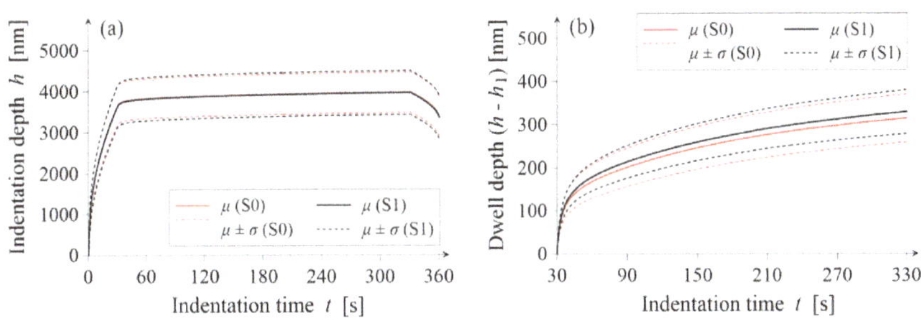

Figure 9. (**a**) Average indenter immersion during all the indentation phases and (**b**) average immersion in the creep phase only. Results for sample S1 versus the reference sample S0.

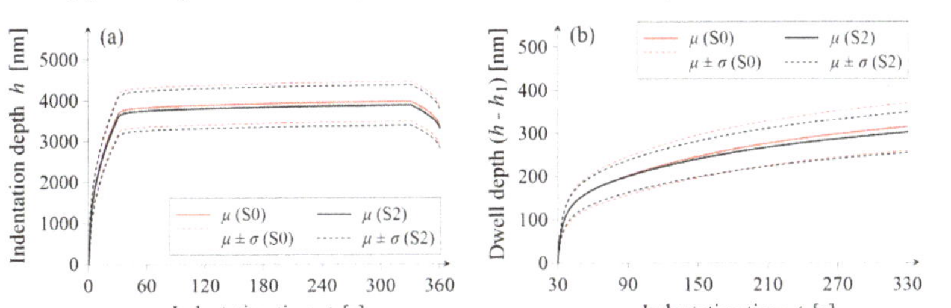

Figure 10. (**a**) Average indenter immersion during all the indentation phases and (**b**) average immersion in the creep phase only. Results for sample S2 versus the reference sample S0.

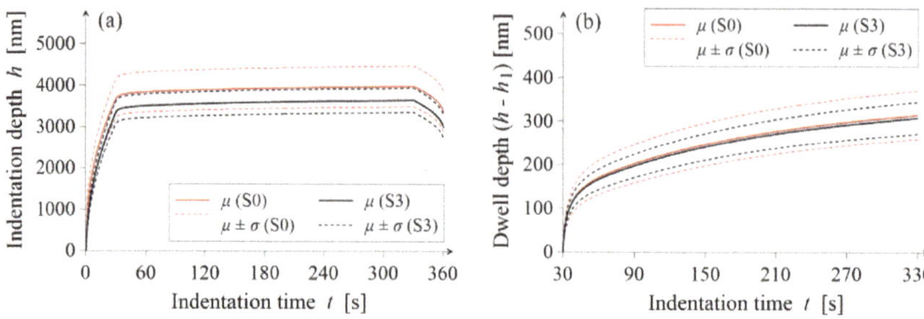

Figure 11. (**a**) Average indenter immersion during all the indentation phases and (b) average immersion in the creep phase only. Results for sample S3 versus the reference sample S0.

To compare the obtained results, Table 6 presents the average values μ and standard deviations σ of the creep coefficient C_{IT} for all the tested series of self-compacting cement matrices.

Table 6. Mean values (μ), standard deviations (σ), and the creep coefficient (C_{IT}).

Concrete Series	S0	S1	S2	S3
μ_{CIT} [%]	8.49	8.96	8.32	9.06
σ_{CIT} [%]	1.46	1.36	1.26	1.03
σ_{CIT} / μ_{CIT}	0.172	0.152	0.151	0.114

In turn, to illustrate the statistical distribution of the obtained results, Figure 12 presents the histograms of the creep coefficient (C_{IT}) for the S0, S1, S2, and S3 series samples.

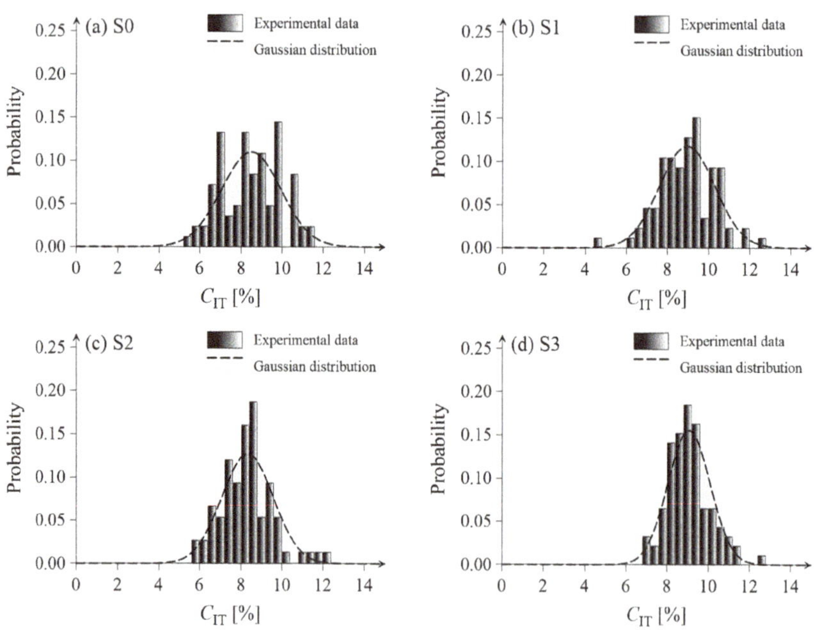

Figure 12. Histograms of the creep coefficient C_{IT} for samples (**a**) S0, (**b**) S1, (**c**) S2, and (**d**) S3.

Based on Figure 12, it can be concluded that the results of the creep coefficient C_{IT} obtained for all the tested series fit well into the Gaussian distribution (except for series S0). This proves that the number of measurements adopted for testing each series (except S0) was sufficient for statistical inference.

4. Discussion

Based on the obtained results (Figures 9–11), it can be concluded that the addition of SiO_2 (S1 series) and Al_2O_3 (S3 series) nanoparticles increased the creep coefficient C_{IT} and the addition of TiO_2 nanoparticles (S2 series) reduced the creep coefficient C_{IT} in comparisonto the reference series S0, which did not contain nanoparticles. It should be noted, however, that the obtained average values of the creep coefficient μc_{IT} for the series S1, S2, and S3 were in each case within the range of the standard deviation σc_{IT} that was obtained for the series S0.

Moreover, the difference between the mean values of the creep coefficient μ_{CIT} for the tested series was not more than 6.7%. Therefore, it can be said that the addition of nanoparticles did not significantly affect the creep of the cement matrix of self-compacting concretes. This is a surprising result with regard to the hardness test results that were obtained for the same series of concrete by Niewiadomski et al. [22]. In this paper, the authors stated that the addition of nanoparticles caused an increase in the hardness of a cement matrix, which was measured at the micro-scale. However, it should be noted that the two tests performed in that sudy were focused on two non-identical properties of hardened cement matrix (creep and hardness), and the parameters of the tests differed qualitatively.

When considering the above research results, it can be said that the impact of nanoparticles on the rheological parameters of a cement matrix is insignificant. In addition, when assuming that nanoparticles do not significantly affect the interfacial transition zone ITZ between the cement matrix and the aggregate (the analysis of the ITZ of the composite was conducted in [24]), it can be concluded that nanoparticles also have a slight effect on the rheological parameters of hardened composites based on a cement matrix, such as self-compacting concrete.

Nevertheless, it is worth emphasizing that the results provided in this article improve the state of knowledge concerning the impact of nanoparticles on the physical and mechanical properties of both hardened self-compacting concrete and its cement matrix.

It is worth considering further directions of research regarding the impact of nanoparticles on the properties of cement matrices. Interference at the micro, or even nano level, in the structure of cement composites can be an important issue that allows concrete to be modified for a specific purpose.

5. Conclusions

This paper evaluates, using the indentation method, the creep of the cement matrix of self-compacting concrete, which is modified by the addition of SiO_2, TiO_2, and Al_2O_3 nanoparticles. It thus expands our knowledge concerning this subject. It turned out that the addition of SiO_2 and Al_2O_3 nanoparticles in an amount of 4.0% of the cement weight increased the creep coefficient C_{IT} of the cement matrix when compared to the reference series, which is an adverse phenomenon. In turn, the use of TiO_2 nanoparticles in an amount of 4.0% of the cement weight resulted in a decrease of the value of the creep coefficient C_{IT},which can be considered beneficial. It is worth emphasizing, however, that the analysis of the obtained results with regard to statistical purposes indicated that the addition of nanoparticles did not significantly affect the creep of the cement matrix of self-compacting concrete. Nevertheless, the methodology used in this study allowed shortening the time needed to evaluate the cement matrix creep phenomenon in comparison to traditional methods.

Author Contributions: Conceptualization, P.N. and D.S.; methodology, D.S.; software, D.S.; validation, P.N. and D.S.; formal analysis, P.N. and D.S.; investigation, P.N.; resources, P.N. and D.S.; data curation, D.S.; writing—original draft preparation, P.N.; writing—review and editing, P.N.; visualization, D.S.; supervision, D.S.; project administration, P.N.; funding acquisition, P.N. All authors have read and agreed to the published version of the manuscript.

Funding: This research received no external funding.

Conflicts of Interest: The authors declare no conflict of interest.

References

1. Sanchez, F.; Sobolev, K. Nanotechnology in concrete—A review. *Constr. Build. Mater.* **2010**, *24*, 2060–2071. [CrossRef]
2. Li, H.; Xiao, H.-G.; Yuan, J.; Ou, J. Microstructure of cement mortar with nano-particles. *Compos. Part B Eng.* **2004**, *35*, 185–189. [CrossRef]
3. Du, H.; Du, S.; Liu, X. Durability performances of concrete with nano-silica. *Constr. Build. Mater.* **2014**, *73*, 705–712. [CrossRef]
4. Ji, T. Preliminary study on the water permeability and microstructure of concrete incorporating nano-SiO2. *Cem. Concr. Res.* **2005**, *35*, 1943–1947. [CrossRef]
5. Li, H.; Xiao, H.-G.; Ou, J.-P. A study on mechanical and pressure-sensitive properties of cement mortar with nanophase materials. *Cem. Concr. Res.* **2004**, *34*, 435–438. [CrossRef]
6. Beigi, M.H.; Berenjian, J.; Omran, O.L.; Sadeghi-Nik, A.; Nikbin, I.M. An experimental survey on combined effects of fibers and nanosilica on the mechanical, rheological, and durability properties of self-compacting concrete. *Mater. Des.* **2013**, *50*, 1019–1029. [CrossRef]
7. Niewiadomski, P.; Hoła, J. Failure process of compressed self-compacting concrete modified with nanoparticles assessed by acoustic emission method. *Autom. Constr.* **2020**, *112*, 103111. [CrossRef]
8. Vandamme, M.; Ulm, F.-J. Nanogranular origin of concrete creep. *Proc. Natl. Acad. Sci. USA* **2009**, *106*, 10552–10557. [CrossRef]
9. Acker, P.; Ulm, F.-J. Creep and shrinkage of concrete: Physical origins and practical measurements. *Nucl. Eng. Des.* **2001**, *203*, 143–158. [CrossRef]
10. Sadowski, Ł.; Stefaniuk, D.; Żak, A.M.; Krakowiak, K.J. Micromechanical properties within the interphase between heterogeneous layers made of cementitious composites. *Constr. Build. Mater.* **2019**, *215*, 1033–1043. [CrossRef]
11. Oliver, W.; Pharr, G. An improved technique for determining hardness and elastic modulus using load and displacement sensing indentation experiments. *J. Mater. Res.* **1992**, *7*, 1564–1583. [CrossRef]
12. Jones, C.; Grasley, Z.C. Short-term creep of cement paste during nanoindentation. *Cem. Concr. Compos.* **2011**, *33*, 12–18. [CrossRef]
13. Frech-Baronet, J.; Sorelli, L.; Charron, J.-P. New evidences on the effect of the internal relative humidity on the creep and relaxation behaviour of a cement paste by micro-indentation techniques. *Cem. Concr. Res.* **2017**, *91*, 39–51. [CrossRef]
14. Němeček, J. Creep effects in nanoindentation of hydrated phases of cement pastes. *Mater. Charact.* **2009**, *60*, 1028–1034. [CrossRef]
15. Vandamme, M.; Ulm, F.-J. Nanoindentation investigation of creep properties of calcium silicate hydrates. *Cem. Concr. Res.* **2013**, *52*, 38–52. [CrossRef]
16. Li, J.; Yao, Y. A study on creep and drying shrinkage of high performance concrete. *Cem. Concr. Res.* **2001**, *31*, 1203–1206. [CrossRef]
17. Kristiawan, S.A.; Nugroho, A.P. Creep Behaviour of Self-compacting Concrete Incorporating High Volume Fly Ash and its Effect on the Long-term Deflection of Reinforced Concrete Beam. *Procedia Eng.* **2017**, *171*, 715–724. [CrossRef]
18. Garas, V.Y.; Kahn, L.F.; Kurtis, K.E. Short-term tensile creep and shrinkage of ultra-high performance concrete. *Cem. Concr. Compos.* **2009**, *31*, 147–152. [CrossRef]
19. Leemann, A.; Lura, P.; Loser, R. Shrinkage and creep of SCC – The influence of paste volume and binder composition. *Constr. Build. Mater.* **2011**, *25*, 2283–2289. [CrossRef]
20. Persson, B. A comparison between mechanical properties of self-compacting concrete and the corresponding properties of normal concrete. *Cem. Concr. Res.* **2001**, *31*, 193–198. [CrossRef]
21. *PN-EN 197-1:2012 Cement—Part 1: Composition, Specifications and Conformity Criteria for Common Cements*; Polish Committee for Standardization: Warsaw, Poland, 2012; Available online: https://infostore.saiglobal.com/preview/is/en/2011/i.s.en197-1-2011.pdf?sku=1492629 (accessed on 6 April 2019).
22. Niewiadomski, P.; Hoła, J.; Cwirzen, A. Study on properties of self-compacting concrete modified with nanoparticles. *Arch. Civ. Mech. Eng.* **2018**, *18*, 877–886. [CrossRef]

23. Sorelli, L.; Constantinides, G.; Ulm, F.-J.; Toutlemonde, F. The nano-mechanical signature of Ultra High Performance Concrete by statistical nanoindentation techniques. *Cem. Concr. Res.* **2008**, *38*, 1447–1456. [CrossRef]
24. Stefaniuk, D.; Niewiadomski, P.; Musiał, M.; Łydżba, D. Elastic properties of self-compacting concrete modified with nanoparticles: Multiscale approach. *Arch. Civ. Mech. Eng.* **2019**, *19*, 1150–1162. [CrossRef]

applied sciences

MDPI

Article

The S-Curve as a Tool for Planning and Controlling of Construction Process—Case Study

Jarosław Konior and Mariusz Szóstak *

Department of Building Engineering, Faculty of Civil Engineering, Wroclaw University of Science and Technology, 50-370 Wrocław, Poland; jaroslaw.konior@pwr.edu.pl

* Correspondence: mariusz.szostak@pwr.edu.pl; Tel.: +48-71-320-23-69

Received: 12 February 2020; Accepted: 13 March 2020; Published: 19 March 2020

Abstract: One of the key tasks of an investor and a contractor at the stage of planning and implementing construction works is to measure the progress of execution with regard to the planned deadlines and costs. During the execution of construction works, the actual progress of the works may differ significantly from the initial plan, and it is unlikely that the construction project will be implemented entirely according to the planned work and expenditure schedule. In order to monitor the process of deviations of the deadline and the budget of the investment task, several rudimentary methods of planning—as well as the cyclical control of the progress of construction projects—are used. An effective tool for measuring the utilization of the financial outlays of a construction project is the presentation of the planned financial flows on a timeline using a cumulative cost chart, the representation of which is the S-curve. The purpose of this paper is to analyze the course of an sample construction project comparing the planned costs of the scheduled works with the actual costs of the performed works, as well as identifying the reasons leading to the failure to meet the planned deadlines and budget of the project implementation. As part of the research conducted at a construction site of a hotel facility, the authors of this paper analyzed each of the 20-month effects of financial expenditures on construction works that were developed and processed by the Bank Investment Supervision (BIS) over a period of three years (between 2017 and 2019). Based on these results, charts and tables of the scheduled and actual cumulative costs of the completed construction project were prepared, the careful analysis of which enables interesting conclusions to be drawn.

Keywords: construction project management; cost; time; Bank Investment Supervision

1. Introduction

Construction project management is a process that includes a number of operations, activities and decisions that are closely related to a project being executed and that aim to create new, or to increase, existing fixed assets in order to achieve utility effects [1]. The utility effect of the construction process may be the construction of a new building, or the renovation or modernization of an existing building. In each construction process, four basic phases are distinguished according to the definition of the building object life cycle: the programming/planning phase, the implementation phase, the operation/use/maintenance phase and the phase of decommissioning or demolition [2]. Appropriate planning of the entire construction process is a very important operation that has a direct impact on whether the investment project implementation is successful [3].

The construction industry is characterized by high complexity of the implemented construction processes. The execution of construction projects is specific and particularly difficult, because each implementation is a unique, complex and dynamic process that consists of a number of interrelated subprocesses in which various participants of the investment process are involved [4].

Each construction project is exposed to various types of risk [5]. The most common risks associated with the implementation of construction projects include, among others, time risk, cost risk, work quality risk, construction risk and technological risk [6]. The construction risks that occur during the execution of works can be classified into five categories: people and their safety, budget, cost, schedule, planning/quality and efficiency [7].

During the implementation of construction projects, a common phenomenon is that the planned budget is exceeded and/or the planned implementation deadlines are not met. This phenomenon occurs in all countries [8,9]. Deviations from the plan may arise as a result of changing weather conditions; changes in the volume of available means of production; untimely deliveries of materials, machinery and equipment; unprofessional project management; mediocre discipline and organization of work; delays in making decisions; incorrect decisions; and so on [10].

An important element of appropriately managing the investment process is the thoughtful planning of construction project costs and the effective control of the progress of works in the area of incurred investment costs during project implementation. When planning the costs of a construction project, the cost of construction works must always be correctly determined, and both the direct costs related to the implementation of works and the indirect costs (i.e., overheads and profit) should be taken into account [11]. Unfortunately, exceeding the planned budget and/or time is a very common attribute of construction projects, and it is unlikely that a construction investment will be carried out completely in accordance with the planned work and expenditure schedule [12].

Therefore, it is important to plan the entire investment process, in particular, by developing a correct Investor's work and expenditure schedule with specific dates for starting and completing the project, with appropriate connections between planned tasks, and also with an outline of the specific duration of the individual tasks and the costs of their implementation. However, it should be borne in mind that during the implementation of a construction project, the actual progress of works may differ significantly from the initial plan, and it is therefore also necessary to specify the principles of control and monitoring for the construction project. Only a properly developed work plan, which includes optimal working sequences and perfect timing for executing each individual activity, can enhance the work efficiency and enable contractors to fulfill the contract at the lowest cost [13].

An effective tool for measuring the utilization of the financial outlays of a construction project is the presentation of the planned financial flows on a timeline using a cumulative cost chart, the representation of which is the S-curve. The S-curve shows the progress of the investment project from the beginning of the construction works through to their completion. The cumulative cost chart for construction projects takes the shape of the letter "S", hence the name of the curve. The variable slope of the cost curve indicates the changing progress of works per unit of time [14].

The S-curve is flatter at the beginning and end of the execution of a construction project, and steeper in the middle. This is due to the fact that a traditional construction project starts quite slowly. At the beginning of the construction process, human resources are organized, construction site development is prepared and simple preparatory works are carried out. After some time, the implementation of work begins to accelerate. Works are carried out on several working fronts using various working teams. Contractors start to undertake more and more tasks simultaneously. The mutual implementation of parallel tasks generates a much greater increase in costs when compared to the initial stage of implementation [15].

As a result of the continuing research, the classic S-curve method is being expanded and constantly modified in the following ways: through the use of the least square method and fuzzy S-curve regression model [16,17], the use of a polynomial function to generalize the S-curve [18], the use of methods of artificial intelligence [19], the use of an S-curve Bayesian model [20] or by dividing the entire duration of a construction project into three periods for the improved accuracy of cost forecasting [21]).

The second large group of methods that are used for controlling and monitoring the progress of the construction project implementation is earned value management. Earned value management involves the control of the investment task through the cyclical comparison of the actually executed scope of work with the planned time and cost of implementation [22]. Project management that uses earned value management is a well-known management system that integrates schedule, costs and technical performance [23,24]. Earned value management allows cost and schedule deviations as well as performance indicators, project cost forecasts and schedule durations to be calculated [25,26]. In literature, there are many studies that present the effective application of earned value management in real-life construction projects [27–30].

Classic earned value management is being expanded and modified by way of, among others, the introduction of a hybrid methodology based on work packages and logical time analysis [31], the introduction of new parameters (e.g., the Schedule Forecast Indicator (SFI) [32]) or by taking into account of the impact of unplanned time and cost deviations on the financial liquidity of a construction project [33].

The purpose of this article is to analyze the course of an sample construction project by comparing the planned costs of the scheduled work with the actual costs of the performed work, and to identify reasons leading to failures to meet planned deadlines and the budget of the project's implementation. The goals of the paper are to demonstrate the reasons for construction project cost overruns and time delays using a representative case study. Both factors are strongly based on the Earned Value Method (EVM) approach, using an indicated project budget breakdown and timeline deviations. In this particular case study the poorly managed project was carried out and the cost spent exceeded the expected and planned figures by over 50%, which is not uncommon in the Polish investment process nowadays.

2. Research and Methodology of Measurements

The following documents were analyzed as part of the research conducted: (1) the basic Investor's work and expenditure schedule of the construction project, which was developed before the commencement of works; and (2) a set of information about the actual progress of the construction process, contained and updated in the monthly reports of the Bank Investment Supervision (BIS) [34].

The Investor's work and expenditure schedule is a document that shows the planned progress of the project over time, taking into account the planned costs. It was developed at the planning stage of the construction project.

Information on the actual progress of the construction process was collected as part of the BIS services during the authors' own research at the construction site of a hotel facility [34]. For non-public investment tasks that are co-financed by two entities—the Investor and the Bank—a third independent entity was appointed: the Banking Supervision Inspector, who performed a monitoring and auditing function. The BIS's tasks included, among others: preliminary reporting (e.g., verification of documentation about a construction project such as permits, administrative decisions, the planned budget, contracts concluded by an Investor, etc.), monthly reporting (e.g., constant monitoring of the project execution, control of the state of the project implementation, settlement verification, analysis of loan tranche disbursement conditions, etc.), and final reporting (i.e., final financial analysis of the project implementation) [35].

In the implementation phase, the Bank Investment Supervision develops documents that enable the mapping and presentation of the actual progress of the construction process. From the monthly reports developed by the BIS, information about, among others, the progress of works carried out in individual implementation periods, the values of works carried out in individual implementation periods and the values of works carried out cumulatively since the beginning of the works were obtained.

As a result of the performed documentation analysis, a collective data summary was prepared. It was presented in a two-dimensional table where each subsequent row of the table contained data on subsequent periods of the construction works. The table distinguished the following information about the project [33,36]:

- the budgeted cost of work scheduled (BCWS) determined for each individual examined period on the basis of the Investor's work and expenditure schedule;
- the cumulative value of the budgeted cost of work scheduled (C_{BCWS}) for each single examined period, calculated as the cumulative value obtained by adding the value of the budgeted cost of

work scheduled from the analyzed period to the value of budgeted cost of work scheduled from the preceding period, determined on the basis of the Investor's work and expenditure schedule;
- the actual cost of work performed (ACWP) determined for each individual examined period on the basis of BIS reports;
- the cumulative value of the actual cost of work performed (C_{ACWP}) for each examined individual period, calculated as a cumulative value obtained by adding the value of the actual cost of work performed from the analyzed period, based on BIS reports, to the value of the actual costs of the work performed from the preceding period;
- the actual percentage advancement of work performed (AC), calculated as a ratio of the value of the cumulative actual cost of work performed to the total actual cost of the construction project;
- the planned percentage advancement of work scheduled (PP), calculated as the ratio of the cumulative value of the budgeted cost of work scheduled to the total budgeted cost of the construction project;
- the actual percentage advancement of work scheduled (AP), calculated as the ratio of the value of the cumulative actual cost of work performed to the total budgeted cost of the construction project;
- the actual schedule variance (ASV), calculated as the difference between the actual duration and the planned duration of the project;
- the actual schedule performance indicator (I_{ASV}), calculated as the ratio of the actual duration of the construction project to the planned duration of the project;
- the at-completion variance (ACV), calculated as the difference between the total actual cost of the construction project and the total budgeted cost of the construction project;
- the performance indicator of the at-completion variance (I_{ACV}), calculated as the ratio of the total actual cost of the construction project to the total budgeted cost of the construction project.

3. Case Study

The subject of the analyzed project is the execution of a hotel with an underground garage (one underground floor), as well as the execution of elements of land development around it. The hotel was located in a Polish city. The erected building, on the ground floor, had a reception, daytime area, offices and hotel administration, and also a catering area with a restaurant and kitchen facilities. On floor +1, there were office rooms, meeting rooms, conference rooms and sanitary facilities with toilets. On the levels +2 to +7 there were 200 hotel rooms. The floor space of the building was approximately 8200 m^2.

The project was carried out according to the general contracting system. The subject of the contract was an experienced construction company concerned the execution of the "shell & core" state. This consisted of the implementation of a comprehensive investment project involving the construction of a hotel with a garage, the execution of land development and the commissioning of the project after obtaining an occupancy permit. It should be emphasized that the original contract with the General Contractor did not take into account the finishing of the rooms, and therefore the scope of works in the finishing part did not include the finishing and equipping of hotel rooms with bathrooms and annexes.

The scope of the commissioned works included the finishing of common parts (e.g., kitchen, restaurants and catering equipment) without providing the technology as described in the tender design. In addition to the contract, the equipping of the sanitary facilities, offices, facilities, reception, fitness area, toilets and technical rooms was required, as well as the provision of the premises' mobile equipment. The deadline for the implementation of the subject of the contract was scheduled from September 2016 to June 2018. A flat-rate remuneration of PLN 36,111,146.13 net + VAT was set for performing the subject of the contract. Figure 1 presents the Investor's work and expenditure schedule of the project, including the planned amounts of construction works to be executed.

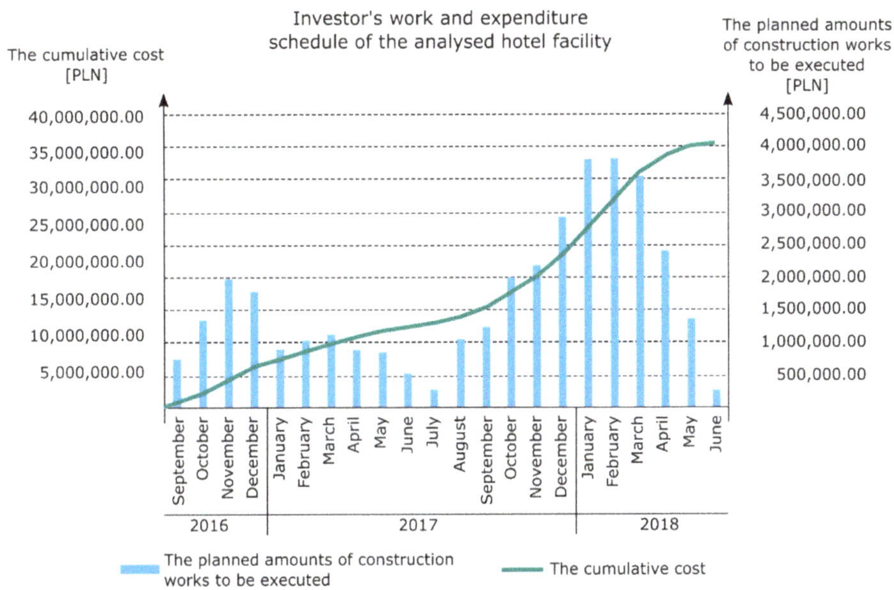

Figure 1. Investor's work and expenditure schedule of the analyzed hotel facility (own elaboration).

The five most important elements of the execution process of the analyzed hotel facility are described below in the chronological order:

- August 2017: The Investor, aiming at a more efficient management of the construction project, signed an annex extending the scope of works of the General Contractor. It included specialist works concerning Leadership in Energy and Environmental Design (LEED) certification, additional fire protection and the execution of power and teletechnical installations in hotel rooms. The value of remuneration increased by 13.0% when compared to the Investor's schedule and reached the level of PLN 40,815,455.00 net.
- December 2017: The General Contractor presented, in consultation with the Investor, an updated work schedule that included subsequent annexes regarding changes in the facility, covering door joinery, automation and the building façade, among others. It should be emphasized that the subsequent annexes to the Agreement introduced new elements to its basic scope that were originally excluded from the scope of the General Contractor. Increasing the scope of the subject of the contract concerned the finishing of the premises, including the installation of shower trays and frames in hotel bathrooms. The value of remuneration in December 2017 increased by 17.2% when compared to the Investor's schedule and reached the level of PLN 42,313,695.00 net. The deadline did not change (curve No. 2 in Figure 2).
- March 2018: The Investor decided to increase the hotel standard (to four stars), which forced the introduction of changes resulting from the difference between the detailed design and the tender design. The changes concerned, among others, sanitary installations, ventilation, electrical installations of rooms, automation and wall arrangement. As a consequence, the value of remuneration increased by 20.8% when compared with the Investor's schedule, and reached the level of PLN 43,635,084.50 net. The annex to the contract extended the duration of the execution until October 2018 (curve No. 3 in Figure 2).
- June 2018: Another annex significantly modified the basic scope of the contract with the General Contractor (as of September 2017). The Investor commissioned the General Contractor an additional scope of works and, in particular, installations of the residence and public areas

(including the hall), catering and multifunctional rooms, and the execution of finishing works involving the equipping of the hotel rooms. The value of remuneration increased by 48.1% when compared to the Investor's schedule and reached the level of PLN 53,473,979.01 net. The implementation time was extended until January 2019 (curve No. 4 in Figure 2).

- July 2018: Due to the increasing scope of work entrusted to the General Contractor and the extension of the deadline, it was necessary to agree and sign off another annex. Therefore, the value of remuneration increased by 59.4% when compared to the Investor's schedule and reached the level of PLN 57,564,756.03 net without the end date changing (curve No. 5 in Figure 2), although in October 2018 the investment completion date was extended to February 2019 (curve No. 6 in Figure 2).
- June 2019: final completion of the investment task (i.e., 12 months later than originally planned (June 2018)). The value of remuneration increased by 62.4% and reached the final value of PLN 58,646, 84.75 net (curve No. 7 in Figure 2).

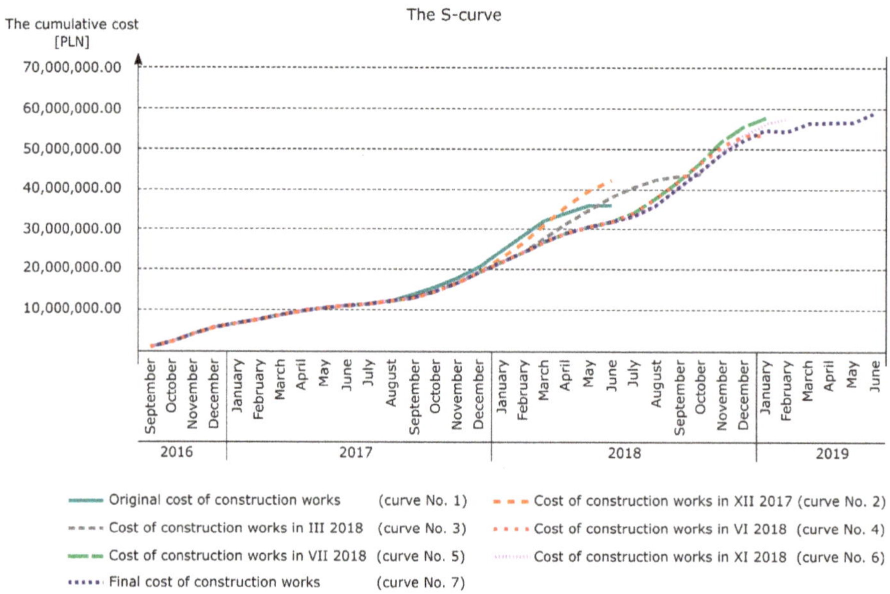

Figure 2. Planned and actual cumulative costs of the completed hotel facility (own elaboration).

Table 1 summarizes the collected data concerning the completed construction project.

Table 1. Results of measurements and cost analyses of the assessed hotel facility (own elaboration).

No.	Period	The Budgeted Cost of Work Scheduled According to the Work and Expenditure Schedule	The Cumulative Value of the Budgeted Cost of Work Scheduled	The Actual Cost of Work Performed	The Cumulative Value of the Actual Cost of Work Performed	The Actual Percentage Advancement of Work Performed	The Planned Percentage Advancement of Work Scheduled	The Actual Percentage Advancement of Work Scheduled
		BCWS	C_{BCWS}	*ACWP*	C_{ACWP}	*AC*	*PP*	*AP*
(1)	(2)	(3)	(4)	(5)	(6)	(7)	(8)	(9)
[–]	[–]	[PLN]	[PLN]	[PLN]	[PLN]	[%]	[%]	[%]
1	Sep.16	743,145.00	743,145.00	743,145.00	743,145.00	1.27	2.06	2.06
2	Oct.16	1,342,960.76	2,086,105.76	1,342,960.76	2,086,105.76	3.56	5.78	5.78
3	Nov.16	1,962,497.14	4,048,602.90	1,962,497.14	4,048,602.90	6.90	11.21	11.21
4	Dec.16	1,786,729.02	5,835,331.92	1,786,729.02	5,835,331.92	9.95	16.16	16.16
5	Jan.17	890,075.82	6,725,407.74	890,075.82	6,725,407.74	11.47	18.62	18.62
6	Feb.17	1,024,911.29	7,750,319.03	1,024,911.28	7,750,319.02	13.22	21.46	21.46
7	Mar.17	1,124,204.85	8,874,523.88	1,124,204.85	8,874,523.87	15.13	24.58	24.58
8	Apr.17	899,863.73	9,774,387.61	899,863.73	9,774,387.60	16.67	27.07	27.07
9	May 17	860,000.82	10,634,388.43	860,000.82	10,634,388.42	18.13	29.45	29.45
10	Jun.17	537,721.00	11,172,109.43	537,721.00	11,172,109.42	19.05	30.94	30.94
11	Jul.17	293,524.63	11,465,634.06	293,524.63	11,465,634.05	19.55	31.75	31.75
12	Aug.17	1,064,465.00	12,530,099.06	1,072,251.29	12,537,885.34	21.38	34.70	34.72
13	Sep.17	1,236,413.52	13,766,512.58	614,825.93	13,152,711.27	22.43	38.12	36.42
14	Oct.17	1,998,683.50	15,765,196.08	1,509,792.56	14,662,503.83	25.00	43.66	40.60
15	Nov.17	2,195,751.00	17,960,947.08	2,231,802.48	16,894,306.31	28.81	49.74	46.78

Table 1. *Cont.*

No.	Period	The Budgeted Cost of Work Scheduled According to the Work and Expenditure Schedule	The Cumulative Value of the Budgeted Cost of Work Scheduled	The Actual Cost of Work Performed	The Cumulative Value of the Actual Cost of Work Performed	The Actual Percentage Advancement of Work Performed	The Planned Percentage Advancement of Work Scheduled	The Actual Percentage Advancement of Work Scheduled
		BCWS	C_{BCWS}	*ACWP*	C_{ACWP}	*AC*	*PP*	*AP*
16	Dec.17	2,925,234.00	20,886,181.08	2,559,405.11	19,453,711.42	33.17	57.84	53.87
17	Jan.18	3,816,618.00	24,702,799.08	2,644,027.80	22,097,739.22	37.68	68.41	61.19
18	Feb.18	3,825,234.00	28,528,033.08	2,202,625.30	24,300,364.52	41.44	79.00	67.29
19	Mar.18	3,540,993.30	32,069,026.38	2,905,521.47	27,205,885.99	46.39	88.81	75.34
20	Apr.18	2,408,845.75	34,477,872.13	1,947,719.29	29,153,605.28	49.71	95.48	80.73
21	May.18	1,358,443.00	35,836,315.13	1,488,799.77	30,642,405.05	52.25	99.24	84.86
22	Jun.18	274,831.00	36,111,146.13	1,191,291.00	31,833,696.05	54.28	100.00	88.15
23	Jul.18	0.00	36,111,146.13	1,622,583.04	33,456,279.09	57.05	100.00	92.65
24	Aug.18	0.00	36,111,146.13	2,415,758.53	35,872,037.62	61.17	100.00	99.34
25	Sep.18	0.00	36,111,146.13	4,252,403.78	40,124,441.40	68.42	100.00	111.11
26	Oct.18	0.00	36,111,146.13	4,722,806.07	44,847,247.47	76.47	100.00	124.19
27	Nov.18	0.00	36,111,146.13	4,470,580.10	49,317,827.57	84.09	100.00	136.57
28	Dec.18	0.00	36,111,146.13	3,013,286.20	52,331,113.77	89.23	100.00	144.92
29	Jan.19	0.00	36,111,146.13	2,226,044.30	54,557,158.07	93.03	100.00	151.08
30	Feb.19	0.00	36,111,146.13	0.00	54,557,158.07	93.03	100.00	151.08
31	Mar.19	0.00	36,111,146.13	2,014,670.58	56,571,828.65	96.46	100.00	156.66
32	Apr.19	0.00	36,111,146.13	0.00	56,571,828.65	96.46	100.00	156.66
33	May.19	0.00	36,111,146.13	0.00	56,571,828.65	96.46	100.00	156.66
34	Jun.19	0.00	36,111,146.13	2,074,556.10	58,646,384.75	100.00	100.00	162.41

Each subsequent annex to the principal contract (as of September 2016) forced the General Contractor to develop an updated work and expenditure schedule. The updated schedule that was used for the implementation of works modified the actual progress of the work performed. Figure 3 shows the variable work progress that resulted from several work and expenditure schedule updates.

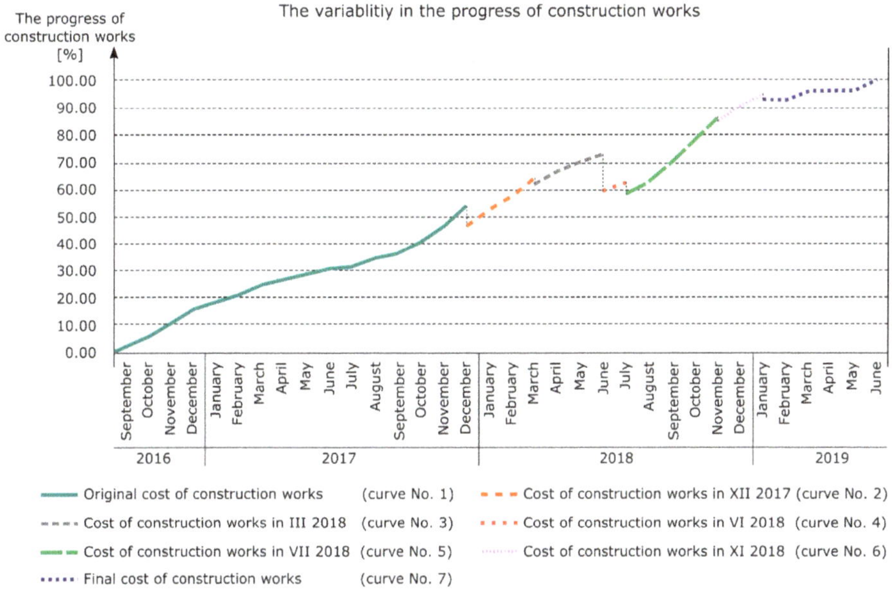

Figure 3. Variability in the progress of construction works as a result of updating the work and expenditure schedule of the hotel facility (own elaboration).

Table 2 summarizes information about individual changes in the remuneration and the completion date of the implemented task.

Table 2. Results of measurements and cost analyses of the assessed hotel facility (own elaboration).

Work and Expenditure Schedule	Duration	Cost of Construction Works	Deviation from Schedule ASV	Deviation from Budget ACV	Schedule Performance Indicator I_{ASV}	Cost Performance Indicator I_{ACV}
(1)	(2)	(3)	(4)	(5)	(6)	(7)
[-]	[months]	[PLN]	[months]	[PLN]	[-]	[-]
Original cost of construction works	22	36,111,146.13	-	-	-	-
Cost of construction works in Dec. 2017	22	42,313,695.00	0	6,202,548.87	1.000	1.172
Cost of construction works in Mar. 2018	26	4, 635,084.50	4	7,523,938.37	1.182	1.208
Cost of construction works in Jun. 2018	29	53,473,979.01	7	17,362,832.88	1.318	1.481
Cost of construction works in Jul. 2018	29	57,564,756.03	7	21,453,609.90	1.318	1.594
Cost of construction works in Nov. 2018	30	57,564,756.03	8	21,453,609.90	1.364	1.594
Final cost of construction works	34	58,646,384.75	12	22,535,238.62	1.545	1.624

4. Results of the Case Study

The hotel facility was completed in June 2019 after 34 months of work. The final cost of the construction works amounted to PLN 58,646,000 and therefore was higher by PLN 22,535,000 from the figure originally assumed (as of September 2016). It should also be noted that the actual date of the commissioning and commencement of operation of the hotel was 12 months later than the planned date, resulting in additional financial losses for the Investor.

Analysis of the actual progress of works clearly indicates how much the actual state of implementation of the project differed from that originally assumed. The original budget of the investment task was underestimated. The changes that occurred during the implementation of the project resulted in failure to meet the parameters assumed by the Investor in mid-2016 (i.e., the time and cost of implementing the project).

The following reasons led to the failure of meeting the planned deadlines and implementation costs:

- A lack of appropriate preparation for the construction project in the investment planning phase. The Investor planned the cost of implementing the construction project based on prices that were valid in 2016 and did not take into account increases in prices during the subsequent periods of implementation, or the financial fluctuations over time determined by discounted techniques.
- The Investor's lack of experience in implementing similar construction projects. The Investor had carried out in the past several smaller projects, but not from the hotel sector.
- Changes in the originally adopted standard of the hotel. The Investor incorrectly analyzed the market demand for hotel buildings and initially adopted a lower standard for the facility than required at this location.
- Changes in the scope of work of the General Contractor. Subsequent scopes of work required the General Contractor to develop additional cost estimates and offers, and this took up additional time. In turn, the increase in the scope of works resulted in a significant increase in employment at the construction site, especially with regard to specialized subcontractors. The extension of the implementation time caused an increase in the indirect costs of the General Contractor.
- The current situation of the market imbalance in the construction industry. In the analyzed project period between 2017 and 2019 there were major problems concerning contracting and subcontracting companies at the initially assumed cost level.

The case study of the analyzed hotel facility enabled the following conclusions to be made regarding the measurement of deviations in the deadlines and budget of the investment task:

- the investment task was completed 12 months later than originally planned;
- the schedule performance efficiency indicator was equal to 1.545, and therefore the project was longer than planned by 54.5%;
- the final cost of the construction works increased by PLN 22,535,338.62 net when compared to the original cost;
- the cost performance efficiency indicator was equal to 1.624, and therefore the project was more expensive than planned by 62.4%;
- analysis of the progress of the hotel facility execution indicates a growing trend of the schedule and cost performance efficiency indicator.

5. Discussion

S-curve provides the basis for monitoring cash flows while planning any construction project. Unfortunately, there is very little likelihood that a project will proceed completely as planned. Small deviations between the plan and the reality can be seen as being within the limits of the norm and usually do not interfere with the purpose. However, greater differences can hinder the goal and require a revision to ensure that project objectives are achieved [12]. The problems of overruns of a planned project budget or rescheduled deadlines are widely recognized in many countries [8,9,37,38].

The simplest scheduling methods utilizing the S-curve are assumed to be deterministic, and do not take into account possible risks and uncertainties. However, there are methods that take into account the use of stochastic curves in probabilistic monitoring and project prediction as an alternative to deterministic curves and traditional forecasting methods. For the generation of stochastic cost curves, a simulation method was adopted based on defining the variability of the duration and the costs of individual activities in the project [39]. The study also used a stochastic model approach to financial management, taking into account the uncertainty of duration and costs at different stages during the project lifecycle [40]. Effective project management requires a reliable knowledge of cash flows at different stages of the project lifecycle. Obtaining this knowledge largely depends on taking into account the precarious environmental conditions of the project, which can be obtained using the method of assessing cash flow based on the project schedule [41]. Uncertainty and imprecision in project planning have been included in cash flow calculation methodologies for projects involving fuzzy activities and/or costs. Cash flow can be represented by the cost area S (as opposed to traditional S-curves) obtained from a combination of cost curves at different risk capability levels. Unfortunately, according to the authors, the proposed concept of the cost area S, in addition to the need to collect a lot of data, also requires the use of advanced software [42].

In the presented case analysis there were continuous time and cost deviations due to poor project management and a lack of application of even the simplest model for monitoring the S-curve. (i.e., using the models indicated in this discussion).

The entire analysis of research conducted by the authors of the paper leads to the main conclusion that the models proposed earlier by various researches of the forecasted S-curve as a rule are not exactly in line with a real state. Some works are too general and too descriptive [15,23]. There are also presented models and methods that are too complicated and thus not very practical or easy to adopt in planning and managing construction projects [3,11,19,40]. In some research, the models seems to be reasonable, however they have not been tested and verified during construction process monitoring [9,15,41]. To make things worse, it is hard to find reliable, proven research data based on solid measures of actually executed construction projects where technical inspections were conducted on construction sites and what was planned, paid and earned was reviewed. Some of the published papers that are accessible have relied on questionnaires, analyses of past documents and assumptions rather than facts [38,39]. However, there are still some strong construction-based papers that present a case study of the application of the S-curve regression method to project control of construction management [17].

One of the authors of the paper represents a strong continuity and solid experience collected from more than 30 years of engineering and construction [7,35]. The main value of the described construction case study is a detailed and systematic analysis of the course of the hotel project that compares the planned costs of the scheduled work with the actual costs of the performed work, as well as the identification of the reasons leading to the failure to meet the planned deadlines and budget of the project's implementation [34]. The paper demonstrates the reasons and the effects of construction project's cost overruns and time delays using a representative case study. Both factors were strongly based on the EVM approach, including an indicated project budget breakdown and timeline deviations. According to the newly published Project Management Institute briefings [1], 70% of construction projects are completed with overruns of cost and time. This statement is rather frightening but credible, and the presented case study highlights the state-of-the-art.

6. Summary and Conclusions

Appropriate planning of the investment process is a very important operation with a direct impact on the successful implementation of a construction project via maintenance of the assumed budget and project completion deadline, while ensuring a consistent quality of on-going construction work.

Correct planning of cash flows is of key importance for Investors and contractors, preferably taking into account financial fluctuations over time determined by discounted techniques.

Planning of costs also has a significant impact on the financial liquidity of construction companies. There is therefore a need to develop simple, fast and effective methods that enable cash flows to be properly planned and controlled. A very helpful tool for planning, monitoring and controlling construction projects is the S-curve. Knowledge of the planned and actual course of cumulative financial expenditures over time and the shape of the S-curve and its deviations permits rational actions to be taken in order to achieve the intended goal and achieve success in construction project implementation.

Cumulative cost S-curves, due to their uniqueness, are unrepeatable. Each construction project is unique because it is situated in a different location and different environment; designed and implemented by different work teams; and carried out using various technical, organizational and technological solutions. Therefore, it is justified to conduct further research into the course of cash flows and cost planning. The development of a methodology for planning cumulative cost curves in construction projects will allow the development of effective methods and verified models for better planning and utilizing financial expenditures during construction works.

Author Contributions: Conceptualization, J.K. and M.S.; Methodology, J.K. and M.S.; Formal analysis, M.S.; Resources, J.K. and M.S.; Writing—Original draft preparation, J.K. and M.S.; Supervision, J.K.; Project administration, M.S.; All authors have read and agreed to the published version of the manuscript.

Funding: This research received no external funding.

Conflicts of Interest: The authors declare no conflict of interest.

References and Note

1. Project Management Institute. *A Guide to the Project Management Body of Knowledge (PMBOK GUIDE)*, 6th ed.; Project Management Institute (PMI): Newtown Square, PA, USA, 2017.
2. Plebankiewicz, E.; Zima, K. Wieczorek Damian Life cycle cost modelling of buildings with consideration of the risk. *Arch. Civ. Eng.* **2016**, *62*, 149–166.
3. Kern, A.P.; Formoso, C.T. A model for integrating cost management and production planning and control in construction. *J. Financ. Manag. Prop. Constr.* **2006**, *11*, 75–90. [CrossRef]
4. Hoła, B.; Sawicki, M.; Skibniewski, M. An IT model of a Knowledge Map which supports management in small and medium-sized companies using selected Polish construction enterprises as an example. *J. Civ. Eng. Manag.* **2015**, *21*, 1014–1026. [CrossRef]
5. Kasprowicz, T. Quantitative Identification of Construction Risk. *Arch. Civ. Eng.* **2017**, *63*, 63–75. [CrossRef]
6. Zavadskas, E.K.; Turskis, Z.; Tamošaitienė, J. Risk assessment of construction projects. *J. Civ. Eng. Manag.* **2010**, *16*, 33–46. [CrossRef]
7. Konior, J. Enterprise's Risk Assessment of Complex Construction Projects. *Arch. Civ. Eng.* **2015**, *61*, 63–74. [CrossRef]
8. Makesh, S.; Mathivanan, M. Analysis on causes of delay in building construction. *Int. J. Innov. Technol. Explor. Eng.* **2019**, *8*, 335–341.
9. Rachid, Z.; Toufik, B.; Mohammed, B. Causes of schedule delays in construction projects in Algeria. *Int. J. Constr. Manag.* **2019**, *19*, 371–381. [CrossRef]
10. Połoński, M. *Management of Construction Investment Process*; Wydawnictwo SGGW (in Polish): Warszawa, Poland, 2018.
11. Leśniak, A.; Zima, K. Cost Calculation of Construction Projects Including Sustainability Factors Using the Case Based Reasoning (CBR) Method. *Sustainability* **2018**, *10*, 1608. [CrossRef]
12. Al-Jibouri, S.H. Monitoring systems and their effectiveness for project cost control in construction. *Int. J. Proj. Manag.* **2003**, *21*, 145–154. [CrossRef]
13. Lo, W.; Chen, Y.-T. Optimization of Contractor's S-Curve. In Proceedings of the 24th International Symposium on Automation & Robotics in Construction (ISARC 2007), 19–21 September 2007; pp. 417–420. Available online: https://www.irbnet.de/daten/iconda/CIB11259.pdf (accessed on 18 March 2020).
14. Chao, L.-C.; Chen, H.-T. Predicting project progress via estimation of S-curve's key geometric feature values. *Autom. Constr.* **2015**, *57*, 33–41. [CrossRef]

15. Cristóbal, J.R.S. The S-curve envelope as a tool for monitoring and control of projects. *Procedia Comput. Sci.* **2017**, *121*, 756–761. [CrossRef]
16. Hsieh, T.Y.; Wang, M.H.L.; Chen, C.W.; Chen, C.Y.; Yu, S.E.; Yang, H.C.; Chen, T.H. A new viewpoint of s-curve regression model and its application to construction management. *Int. J. Artif. Intell. Tools* **2006**, *15*, 131–142. [CrossRef]
17. Hsieh, T.-Y.; Hsiao-Lung Wang, M.; Chen, C.-W. A Case Study of S-Curve Regression Method to Project Control of Construction Management via T-S Fuzzy Model. *J. Mar. Sci. Technol.* **2004**, *12*, 209–216.
18. Chao, L.-C.; Chien, C.-F. Estimating Project S-Curves Using Polynomial Function and Neural Networks. *J. Constr. Eng. Manag.* **2009**, *135*, 169–177. [CrossRef]
19. Chao, L.-C.; Chien, C.-F. A Model for Updating Project S-curve by Using Neural Networks and Matching Progress. *Autom. Constr.* **2010**, *19*, 84–91. [CrossRef]
20. Kim, B.C.; Reinschmidt, K. An S-curve Bayesian model for forecasting probability distributions on project duration and cost at completion. In Proceedings of the CME 25 Conference of Construction Management and Economics, Reading, UK, 16–18 July 2007; pp. 1449–1459.
21. Cheng, Y.M.; Yu, C.H.; Wang, H.T. Short-interval dynamic forecasting for actual S-curve in the construction phase. *J. Constr. Eng. Manag.* **2011**, *137*, 933–941. [CrossRef]
22. Chen, H.L.; Chen, W.T.; Lin, Y.L. Earned value project management: Improving the predictive power of planned value. *Int. J. Proj. Manag.* **2016**, *34*, 22–29. [CrossRef]
23. De Marco, A.; Narbaev, T. Earned value-based performance monitoring of facility construction projects. *J. Facil. Manag.* **2013**, *11*, 69–80. [CrossRef]
24. Dziadosz, A.; Kapliński, O.; Rejment, M. Usefulness and fields of the application of the Earned Value Management in the implementation of construction projects. *Bud. Archit.* **2014**, *13*, 357–364.
25. Waris, M.; Khamidi, M.F.; Idrus, A. The Cost Monitoring of Construction Projects through Earned Value Analysis. *J. Constr. Eng. Proj. Manag.* **2012**, *2*, 42–45. [CrossRef]
26. Bhosekar, S.K.; Vyas, G. Cost Controlling Using Earned Value Analysis in Construction Industries. *Int. J. Eng. Innov. Tech.* **2012**, *1*, 324–332.
27. Vandevoorde, S.; Vanhoucke, M. A comparison of different project duration forecasting methods using earned value metrics. *Int. J. Proj. Manag.* **2006**, *24*, 289–302. [CrossRef]
28. Kwon, O.; Kim, S.; Paek, J.; Eom, S.-J. Application of Earned Value in the Korean Construction Industry. *J. Asian Archit. Build. Eng.* **2008**, *7*, 69–76. [CrossRef]
29. Khamidi, M.F.; Ali, W.; Idrus, A. Application of Earned Value Management System on an Infrastructure Project: A Malaysian Case Study. *Int. Conf. Manag. Sci. Eng.* **2011**, *8*, 1–5.
30. Połoński, M.; Komendarek, P. Earned value method for operational cost control of civil structure. *Quantit. Meth. Econ.* **2011**, *12*, 279–290.
31. Howes, R. Improving the performance of Earned Value Analysis as a construction project management tool. *Eng. Constr. Archit. Manag.* **2000**, *7*, 399–411. [CrossRef]
32. Czemplik, A. Application of earned value method to progress control of construction projects. *Procedia Eng.* **2014**, *91*, 424–428. [CrossRef]
33. Przywara, D.; Rak, A. The time-cost analysis of schedule monitoring using the earned value method. *Tech. Trans.* **2017**, *5*, 57–66.
34. Bauer-Celny, A.; Konior, J.; Szóstak, M. PM Group Polska sp. z o.o. 2006–2015 and 3EPCM sp. z o.o. 2016–2019, "Initial, Monthly, Close-out Reporting of Bank Investment Supervision".
35. Konior, J. Monitoring of Construction Projects Feasibility by Bank Investment Supervision Approach. *Civ. Eng. Archit.* **2019**, *7*, 31–35. [CrossRef]
36. Kozień, E. Application of approximation technique to on-line updating of the actual cost curve in the earned value method. *Czas. Tech.* **2017**, *9*, 181–195.
37. Gardezi, S.S.S.; Manarvi, I.A.; Gardezi, S.J.S. Time Extension Factors in Construction Industry of Pakistan. *Procedia Eng.* **2014**, *77*, 196–204. [CrossRef]
38. Gebrehiwet, T.; Luo, H. Analysis of Delay Impact on Construction Project Based on RII and Correlation Coefficient: Empirical Study. *Procedia Eng.* **2017**, *196*, 366–374. [CrossRef]
39. Barraza, G.A.; Back, W.E.; Mata, F. Probabilistic Forecasting of Project Performance Using Stochastic S Curves. *J. Constr. Eng. Manag.* **2004**, *130*, 25–32. [CrossRef]

40. Yao, J.-S.; Chen, M.-S.; Lu, H.-F. A fuzzy stochastic single-period model for cash management. *Eur. J. Oper. Res.* **2006**, *170*, 72–90. [CrossRef]
41. Mohagheghi, V.; Meysam Mousavi, S.; Vahdani, B. An Assessment Method for Project Cash Flow under Interval-Valued Fuzzy Environment. *J. Optim. Ind. Eng.* **2017**, *22*, 79–80.
42. Maravas, A.; Pantouvakis, J.-P. Project cash flow analysis in the presence of uncertainty in activity duration and cost. *Int. J. Proj. Manag.* **2012**, *30*, 374–384. [CrossRef]

applied sciences

MDPI

Article

A Proposed Soft Computing Model for Ultimate Strength Estimation of FRP-Confined Concrete Cylinders

Reza Kamgar [1], Hosein Naderpour [2], Houman Ebrahimpour Komeleh [3], Anna Jakubczyk-Gałczyńska [4,*] and Robert Jankowski [4]

1 Department of Civil Engineering, Shahrekord University, Shahrekord 8818634141, Iran; kamgar@sku.ac.ir
2 Faculty of Civil Engineering, Semnan University, Semnan 98 23, Iran; naderpour@semnan.ac.ir
3 Department of Civil Engineering, Shahid Bahonar University of Kerman, Kerman 7616913439, Iran; hek2006@gmail.com
4 Faculty of Civil and Environmental Engineering, Gdansk University of Technology, 80-233 Gdansk, Poland; jankowr@pg.edu.pl
* Correspondence: annjakub@pg.edu.pl

Received: 19 January 2020; Accepted: 28 February 2020; Published: 4 March 2020

Abstract: In this paper, the feed-forward backpropagation neural network (FFBPNN) is used to propose a new formulation for predicting the compressive strength of fiber-reinforced polymer (FRP)-confined concrete cylinders. A set of experimental data has been considered in the analysis. The data include information about the dimensions of the concrete cylinders (diameter, length) and the total thickness of FRP layers, unconfined ultimate concrete strength, ultimate confinement pressure, ultimate tensile strength of the FRP laminates and the ultimate concrete strength of the concrete cylinders. The confined ultimate concrete strength is considered as the output data, while other parameters are considered as the input data. These parameters are mostly used in existing FRP-confined concrete models. Soft computing techniques are used to estimate the compressive strength of FRP-confined concrete cylinders. Finally, a new formulation is proposed. The results of the proposed formula are compared to the existing methods. To verify the proposed method, results are compared with other methods. The results show that the described method can forecast the compressive strength of FRP-confined concrete cylinders with high precision in comparison with the existing formulas. Moreover, the mean percentage of error for the proposed method is very low (3.49%). Furthermore, the proposed formula can estimate the ultimate compressive capacity of FRP-confined concrete cylinders with a different type of FRP and arbitrary thickness in the initial design of practical projects.

Keywords: FRP; soft computing; compressive strength; confined concrete; artificial neural network

1. Introduction

A combination of high-strength fibers and matrix leads to the construction of a fiber-reinforced polymer (FRP). The primary role of the matrix is to bind these fibers together to construct structural shapes. Four common types of fibers (i.e., aramid, carbon, glass, and high-strength steel) and also two standard matrices exist (i.e., epoxies and esters) [1,2]. A new area has been opened in the civil engineering field due to the beneficial properties of FRP in the repair and rehabilitation of existing structures. The FRP can create a continuous confinement action for the concrete member, and can also increase the corrosion resistance of members [3]. Hereby, FRPs are popularly used to repair or retrofit the reinforcing frame members [4–10]. Studies on the behavior of FRP and FRP-confined concrete have advanced rapidly in recent years [11]. There are a lot of publications proposing a formula for FRP-confined concrete [12–19]. These proposed formulas are usually based on the Richard et al. method [20].

Nowadays, the use of artificial neural networks, Bayesian networks and neuro-fuzzy systems has a special place in engineering solutions, including FRP-strengthened concrete structures, structural optimization, water resource management, vibration control, bridge engineering, etc. [11,21–37]. In the study of vibrations in buildings, e.g., caused by earthquakes, the search for alternative solutions is also underway (see [38–41], for example).

An artificial neural network (ANN) was used by Lee and Lee [42] to estimate the shear strength of FRP-reinforced concrete flexural members. Sobhani et al. [43] used ANN, adaptive neuro-fuzzy inference system (ANFIS) and regression analysis to predict the compressive strength of no-slump concrete. Cheng and Cao [44] predicted the shear strength of reinforced concrete deep beams using evolutionary multivariate adaptive regression splines. In addition, the M5 model tree, used by Behnood et al. [45], is capable of predicting the elastic modulus of recycled aggregate concrete. Ebrahimpour Komleh and Maghsoudi [46] proposed a new formulation to estimate the curvature ductility factor for FRP-reinforced high-strength concrete beams using ANFIS and multiple regression methods. The ANFIS model was also used by Gu and Oyadiji [47] to control a multi-degree of freedom structures equipped with an MR damper. The ANFIS and ANN models were applied by Amini and Moeini [48] to compare results obtained for the shear strength of reinforced concrete beams with building codes. The strength of FRP connections using the backpropagation neural network was studied by Mashrei et al. [49]. The deflection of high-strength self-compacting concrete deep beams was studied by Mohammadhassani et al. applying ANFIS [50]. Nehdi and Nikopour [51] used the genetic algorithm to predict the shear capacity of reinforced concrete beams reinforced with FRP sheets.

Currently, seawater and sea sand concrete is also becoming popular due to the shortage of resources and, therefore, many researchers have focused their studies on these types of materials [52–54]. Some mechanical properties of FRP-confined concrete columns made of sea sand and seawater were studied by Li et al. [52]. They presented some theoretical models for hoop stress and strain relations and axial compression–strain relations. Zhou et al. [54] experimentally considered the effects of a chloride environment on the mechanical performance and durability of FRP-confined concrete columns made of seawater.

In this paper, the feed-forward backpropagation neural network (FFBPNN) method has been used to estimate the ultimate compressive capacity of FRP-confined concrete cylinders. For this purpose, a set of previously published and available experimental data (281 instances) for concrete made of ordinary sand has been collected for training and testing. Finally, a new formulation has been proposed to estimate the ultimate compressive capacity of FRP-confined concrete cylinders. It should be noted that the correlation coefficient of the proposed formula is equal to 0.9809, which shows a good agreement with the actual values. A comparison has been performed between the results obtained by FFBPNN and the results of the other existing models to demonstrate the ability of the proposed method. The results show that the values of the mean percentage of error (3.49%), root mean square error (3.99), and average absolute error (0.035) for the proposed method are less than other studied methods. It means that, for the proposed formula, more than 96% of the simulated results are entirely consistent with the experimental results, and also that the proposed method is very accurate compared to other existing methods. Furthermore, it is shown that the FFBPNN is a formula that can be used for all types of FRP (carbon, aramid, and glass). The proposed method can be easily employed using a calculator with high precision while, in the case of neuro-fuzzy, neural network and other known methods, a computer and sophisticated software are usually needed.

2. Research Objectives

Generally, ANNs have been used in applied science and engineering problems, because of their positive features. These features can be summarized as: (I) ability to handle the uncertainties, (II) ability to find the existing sensitivity and, finally, (III) proposing a mathematical relationship between input and output data. This research work addresses the following main objectives. First, the feed-forward backpropagation neural network is used to predict the compressive strength of FRP-confined concrete cylinders from a set of experimental data. For this purpose, a database of experimental data has been established based on various publications. Based on these data, the main effective parameters that have an influence on the compressive strength of FRP-confined concrete cylinders (FRPCCC) are assessed. Finally, using the feed-forward backpropagation neural network, a new formulation is proposed, and the effects of the presented formula are compared with existing models.

3. Overview of Existing Models

Some published publications offer a formula to forecast the compressive strength of FRPCCC (f'_{cc}). In these papers, certain parameters are adopted as the input parameters. These parameters include the diameter of the concrete cylinder (d), length of the concrete cylinder (L), unconfined ultimate concrete strength (f'_{co}), the thickness of FRP layer (t), ultimate confinement pressure (f_l) and ultimate tensile strength of the FRP laminate (f_f). Table 1 shows the existing formula to compute the compressive strength of FRPCCC.

Table 1. Some of the existing formulas for predicting the compressive strength of fiber-reinforced polymer-confined concrete cylinders (FRPCCC).

Author	Year	Formula	Note
Mansouri et al. [55]	2016	Neuro-fuzzy, multivariate adaptive regression splines, neural network, and M5 model tree techniques (without any proposed formula)	Nonlinear
Naderpour et al. [56]	2010	$f'_{cc} = \left(f'_{cc}\right)_{chart} \times C(d) \times C(L) \times C(f'_{co}) \times C(t) \times C(f_l)$ $C(d) = -0.490\left(\frac{d}{140}\right) + 1.494$ $C(L) = 0.159\ \ln\left(\frac{L}{300}\right) + 1.009$ $C(f'_{co}) = 1.082\left(\frac{f'_{co}}{35}\right)^4 - 5.071\left(\frac{f'_{co}}{35}\right)^3 + 8.209\left(\frac{f'_{co}}{35}\right)^2 - 5.025\left(\frac{f'_{co}}{35}\right) + 1.798$ $C(t) = -0.064\left(\frac{t}{1.2}\right)^2 + 0.669\left(\frac{t}{1.2}\right) + 0.387$ $C(f_l) = -0.213\left(\frac{f_l}{1500}\right)^4 + 0.901\left(\frac{f_l}{1500}\right)^3 - 1.008\left(\frac{f_l}{1500}\right)^2 + 0.723\left(\frac{f_l}{1500}\right) + 0.604$	Nonlinear
Vintzileou and Panagiotidou [18]	2008	$f'_{cc}/f'_{co} = 1 + 2.8(f_l/f'_{co})$	Linear
Berthet et al. [12]	2006	$f'_{cc}/f'_{co} = 1 + 3.45(f_l/f'_{co})$ $20 \leq f'_{co} \leq 50\ (MPa)$ $f'_{cc}/f'_{co} = 1 + 0.95(f_l/f'^{1.25}_{co})$ $50 \leq f'_{co} \leq 200\ (MPa)$	Linear-Nonlinear
Matthys et al. [15]	2005	$f'_{cc}/f'_{co} = 1 + 2.3(f_l/f'_{co})^{0.85}$	Nonlinear
Matthys et al. [15]	2005	$f'_{cc}/f'_{co} = 1 + 2.3(f_l/f'_{co})^{0.85}$	Nonlinear
Lam and Teng [13]	2002	$f'_{cc}/f'_{co} = 1 + 2(f_l/f'_{co})$	Linear
Xiao and Wu [19]	2000	$f'_{cc}/f'_{co} = 1.1 + (4.1 - 0.75f'^{2}_{co}/E_l)(f_l/f'_{co})$	Nonlinear
Miyauchi et al. [16]	2000	$f'_{cc}/f'_{co} = 1 + 3.485(f_l/f'_{co})$	Linear
Saafi et al. [17]	1999	$f'_{cc}/f'_{co} = 1 + 2.2(f_l/f'_{co})^{0.84}$	Nonlinear
Miyauchi et al. [16]	2000	$f'_{cc}/f'_{co} = 1 + 2.98(f_l/f'_{co})$	Linear

Table 1. *Cont.*

Author	Year	Formula	Note
Spoelstra and Monti [57]	1999	$f'_{cc}/f'_{co} = 0.2 + 3(f_l/f'_{co})^{0.5}$	Second-order
Toutanji [58]	1999	$f'_{cc}/f'_{co} = 1 + 3.5(f_l/f'_{co})^{0.85}$	Nonlinear
Samaan et al. [59]	1998	$f'_{cc}/f'_{co} = 1 + 6(f_l^{0.7}/f'_{co})$	Nonlinear
Kono et al. [60]	1998	$f'_{cc}/f'_{co} = 1 + 0.0572 f_l$	Linear
Karbhari and Gao [61]	1997	$f'_{cc}/f'_{co} = 1 + 2.1(f_l/f'_{co})^{0.87}$	Nonlinear
Mander et al. [62]	1988	$f'_{cc}/f'_{co} = -1.254 - 2(f_l/f'_{co}) + 2.254(1 + 7.94 f_l/f'_{co})^{0.5}$	Second-order
Fardis and Khalili [63]	1981	$f'_{cc}/f'_{co} = 1 + 2.05(f_l/f'_{co})$	Linear
Fardis and Khalili [64]	1982	$f'_{cc}/f'_{co} = 1 + 3.7(f_l/f'_{co})^{0.85}$	Nonlinear
Richart et al. [20]	1928	$f'_{cc}/f'_{co} = 1 + 4.1(f_l/f'_{co})$	Linear

It should be noted that when a concrete cylinder is subjected to the axial compression force, the compressive strength is less than its value for the FRPCCC (see Figure 1). It means that $P_1 < P_2$.

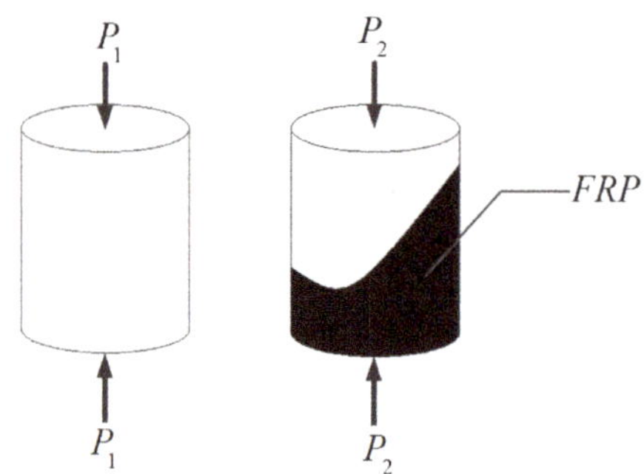

Figure 1. Two specimens (i.e., concrete cylinder and FRPCCC) subjected to the compression (an axial force).

4. Proposing a New Formulation to Predict the Compressive Strength of FRP-Confined Concrete Cylinder

In this paper, firstly, a set of experimental data is collected from the published literature [17,58,60,65–78] (see Table A1 in Appendix A). Then, the collected data are divided into input and output parameters (see Table 2).

Table 2. Input and output parameters.

Type	Parameters	Expression
Input	d (mm)	The diameter of the concrete cylinder
	L (mm)	length of the concrete cylinder
	f'_{co} (MPa)	Unconfined ultimate concrete strength
	t (mm)	The thickness of the FRP layer
	f_l (MPa)	Ultimate confinement pressure
	f_f (MPa)	The ultimate tensile strength of the CFRP laminate
Output	f'_{cc} (MPa)	Confined ultimate concrete strength

The values for minimum, maximum, mean, standard deviation, and coefficient of variation for the collected data are depicted in Table 3.

Table 3. Statistical properties for experimental data collected from the published literature.

Quantity	d (mm)	L (mm)	t (mm)	f'_{co} (MPa)	f_l (MPa)	f_f (MPa)	f'_{cc} (MPa)
Mean	133.854	272.014	0.835	42.642	15.857	2123.174	80.448
Minimum	51	102	0.089	19.4	2.33	229.762	33.8
Maximum	219	438	5.9	103	94.57	3820.359	303.6
standard deviation	27.283	58.250	1.133	17.110	12.463	1112.343	29.173
coefficient of variation	0.204	0.214	1.357	0.401	0.786	0.524	0.363

4.1. The Artificial Neural Network Model

ANNs are among the computational software methods used. The neural networks can find the existing patterns between the input and output data of experiments or simulations via training [79]. It is noteworthy that layers, neurons and weights can compose the neural networks. Here, the primary role of the weights is to relate every neuron in each layer to the neurons in other layers. Every neuron is associated with neurons in other layers by the weights. Every layer processes the input data and transfers them to the next layer. Additionally, an input layer, two or more hidden layers and an output layer compose the feed-forward neural network. A three-layer neural network is depicted in Figure 2. As mentioned in Section 3, the number of collected data is 281. These data are used for the learning, validating, and testing of ANNs. In the neural network modeling, log-sigmoid transfer functions are used and one hidden layer is selected. Firstly, all selected data are normalized based on the following equation:

$$f_{scaled} = (0.9 - 0.1)\left(\frac{f - f_{\min}}{f_{\max} - f_{\min}}\right) + 0.1 \quad 0.1 \leq f_{scaled} \leq 0.9 \tag{1}$$

where f, $f_{\min}$, $f_{\max}$ and f_{scaled} are the selected parameters, their minimum and maximum values are based on Table 3 and the value of the scaled parameters, respectively. Based on Equation (1), the scaled parameters place in the range between 0.1 and 0.9, as recognized by the log-sigmoid transfer functions.

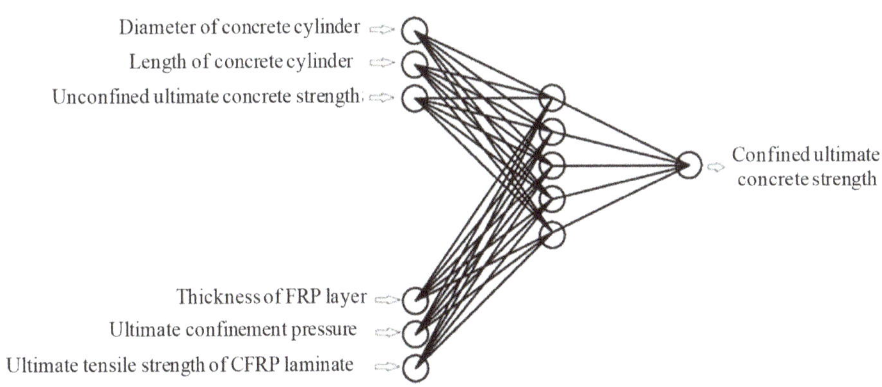

Figure 2. A three-layer artificial neural network.

The Levenberg–Marquardt algorithm is used to train randomly divided input and output vectors, which are called training (also learning), validating (also verifying) and testing datasets. Since improving the performance of the ANN model can be done by finding the optimal distribution of the datasets, various sets were analyzed. Finally, the best division was chosen, in which 70% of all data were training sets, while 15% of all data were validating and testing sets, respectively.

For this purpose, a 6:n:1 network is considered with six inputs, n hidden neurons and one output, respectively (see Figure 2). Moreover, the flowchart of the utilized ANN is depicted in Figure 3.

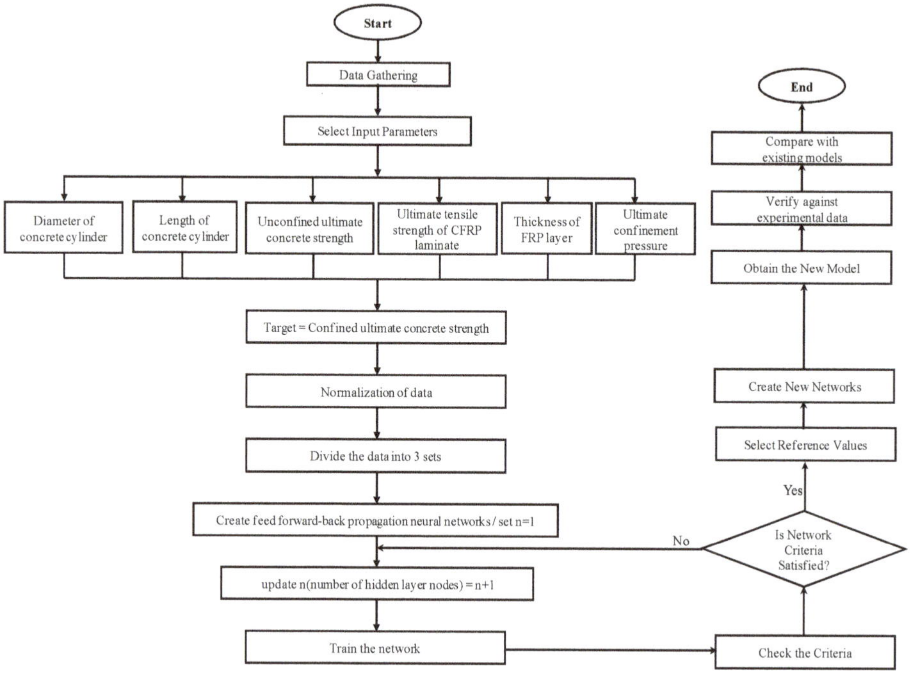

Figure 3. Flowchart of the utilized artificial neural network (ANN) for predicting the confined ultimate concrete strength.

The mean squared error (MSE) is considered as a criterion to stop the training of the networks. The MSE is defined as the average squared difference and is an important value that indicates an error between the network output and the actual value obtained from research. Therefore, when the quantity for the desired network has a minimum value, this network has a better performance. In addition, in a network, the correlation between outputs and targets is measured by regression values (R-values). The R-value is a parameter to measure the correlation between targets and outputs. These two criteria are selected to recognize which network has a better performance.

Figure 4 shows the regression values of the networks versus the different numbers of neurons in hidden layers. Furthermore, Figure 5 presents the maximum absolute value for the error of each network. From the above description and considering Figures 4 and 5, it can be concluded that a network with 15 hidden neurons had the best performance.

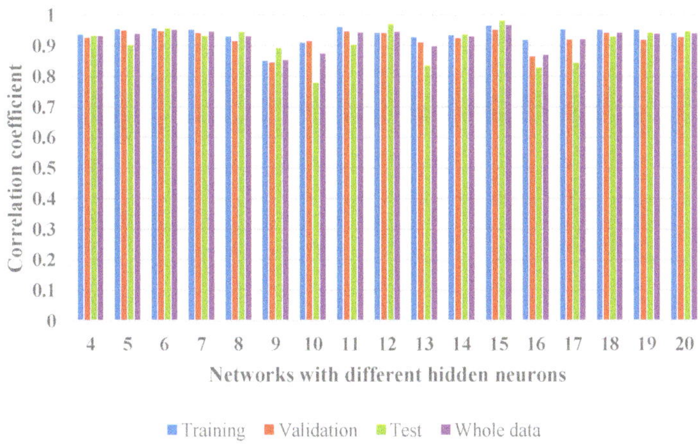

Figure 4. The correlation coefficient with different values of ANN 6:*n*:1.

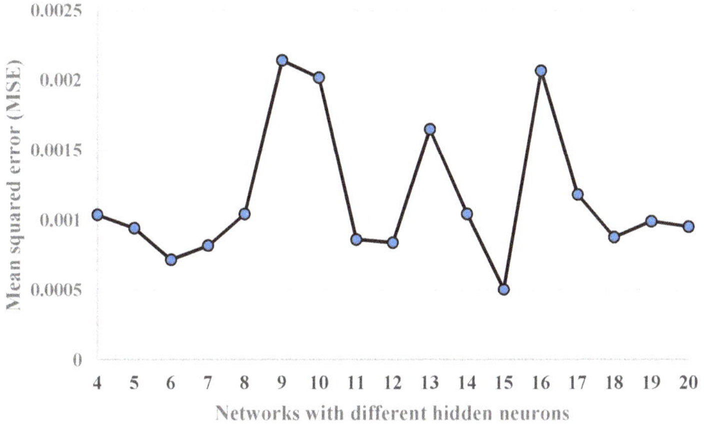

Figure 5. Mean squared error (MSE) versus some hidden-layer neurons.

After selecting a desirable network (6:15:1), the results for the training of this network are shown in Figures 6–8. It can be seen from Figure 6 that the network is well established and learned, since the values for MSE of the network begin at a large value and stop at a smaller one.

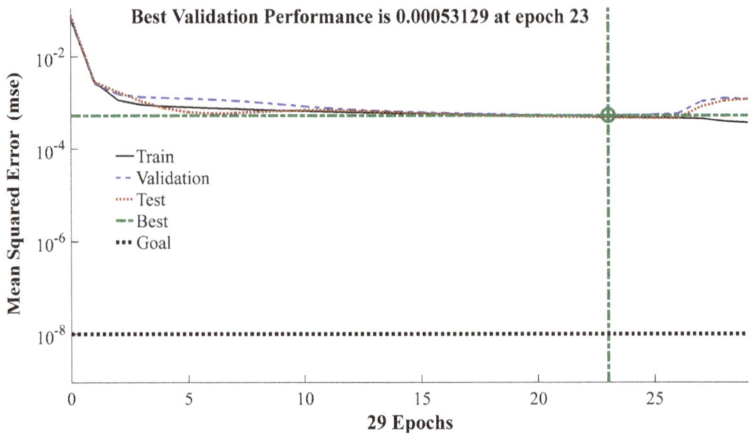

Figure 6. Performance of ANN 6:15:1.

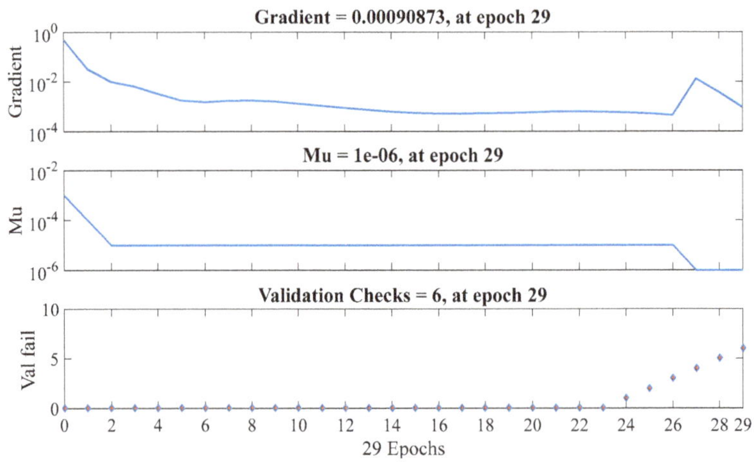

Figure 7. Training state of ANN 6:15:1.

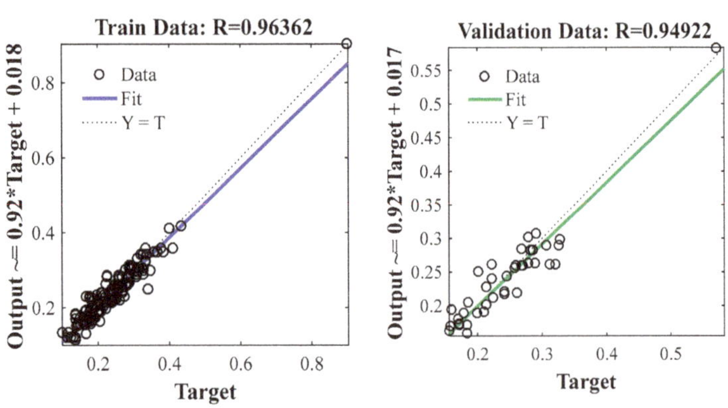

Figure 8. *Cont.*

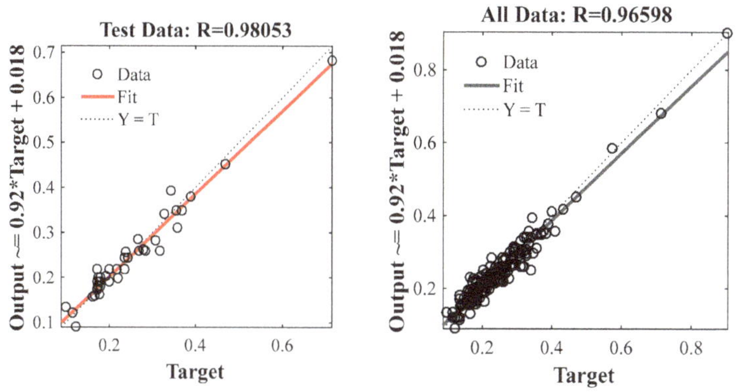

Figure 8. Regressions of training, validating, and testing datasets simulated by ANN 6:15:1.

It should be noted that the ANN technique cannot propose a formulation to predict the compressive strength of FRPCCC. Therefore, in the next section of this paper, the K-fold cross-validation technique is used to obtain a new formulation. Then, the efficiency of the proposed formula is examined.

4.2. Using a Model with a K-Fold Cross-Validation Technique in FFBPNN

In this section of the paper, a K-fold cross-validation (KFCV) technique is applied for the optimization and evaluation of the perfected ANN [80,81]. In the KFCV technique, the data are divided randomly into K folds. Then, the K-1 folds are used for training, and the last fold is used to test the neural network. In the parametric study conducted, the values for K, changing from two to five and K = 4, are considered. The process of learning and testing is conducted for all the K sections. Therefore, all the K sections contribute to the learning and testing of the ANN. This process is iterated three times for the reduction and variation of KFCV and similar distribution of data in each K. The performance of the neural network for each iteration can be computed by the percentage of correct predictions in the neural network for K folds.

In every epoch, the performance evaluation of the neural network is calculated. The curve is the correct classification factor (CCF), it is drawn for three iterations and, finally, it is averaged. In the CCF curve, after a specified epoch, the curve is saturated. Then, the optimal epoch is defined using 10% of the curve plateau. In this study, a neural network with three layers is selected for the sake of simplicity. For optimization of the ANN structure, some neurons in the hidden layer are optimized. For this purpose, the selecting criteria are considered to be the area under the CCF curve (AUCCF). Therefore, the AUCCF is measured until it reaches the optimal epoch. Hence, different neurons, from two to 13 neurons in the hidden layer, are selected and the KFCV process is repeated for the structures. Finally, the structure with the maximum efficiency can be determined by drawing the CCF and calculating the AUCCF. Figure 9 shows the AUCCF curve. As can be seen from this figure, the 6:11:1 structure with 11 neurons in the hidden layer has the highest performance with 86.6%. Figure 10 shows the CCF curve for the optimized ANN structure. As shown in this figure, the optimum epoch is 224.

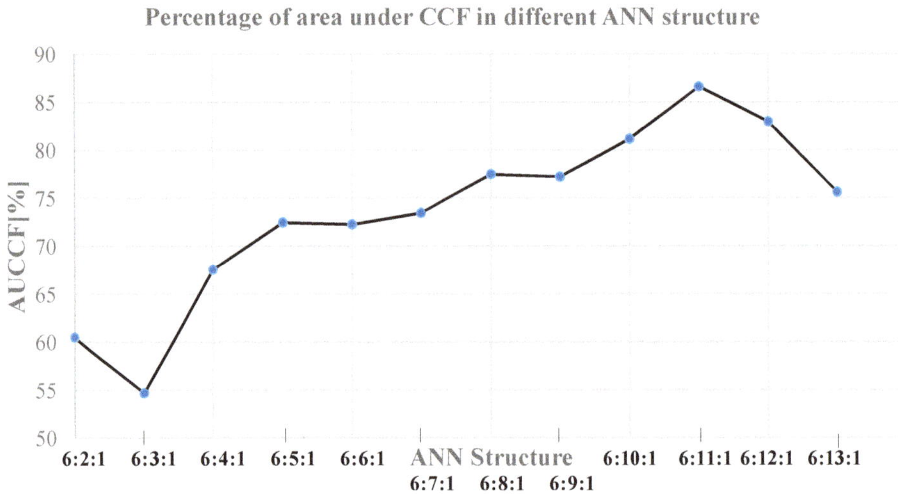

Figure 9. Area under the correct classification factor (AUCCF) curve of the neural network with different structures.

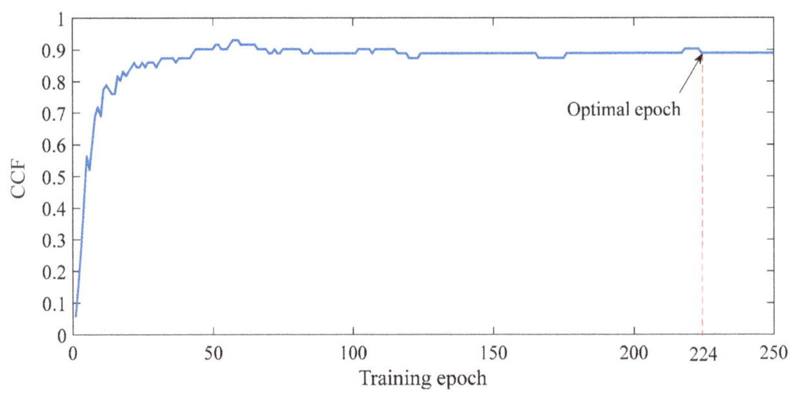

Figure 10. The CCF curve of optimized ANN structure.

Data that are predicted by the optimized ANN neural network and the training data are plotted in Figure 11. As shown in this figure, the correlation coefficient is equal to 0.9809, which confirms the performance of the optimized ANN structure.

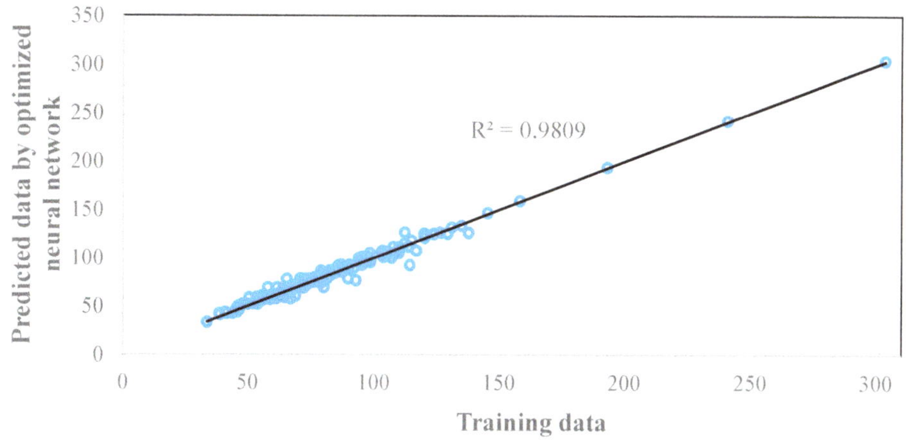

Figure 11. The correlation coefficient of the predicted data by optimized ANN structure and training data.

The Tansig and Pureline activation functions are selected for the hidden layer and the output layer, respectively. Considering the optimum structure of the neural network, weights, biases, and activation functions, a relation, such as Equation (2), could be extracted:

$$Output = \left(\frac{2}{1 + exp^{(-2\times(Input\times Iw + b_1))}} - 1 \right) \times Lw + b_2 \tag{2}$$

Input, *Output*, *IW*, *LW*, $b_{1,}$ and b_2 in Equation (2) are constant coefficients, which are defined as follows:

$$IW = \begin{bmatrix} -1.2007 & -0.6174 & 2.6247 & 0.3393 & -0.7012 & 1.6310 \\ 13.2928 & -16.8285 & 8.4865 & 1.3056 & -0.4213 & -7.0180 \\ -1.0486 & -0.0470 & -2.9279 & 0.5110 & -2.2153 & -0.2471 \\ -6.1496 & 3.0775 & 7.2171 & -3.7018 & 16.1829 & -15.3115 \\ 3.2838 & -0.7242 & -0.5853 & 0.0652 & 0.5304 & -1.0226 \\ -0.9033 & 0.8337 & 3.1251 & 2.2056 & 1.3875 & 2.3565 \\ -0.9161 & 0.6510 & 1.1284 & -0.1900 & 0.5996 & 0.5913 \\ 5.4701 & 1.9650 & 1.4193 & -0.8012 & -0.3120 & 1.6189 \\ -1.3924 & -1.9230 & 6.6908 & -12.3073 & 2.1772 & 1.0588 \\ -12.8275 & 14.9574 & 11.9745 & 3.8238 & -21.2448 & 23.5244 \\ 12.3926 & 10.8354 & -2.5711 & -8.0694 & 11.5337 & 0.1035 \end{bmatrix}^T$$
$$LW = \begin{bmatrix} -7.7570 & -0.1861 & 0.2698 & -0.2166 & 1.1077 & 1.2329 & 1.6244 & -0.7781 & -3.3951 & 0.1298 & -1.1472 \end{bmatrix}^T$$
$$b_1 = \begin{bmatrix} -1.7314 & 4.5025 & -3.2840 & 26.0974 & 0.7293 & 1.5696 & 1.5758 & -0.6336 & 16.9266 & 2.9547 & 17.0029 \end{bmatrix}$$
$$b_2 = -3.9606$$
$$Input = \begin{bmatrix} d & L & t & f'_{co} & f_l & f_f \end{bmatrix}$$
$$Output = [f'_{cc}] \tag{3}$$

5. Comparison of the Proposed Strength Model with Existing Empirical Ones

Five known models are selected [12,13,15,16,18,56] to verify the proposed formula. It must be noted that no formula has been proposed in the most recent available publication [55]. The formula proposed in this paper can be implemented in a calculator, while, in the case of the neuro-fuzzy, neural network, multivariate adaptive regression splines and M5 model tree techniques (all considered in [55]), a computer and professional programs should be used.

Figure 12 shows the values of the compressive strength of the FRPCCC obtained by the proposed and existing formula versus the experimental values. Table A1 in the Appendix section shows the experimental data that have been used to judge the ability of different methods. In fact, for all formulas, the same data are applied to forecast the compressive strengths of the FRPCCC. Figure 12 shows that the presented formula can estimate the compressive strengths of the FRPCCC with a higher precision compared to the existing formulas.

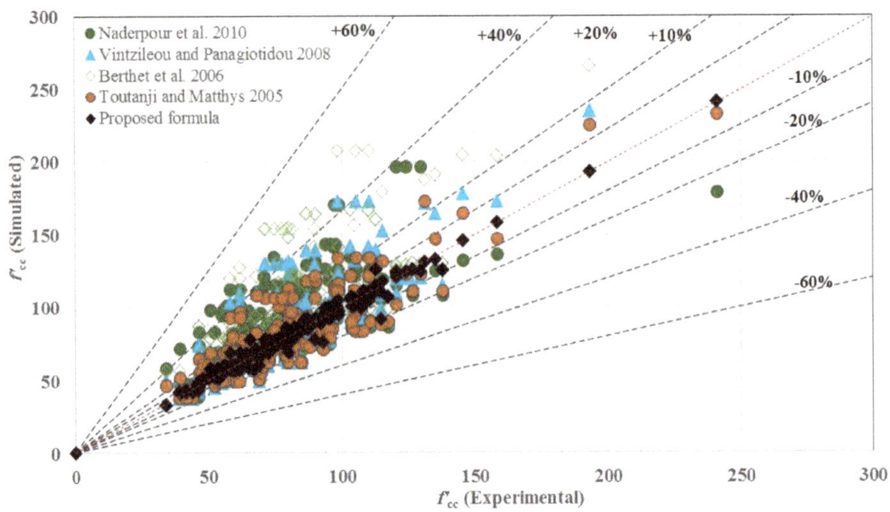

Figure 12. Comparison between simulated and experimental results for the compressive strength of FRPCCC.

The mean percentage of error, correlation coefficient, root mean square error (RMSE), and average absolute error (AAE) for the studied methods are shown in Table 4 to verify the efficiency of the proposed method. Based on this table, it should be noted that the mean percentage of error and the correlation coefficient for the proposed method are equal to 3.49% and 0.9809, respectively. Meanwhile, the corresponding values for other existing methods are equal to over 13% and 0.41, respectively. This means that, for the proposed formula, more than 96% of the simulated results are entirely consistent with the experimental ones. Furthermore, the minimum values of RMSE and AAE are obtained for the proposed formula. Therefore, it should be pointed out that the proposed formula is very accurate compared to other existing ones, for which the accuracy is lower than 85%.

Table 4. Comparison between different studied models.

Method Error	Miyauchi et al. [16]	Lam and Teng [13]	Matthys et al. [15]	Berthet et al. [12]	Vintzileou and Panagiotidou [18]	Naderpour et al. [56]	Proposed Formula
Mean percentage of error	17.77%	16.95%	13.14%	17.59%	15.42%	8.44%	3.49%
RMSE	31.83	14.96	13.95	31.20	20.64	12.86	3.99
AAE	0.28	0.15	0.14	0.25	0.18	0.11	0.035
correlation coefficients	0.6813	0.7897	0.8064	0.6835	0.7288	0.7686	0.9809

Based on Figures 11 and 12, as well as Table 4, it is evident that the proposed formula has a good agreement with the actual values. Therefore, it can be used in the practical projects to evaluate the amount of column compressive capacity reinforced by FPR sheets in the initial design. It should be noted that the collected data (see Appendix A) are for different types of FRP sheets (carbon, aramid, and glass) and the FFBPNN method has been trained and tested with these data. Therefore, the proposed formula can estimate the ultimate compressive capacity of FRP-confined concrete cylinders with a different type of FRP and arbitrary thickness.

6. Concluding Remarks

A soft computing model for the ultimate strength estimation of FRPCCC has been proposed in this paper. A set of experimental data from the published literature has been collected and divided into input and output parameters. Firstly, the ANN model has been created and analyzed. The mean squared error and R-values have been used to verify the efficiency of the network.

The results of the analysis indicate that a network with 15 hidden neurons has the best performance. However, it should be noted that the basic ANN technique cannot propose a formulation to forecast the compressive strength of FRPCCC. Therefore, in the next step of the study, the author's improvement approach has been presented. A model with a K-fold cross-validation technique in the feed-forward backpropagation neural network has been presented. The correlation coefficient, root mean square error, mean percentage of error and average absolute error have been used to check its efficiency. The structure with 11 neurons in the hidden layer has been found to give the best performance. Finally, a comparison between the proposed formula and existing empirical ones has been conducted. To verify the proposed formula, five known models described in this paper have been selected. The results of the study show that the proposed method can estimate the compressive strengths of the FRPCCC with higher precision compared to the existing formulas. Moreover, it can be used to predict the compressive strength of FRPCCC with different types and arbitrary thicknesses of FRP (carbon, aramid, and glass). It should be noted that the mean percentage of error and the correlation coefficient for the proposed method are equal to 3.49% and 0.9809, respectively. Meanwhile, the corresponding values for other existing methods are equal to over 13% and 0.41, respectively. It means that, for the proposed formula, more than 96% of the simulated results are entirely consistent with the experimental results. Furthermore, the minimum values of RMSE and AAE have been obtained for the proposed formula. Therefore, it should be pointed out that the proposed formula is very accurate compared to other existing methods, for which the accuracy is usually lower than 85%. It should also be added that the proposed method can be easily employed using a calculator with high precision while, in the case of the neuro-fuzzy network, neural network and other known methods, a computer and sophisticated software is usually needed. Therefore, our model can be used to estimate the ultimate compressive capacity of FRP-confined concrete cylinders in the initial design of practical projects.

Finally, it should be noted that there is a lack of experimental tests on concrete cylinders made of seawater and sea sand retrofitted with FRP sheets in order to propose a formula that covers the entire region. This should be a focus in future studies.

Author Contributions: Conceptualization, R.K. and H.N.; methodology, R.K., H.N., H.E.K., A.J.-G. and R.J.; software, R.K., H.N. and H.E.K.; validation, R.K., H.N., H.E.K., A.J.-G. and R.J.; formal analysis, R.K., H.N. and H.E.K.; investigation, R.K. and H.N.; writing—original draft preparation, R.K., H.N., and H.E.K.; writing—review and editing, A.J.-G. and R.J. All authors have read and agreed to the published version of the manuscript.

Funding: This research received no external funding.

Conflicts of Interest: The authors declare no conflict of interest.

Appendix A

The collected data are indicated in Table A1.

Table A1. The collected data from experimental studies.

No.	Ref.	Fiber Type	d (mm)	L (mm)	t (mm)	f'_{co} (MPa)	f_l (MPa)	f_f (MPa)	f'_{cc} (MPa)
1	[72]	CFRP	150	300	0.11	45.2	5.2	3481	59.4
2	[72]	CFRP	150	300	0.22	45.2	10.08	3481	79.4
3	[72]	CFRP	150	300	0.11	31.2	5.04	3481	52.4
4	[72]	CFRP	150	300	0.22	31.2	10.14	3481	67.4
5	[72]	CFRP	150	300	0.33	31.2	15.31	3481	81.7
6	[72]	CFRP	100	200	0.11	51.9	7.7	3481	75.2
7	[72]	CFRP	100	200	0.22	51.9	15.15	3481	104.6
8	[72]	CFRP	100	200	0.11	33.7	7.57	3481	69.6
9	[72]	CFRP	100	200	0.22	33.7	15.39	3481	88
10	[72]	CFRP	150	300	0.11	45.2	5.2	3481	59.4
11	[60]	CFRP	100	200	0.167	34.3	12.77	3820	57.4
12	[60]	CFRP	100	200	0.167	34.3	12.85	3820	64.9
13	[60]	CFRP	100	200	0.167	32.3	12.85	3820	58.2
14	[60]	CFRP	100	200	0.167	32.3	12.77	3820	61.8
15	[60]	CFRP	100	200	0.167	32.3	12.67	3820	57.7
16	[60]	CFRP	100	200	0.334	32.3	27.69	3820	61.8
17	[60]	CFRP	100	200	0.334	32.3	35.55	3820	80.2
18	[60]	CFRP	100	200	0.334	32.3	16.4	3820	58.2
19	[60]	CFRP	100	200	0.501	32.3	38.33	3820	86.9
20	[60]	CFRP	100	200	0.501	32.3	38.27	3820	90.1
21	[60]	CFRP	100	200	0.167	34.8	12.8	3820	57.8
22	[60]	CFRP	100	200	0.167	34.8	12.67	3820	55.6
23	[60]	CFRP	100	200	0.167	34.8	12.66	3820	50.7
24	[60]	CFRP	100	200	0.334	34.8	25.4	3820	82.7
25	[60]	CFRP	100	200	0.334	34.8	25.46	3820	81.4
26	[60]	CFRP	100	200	0.501	34.8	38.38	3820	103.3
27	[60]	CFRP	100	200	0.501	34.8	38.24	3820	110.1
28	[69]	CFRP	150	300	0.117	34.9	4.08	2600	46.1
29	[69]	CFRP	150	300	0.235	34.9	3.44	1100	45.8
30	[77]	CFRP	153	306	0.36	19.4	10.77	2275	33.8
31	[77]	CFRP	153	306	0.66	19.4	19.71	2275	46.4
32	[77]	CFRP	153	306	0.9	19.4	26.87	2275	62.6
33	[77]	CFRP	153	306	1.08	19.4	32.2	2275	75.7
34	[77]	CFRP	153	306	1.25	19.4	37.32	2275	80.2
35	[77]	CFRP	153	306	0.36	49	10.68	2275	59.1
36	[77]	CFRP	153	306	0.66	49	19.77	2275	76.5
37	[77]	CFRP	153	306	0.9	49	26.85	2275	98.8
38	[77]	CFRP	153	306	1.08	49	32.3	2275	112.7
39	[75]	CFRP	100	200	0.6	42	15.21	1265	73.5
40	[75]	CFRP	100	200	0.6	42	15.21	1265	73.5
41	[75]	CFRP	100	200	0.6	42	15.15	1265	67.62
42	[75]	AFRP	150	300	1.26	43	3.82	230	47.3
43	[75]	AFRP	150	300	2.52	43	7.76	230	58.91
44	[75]	AFRP	150	300	3.78	43	11.66	230	70.95
45	[75]	AFRP	150	300	5.04	43	15.46	230	74.39
46	[70]	GFRP	100	200	0.35	32	10.63	1520	54
47	[70]	GFRP	100	200	0.35	32	10.69	1520	48
48	[70]	GFRP	100	200	0.35	32	10.63	1520	54
49	[70]	GFRP	100	200	0.35	32	10.61	1520	50
50	[70]	CFRP	100	200	0.16	37	12.2	3790	60
51	[70]	CFRP	100	200	0.16	37	12.15	3790	62
52	[70]	CFRP	100	200	0.16	37	12.13	3790	59
53	[70]	CFRP	100	200	0.16	37	12.11	3790	57
54	[76]	CFRP	150	300	0.169	25.15	4.63	2024	44.13
55	[76]	CFRP	150	300	0.169	25.15	4.61	2024	41.56
56	[76]	CFRP	150	300	0.169	25.15	4.55	2024	38.75
57	[76]	CFRP	150	300	0.338	25.15	9.11	2024	60.09
58	[76]	CFRP	150	300	0.338	25.15	9.14	2024	55.93
59	[76]	CFRP	150	300	0.338	25.15	9.19	2024	61.61
60	[76]	CFRP	150	300	0.507	25.15	13.64	2024	67

Table A1. *Cont.*

No.	Ref.	Fiber Type	d (mm)	L (mm)	t (mm)	f'_{co} (MPa)	f_l (MPa)	f_f (MPa)	f'_{cc} (MPa)
61	[76]	CFRP	150	300	0.507	25.15	13.72	2024	67.27
62	[76]	CFRP	150	300	0.507	25.15	13.72	2024	70.18
63	[76]	CFRP	150	300	0.169	47.44	4.64	2024	72.26
64	[76]	CFRP	150	300	0.169	47.44	4.61	2024	64.4
65	[76]	CFRP	150	300	0.169	47.44	4.57	2024	66.19
66	[76]	CFRP	150	300	0.338	47.44	9.12	2024	82.36
67	[76]	CFRP	150	300	0.338	47.44	9.12	2024	82.35
68	[76]	CFRP	150	300	0.338	47.44	9.08	2024	79.11
69	[76]	CFRP	150	300	0.507	47.44	13.79	2024	96.29
70	[76]	CFRP	150	300	0.507	47.44	13.74	2024	95.22
71	[76]	CFRP	150	300	0.507	47.44	13.77	2024	103.9
72	[76]	CFRP	150	300	0.169	51.84	4.62	2024	78.65
73	[76]	CFRP	150	300	0.169	51.84	4.54	2024	79.18
74	[76]	CFRP	150	300	0.169	51.84	4.57	2024	72.76
75	[76]	CFRP	150	300	0.338	51.84	9.23	2024	95.4
76	[76]	CFRP	150	300	0.338	51.84	9.16	2024	90.3
77	[76]	CFRP	150	300	0.338	51.84	9.02	2024	90.65
78	[76]	CFRP	150	300	0.507	51.84	13.77	2024	110.5
79	[76]	CFRP	150	300	0.507	51.84	13.64	2024	103.6
80	[76]	CFRP	150	300	0.507	51.84	13.65	2024	117.2
81	[76]	CFRP	150	300	0.845	51.84	22.78	2024	112.6
82	[76]	CFRP	150	300	0.845	51.84	22.87	2024	126.6
83	[76]	CFRP	150	300	0.845	51.84	22.67	2024	137.9
84	[76]	CFRP	150	300	0.169	70.48	4.53	2024	87.29
85	[76]	CFRP	150	300	0.169	70.48	4.53	2024	84.03
86	[76]	CFRP	150	300	0.169	70.48	4.53	2024	83.22
87	[76]	CFRP	150	300	0.338	70.48	9.19	2024	94.06
88	[76]	CFRP	150	300	0.338	70.48	9.14	2024	98.13
89	[76]	CFRP	150	300	0.338	70.48	9.22	2024	107.2
90	[76]	CFRP	150	300	0.507	70.48	13.7	2024	114.1
91	[76]	CFRP	150	300	0.507	70.48	13.63	2024	108
92	[76]	CFRP	150	300	0.507	70.48	13.48	2024	110.3
93	[76]	CFRP	150	300	0.169	82.13	4.75	2024	94.08
94	[76]	CFRP	150	300	0.169	82.13	5.2	2024	97.6
95	[76]	CFRP	150	300	0.169	82.13	4.98	2024	95.83
96	[76]	CFRP	150	300	0.338	82.13	10.15	2024	97.43
97	[76]	CFRP	150	300	0.338	82.13	9.14	2024	98.85
98	[76]	CFRP	150	300	0.338	82.13	9.92	2024	98.24
99	[76]	CFRP	150	300	0.507	82.13	13.59	2024	124.2
100	[76]	CFRP	150	300	0.507	82.13	13.76	2024	129.5
101	[76]	CFRP	150	300	0.507	82.13	13.42	2024	120.3
102	[65]	GFRP	102	203	1	38.99	40.75	2078	115.3
103	[65]	GFRP	102	203	1	50.51	40.75	2078	135.1
104	[65]	GFRP	102	203	1	64.2	40.75	2078	145.59
105	[73]	GFRP	150	300	0.3	36.3	2.33	583	46
106	[73]	GFRP	150	300	0.3	36.3	2.33	583	41.2
107	[73]	GFRP	150	300	0.6	36.3	4.67	584	60.52
108	[73]	GFRP	150	300	0.6	36.3	4.67	584	59.23
109	[73]	GFRP	150	300	0.6	36.3	4.67	584	59.77
110	[73]	GFRP	150	300	0.6	36.3	4.67	584	60.16
111	[73]	GFRP	150	300	0.6	36.3	4.67	584	69.02
112	[73]	GFRP	150	300	0.6	36.3	4.67	584	55.75
113	[73]	GFRP	150	300	0.6	36.3	4.67	584	56.41
114	[73]	GFRP	150	300	1.2	36.3	9.33	583	84.88
115	[73]	GFRP	150	300	1.2	36.3	9.33	583	84.33
116	[73]	GFRP	150	300	1.2	36.3	9.33	583	79.64
117	[73]	AFRP	150	300	2.4	36.3	18.67	583	106.87
118	[73]	AFRP	150	300	2.4	36.3	18.67	583	104.94
119	[73]	AFRP	150	300	2.4	36.3	18.67	583	107.91
120	[67]	CFRP	51	102	0.089	41	12.22	3501	86

Table A1. *Cont.*

No.	Ref.	Fiber Type	d (mm)	L (mm)	t (mm)	f'_{co} (MPa)	f_l (MPa)	f_f (MPa)	f'_{cc} (MPa)
121	[67]	CFRP	51	102	0.179	41	24.57	3500	120.5
122	[67]	CFRP	51	102	0.344	41	47.22	3500	158.4
123	[67]	CFRP	51	102	0.689	41	94.57	3500	241
124	[67]	CFRP	51	102	0.179	103	24.57	3500	131.1
125	[67]	CFRP	51	102	0.344	103	47.22	3500	193.2
126	[67]	CFRP	51	102	0.689	103	94.57	3500	303.6
127	[74]	CFRP	153	305	0.36	39.7	5.98	1271	55.98
128	[78]	CFRP	100	200	0.1675	30.2	9.14	2728	46.6
129	[78]	CFRP	100	200	0.5025	30.2	27.42	2728	87.2
130	[78]	CFRP	100	200	0.67	30.2	36.56	2728	104.6
131	[78]	CFRP	100	200	0.14	30.2	4.38	1564	41.7
132	[78]	CFRP	100	200	0.28	30.2	8.75	1563	56
133	[78]	CFRP	100	200	0.42	30.2	13.13	1563	63.3
134	[78]	AFRP	100	200	0.145	30.2	7.7	2655	39
135	[78]	AFRP	100	200	0.29	30.2	15.39	2653	68.5
136	[78]	AFRP	100	200	0.435	30.2	23.09	2654	92.1
137	[72]	CFRP	150	300	0.11	45.2	5.11	3484	59.4
138	[72]	CFRP	150	300	0.22	45.2	10.21	3481	79.4
139	[72]	CFRP	150	300	0.11	31.2	5.11	3484	52.4
140	[72]	CFRP	150	300	0.22	31.2	10.21	3481	67.4
141	[72]	CFRP	150	300	0.33	31.2	15.32	3482	81.7
142	[72]	CFRP	100	200	0.11	51.9	7.66	3482	75.2
143	[72]	CFRP	100	200	0.22	51.9	15.32	3482	104.6
144	[72]	CFRP	100	200	0.11	33.7	7.66	3482	69.6
145	[72]	CFRP	100	200	0.22	33.7	15.32	3482	88
146	[72]	CFRP	150	300	0.11	45.2	5.11	3484	59.4
147	[60]	CFRP	100	200	0.167	34.3	12.76	3820	57.4
148	[60]	CFRP	100	200	0.167	34.3	12.76	3820	64.9
149	[60]	CFRP	100	200	0.167	32.3	12.76	3820	58.2
150	[60]	CFRP	100	200	0.167	32.3	12.76	3820	61.8
151	[60]	CFRP	100	200	0.167	32.3	12.76	3820	57.7
152	[60]	CFRP	100	200	0.334	32.3	25.52	3820	58.2
153	[60]	CFRP	100	200	0.334	32.3	25.52	3820	61.8
154	[60]	CFRP	100	200	0.334	32.3	25.52	3820	80.2
155	[60]	CFRP	100	200	0.501	32.3	38.28	3820	86.9
156	[60]	CFRP	100	200	0.501	32.3	38.28	3820	90.1
157	[60]	CFRP	100	200	0.167	34.8	12.76	3820	57.8
158	[60]	CFRP	100	200	0.167	34.8	12.76	3820	55.6
159	[60]	CFRP	100	200	0.167	34.8	12.76	3820	50.7
160	[60]	CFRP	100	200	0.334	34.8	25.52	3820	82.7
161	[60]	CFRP	100	200	0.334	34.8	25.52	3820	81.4
162	[60]	CFRP	100	200	0.501	34.8	38.28	3820	103.3
163	[60]	CFRP	100	200	0.501	34.8	38.28	3820	110.1
164	[58]	GFRP	76	305	0.236	30.93	9.43	1518	60.82
165	[58]	CFRP	76	305	0.22	30.93	20.18	3486	95.02
166	[58]	CFRP	76	305	0.33	30.93	25.53	2940	94.01
167	[69]	CFRP	150	300	0.117	34.9	4.06	2603	46.1
168	[69]	CFRP	150	300	0.235	34.9	3.45	1101	45.8
169	[77]	CFRP	153	305	0.36	19.4	10.74	2282	33.8
170	[77]	CFRP	153	305	0.66	19.4	19.69	2282	46.4
171	[77]	CFRP	153	305	0.9	19.4	26.85	2282	62.6
172	[77]	CFRP	153	305	1.08	19.4	32.22	2282	75.7
173	[77]	CFRP	153	305	1.25	19.4	37.3	2283	80.2
174	[77]	CFRP	153	305	0.36	49	10.74	2282	59.1
175	[77]	CFRP	153	305	0.66	49	19.69	2282	76.5
176	[77]	CFRP	153	305	0.9	49	26.85	2282	98.8
177	[77]	CFRP	153	305	1.08	49	32.22	2282	112.7
178	[75]	CFRP	100	200	0.6	42	15.18	1265	73.5
179	[75]	CFRP	100	200	0.6	42	15.18	1265	73.5
180	[75]	CFRP	100	200	0.6	42	15.18	1265	67.62

Table A1. *Cont.*

No.	Ref.	Fiber Type	d (mm)	L (mm)	t (mm)	f'_{co} (MPa)	f_l (MPa)	f_f (MPa)	f'_{cc} (MPa)
181	[75]	AFRP	150	300	1.26	43	3.86	230	47.3
182	[75]	AFRP	150	300	2.52	43	7.73	230	58.91
183	[75]	AFRP	150	300	3.78	43	11.59	230	70.95
184	[75]	AFRP	150	300	5.04	43	15.46	230	74.39
185	[70]	GFRP	100	200	0.35	32	10.64	1520	54
186	[70]	GFRP	100	200	0.35	32	10.64	1520	48
187	[70]	GFRP	100	200	0.35	32	10.64	1520	54
188	[70]	GFRP	100	200	0.35	32	10.64	1520	50
189	[70]	CFRP	100	200	0.16	37	12.13	3791	60
190	[70]	CFRP	100	200	0.16	37	12.13	3791	62
191	[70]	CFRP	100	200	0.16	37	12.13	3791	59
192	[70]	CFRP	100	200	0.16	37	12.13	3791	57
193	[76]	CFRP	150	300	0.169	25.15	4.56	2024	44.13
194	[76]	CFRP	150	300	0.169	25.15	4.56	2024	41.56
195	[76]	CFRP	150	300	0.169	25.15	4.56	2024	38.75
196	[76]	CFRP	150	300	0.338	25.15	9.12	2024	60.09
197	[76]	CFRP	150	300	0.338	25.15	9.12	2024	55.93
198	[76]	CFRP	150	300	0.338	25.15	9.12	2024	61.61
199	[76]	CFRP	150	300	0.507	25.15	13.68	2024	67
200	[76]	CFRP	150	300	0.507	25.15	13.68	2024	67.27
201	[76]	CFRP	150	300	0.507	25.15	13.68	2024	70.18
202	[76]	CFRP	150	300	0.169	47.44	4.56	2024	72.26
203	[76]	CFRP	150	300	0.169	47.44	4.56	2024	64.4
204	[76]	CFRP	150	300	0.169	47.44	4.56	2024	66.19
205	[76]	CFRP	150	300	0.338	47.44	9.12	2024	82.36
206	[76]	CFRP	150	300	0.338	47.44	9.12	2024	82.35
207	[76]	CFRP	150	300	0.338	47.44	9.12	2024	79.11
208	[76]	CFRP	150	300	0.507	47.44	13.68	2024	96.29
209	[76]	CFRP	150	300	0.507	47.44	13.68	2024	95.22
210	[76]	CFRP	150	300	0.507	47.44	13.68	2024	103.97
211	[76]	CFRP	150	300	0.169	51.84	4.56	2024	78.65
212	[76]	CFRP	150	300	0.169	51.84	4.56	2024	79.18
213	[76]	CFRP	150	300	0.169	51.84	4.56	2024	72.76
214	[76]	CFRP	150	300	0.338	51.84	9.12	2024	95.4
215	[76]	CFRP	150	300	0.338	51.84	9.12	2024	90.3
216	[76]	CFRP	150	300	0.338	51.84	9.12	2024	90.65
217	[76]	CFRP	150	300	0.507	51.84	13.68	2024	110.54
218	[76]	CFRP	150	300	0.507	51.84	13.68	2024	103.62
219	[76]	CFRP	150	300	0.507	51.84	13.68	2024	117.23
220	[76]	CFRP	150	300	0.845	51.84	22.8	2024	112.66
221	[76]	CFRP	150	300	0.845	51.84	22.8	2024	126.69
222	[76]	CFRP	150	300	0.845	51.84	22.8	2024	137.93
223	[76]	CFRP	150	300	0.169	70.48	4.56	2024	87.29
224	[76]	CFRP	150	300	0.169	70.48	4.56	2024	84.03
225	[76]	CFRP	150	300	0.169	70.48	4.56	2024	83.22
226	[76]	CFRP	150	300	0.338	70.48	9.12	2024	94.06
227	[76]	CFRP	150	300	0.338	70.48	9.12	2024	98.13
228	[76]	CFRP	150	300	0.338	70.48	9.12	2024	107.2
229	[76]	CFRP	150	300	0.507	70.48	13.68	2024	114.12
230	[76]	CFRP	150	300	0.507	70.48	13.68	2024	108.07
231	[76]	CFRP	150	300	0.507	70.48	13.68	2024	110.38
232	[76]	CFRP	150	300	0.169	82.13	4.56	2024	94.08
233	[76]	CFRP	150	300	0.169	82.13	4.56	2024	97.6
234	[76]	CFRP	150	300	0.169	82.13	4.56	2024	95.83
235	[76]	CFRP	150	300	0.338	82.13	9.12	2024	97.43
236	[76]	CFRP	150	300	0.338	82.13	9.12	2024	98.85
237	[76]	CFRP	150	300	0.338	82.13	9.12	2024	98.24
238	[76]	CFRP	150	300	0.507	82.13	13.68	2024	124.2
239	[76]	CFRP	150	300	0.507	82.13	13.68	2024	129.58
240	[76]	CFRP	150	300	0.507	82.13	13.68	2024	120.36

Table A1. *Cont.*

No.	Ref.	Fiber Type	d (mm)	L (mm)	t (mm)	f'_{co} (MPa)	f_l (MPa)	f_f (MPa)	f'_{cc} (MPa)
241	[17]	GFRP	152	435	0.8	35	4.72	448	52.8
242	[17]	GFRP	152	435	1.6	35	10.6	504	66
243	[17]	GFRP	152	435	2.4	35	17.64	559	83
244	[17]	CFRP	152	435	0.11	35	4.76	3289	55
245	[17]	CFRP	152	435	0.23	35	10.72	3542	68
246	[17]	CFRP	152	435	0.55	35	26.71	3691	97
247	[71]	GFRP	153	305	1.44	30.86	9.9	526	53.66
248	[71]	GFRP	153	305	1.44	30.86	9.9	526	56.5
249	[71]	GFRP	153	305	1.44	29.64	9.9	526	67.12
250	[71]	GFRP	153	305	1.44	29.64	9.9	526	55.29
251	[71]	GFRP	153	305	1.44	29.64	9.9	526	60.23
252	[71]	GFRP	153	305	1.44	31.97	9.9	526	59.06
253	[71]	GFRP	153	305	1.44	31.97	9.9	526	60.79
254	[71]	GFRP	153	305	2.2	30.86	16.71	581	72.92
255	[71]	GFRP	153	305	2.2	30.86	16.71	581	65.67
256	[71]	GFRP	153	305	2.2	30.86	16.71	581	77.99
257	[71]	GFRP	153	305	2.2	29.64	16.71	581	74.56
258	[71]	GFRP	153	305	2.2	29.64	16.71	581	93.02
259	[71]	GFRP	153	305	2.2	29.64	16.71	581	71.74
260	[71]	GFRP	153	305	2.2	31.97	16.71	581	77.35
261	[71]	GFRP	153	305	2.2	31.97	16.71	581	77.08
262	[71]	GFRP	153	305	2.97	30.86	24.97	643	85.72
263	[71]	GFRP	153	305	2.97	30.86	24.97	643	86.76
264	[71]	GFRP	153	305	2.97	29.64	24.97	643	86.22
265	[71]	GFRP	153	305	2.97	29.64	24.97	643	114.66
266	[71]	GFRP	153	305	2.97	29.64	24.97	643	87.44
267	[71]	GFRP	153	305	2.97	31.97	24.97	643	86.11
268	[71]	GFRP	153	305	2.97	31.97	24.97	643	83.99
269	[68]	GFRP	150	300	4.28	25.61	37.21	652	71
270	[68]	GFRP	150	300	4.28	25.61	37.21	652	71.3
271	[68]	GFRP	150	300	4.28	25.61	37.21	652	74.7
272	[68]	GFRP	150	300	4.28	25.61	37.21	652	79.2
273	[68]	GFRP	150	300	4.28	25.61	37.21	652	81.5
274	[68]	GFRP	150	300	4.28	25.61	37.21	652	77.5
275	[68]	GFRP	150	300	4.28	25.61	37.21	652	89.9
276	[68]	GFRP	150	300	5.9	25.61	52.71	670	98.5
277	[68]	GFRP	150	300	5.9	25.61	52.71	670	110.3
278	[68]	GFRP	150	300	5.9	25.61	52.71	670	105.2
279	[66]	GFRP	168	336	3.73	58	24.33	548	90
280	[66]	GFRP	219	438	3.7	58	18.52	548	68
281	[66]	GFRP	100	200	3.08	37	24.52	398	81

References

1. Kamgar, R.; Bagherinejad, M.H.; Heidarzadeh, H. A new formulation for prediction of the shear capacity of FRP in strengthened reinforced concrete beams. *Soft Comput.* **2019**. [CrossRef]
2. Wu, H.C.; Eamon, C.D. *Strengthening of Concrete Structures Using Fiber Reinforced Polymers (FRP): Design, Construction and Practical Applications*; Woodhead Publishing: Cambridge, MA, USA, 2017.
3. De Lorenzis, L.; Tepfers, R. Comparative study of models on confinement of concrete cylinders with fiber-reinforced polymer composites. *J. Compos. Constr.* **2003**, *7*, 219–237. [CrossRef]
4. Colomb, F.; Tobbi, H.; Ferrier, E.; Hamelin, P. Seismic retrofit of reinforced concrete short columns by CFRP materials. *Compos. Struct.* **2008**, *82*, 475–487. [CrossRef]
5. Niroomandi, A.; Maheri, A.; Maheri, M.R.; Mahini, S.S. Seismic performance of ordinary RC frames retrofitted at joints by FRP sheets. *Eng. Struct.* **2010**, *32*, 2326–2336. [CrossRef]
6. Ozcan, O.; Binici, B.; Ozcebe, G. Improving seismic performance of deficient reinforced concrete columns using carbon fiber-reinforced polymers. *Eng. Struct.* **2008**, *30*, 1632–1646. [CrossRef]
7. Promis, G.; Ferrier, E.; Hamelin, P. Effect of external FRP retrofitting on reinforced concrete short columns for seismic strengthening. *Compos. Struct.* **2009**, *88*, 367–379. [CrossRef]

8. Satasivam, S.; Bai, Y. Mechanical performance of modular FRP-steel composite beams for building construction. *Mater. Struct.* **2016**, *49*, 4113–4129. [CrossRef]
9. Shateri, M.; Ghaib, M.; Svecova, D.; Thomson, D. On acoustic emission for damage detection and failure prediction in fiber reinforced polymer rods using pattern recognition analysis. *Smart Mater. Struct.* **2017**, *26*, 065023. [CrossRef]
10. Zafar, A.; Andrawes, B. Incremental dynamic analysis of concrete moment resisting frames reinforced with shape memory composite bars. *Smart Mater. Struct.* **2012**, *21*, 025013. [CrossRef]
11. Cevik, A. Modeling strength enhancement of FRP confined concrete cylinders using soft computing. *Expert Syst. Appl.* **2011**, *38*, 5662–5673. [CrossRef]
12. Berthet, J.; Ferrier, E.; Hamelin, P. Compressive behavior of concrete externally confined by composite jackets: Part B: Modeling. *Constr. Build. Mater.* **2006**, *20*, 338–347. [CrossRef]
13. Lam, L.; Teng, J. Strength models for fiber-reinforced plastic-confined concrete. *J. Struct. Eng.* **2002**, *128*, 612–623. [CrossRef]
14. Li, Y.F.; Lin, C.T.; Sung, Y.Y. A constitutive model for concrete confined with carbon fiber reinforced plastics. *Mech. Mater.* **2003**, *35*, 603–619. [CrossRef]
15. Matthys, S.; Toutanji, H.; Audenaert, K.; Taerwe, L. Axial load behavior of large-scale columns confined with fiber-reinforced polymer composites. *ACI Struct. J.* **2005**, *102*, 258.
16. Miyauchi, K.; Inoue, S.; Kuroda, T.; Kobayashi, A. Strengthening effects of concrete column with carbon fiber sheet. *Trans. Jpn. Concr. Inst.* **2000**, *21*, 143–150.
17. Saafi, M.; Toutanji, H.; Li, Z. Behavior of concrete columns confined with fiber reinforced polymer tubes. *Mater. J.* **1999**, *96*, 500–509.
18. Vintzileou, E.; Panagiotidou, E. An empirical model for predicting the mechanical properties of FRP-confined concrete. *Constr. Build. Mater.* **2008**, *22*, 841–854. [CrossRef]
19. Xiao, Y.; Wu, H. Compressive behavior of concrete confined by carbon fiber composite jackets. *J. Mater. Civ. Eng.* **2000**, *12*, 139–146. [CrossRef]
20. Richart, F.E.; Brandtzaeg, A.; Brown, R.L. *A Study of the Failure of Concrete under Combined Compressive Stresses*; Bulletin No. 185; University of Illinois at Urbana Champaign, College of Engineering, Engineering Experimental Station: Champaign, IL, USA, 1928.
21. Cevik, A.; Cabalar, A.F. A genetic-programming-based formulation for the strength enhancement of fiber-reinforced-polymer-confined concrete cylinders. *J. Appl. Polym. Sci.* **2008**, *110*, 3087–3095. [CrossRef]
22. Cevik, A.; Göğüş, M.T.; Güzelbey, İ.H.; Filiz, H. Soft computing based formulation for strength enhancement of CFRP confined concrete cylinders. *Adv. Eng. Softw.* **2010**, *41*, 527–536. [CrossRef]
23. Cevik, A.; Guzelbey, I.H. Neural network modeling of strength enhancement for CFRP confined concrete cylinders. *Build. Environ.* **2008**, *43*, 751–763. [CrossRef]
24. Elsanadedy, H.M.; Al-Salloum, Y.A.; Abbas, H.; Alsayed, S.H. Prediction of strength parameters of FRP-confined concrete. *Compos. Part B Eng.* **2012**, *43*, 228–239. [CrossRef]
25. Mohammad Rezapour Tabari, M.; Nasr Azadani, M.; Kamgar, R. Development of operation multi-objective model of dam reservoir under conditions of temperature variation and loading using NSGA-II and DANN models: A case study of Karaj /Amir Kabir dam. *Soft Comput.* **2020**. [CrossRef]
26. Ghaleini, E.N.; Koopialipoor, M.; Momenzadeh, M.; Sarafraz, M.E.; Mohamad, E.T.; Gordan, B. A combination of artificial bee colony and neural network for approximating the safety factor of retaining walls. *Eng. Comput.* **2019**, *35*, 647–658. [CrossRef]
27. Hüsken, M.; Jin, Y.; Sendhoff, B. Structure optimization of neural networks for evolutionary design optimization. *Soft Comput.* **2005**, *9*, 21–28. [CrossRef]
28. Jalal, M.; Ramezanianpour, A.A. Strength enhancement modeling of concrete cylinders confined with CFRP composites using artificial neural networks. *Compos. Part B Eng.* **2012**, *43*, 2990–3000. [CrossRef]
29. Kamgar, R.; Hatefi, S.M.; Majidi, N. A fuzzy inference system in constructional engineering projects to evaluate the design codes for RC buildings. *Civ. Eng. J.* **2018**, *4*, 2155–2172. [CrossRef]
30. Keshtegar, B.; Ozbakkaloglu, T.; Gholampour, A. Modeling the behavior of FRP-confined concrete using dynamic harmony search algorithm. *Eng. Comput.* **2017**, *33*, 415–430. [CrossRef]
31. Moosazadeh, S.; Namazi, E.; Aghababaei, H.; Marto, A.; Mohamad, H.; Hajihassani, M. Prediction of building damage induced by tunnelling through an optimized artificial neural network. *Eng. Comput.* **2019**, *35*, 579–591. [CrossRef]

32. Najafzadeh, M.; Azamathulla, H.M. Neuro-fuzzy GMDH to predict the scour pile groups due to waves. *J. Comput. Civ. Eng.* **2013**, *29*, 04014068. [CrossRef]
33. Najafzadeh, M.; Etemad-Shahidi, A.; Lim, S.Y. Scour prediction in long contractions using ANFIS and SVM. *Ocean Eng.* **2016**, *111*, 128–135. [CrossRef]
34. Najafzadeh, M.; Lim, S.Y. Application of improved neuro-fuzzy GMDH to predict scour depth at sluice gates. *Earth Sci. Inform.* **2015**, *8*, 187–196. [CrossRef]
35. Najafzadeh, M.; Sattar, A.M. Neuro-fuzzy GMDH approach to predict longitudinal dispersion in water networks. *Water Resour. Manag.* **2015**, *29*, 2205–2219. [CrossRef]
36. Najafzadeh, M.; Tafarojnoruz, A. Evaluation of neuro-fuzzy GMDH-based particle swarm optimization to predict longitudinal dispersion coefficient in rivers. *Environ. Earth Sci.* **2016**, *75*, 157–169. [CrossRef]
37. Siemaszko, A.; Jakubczyk-Gałczyńska, A.; Jankowski, R. The idea of using Bayesian networks in forecasting impact of traffic-induced vibrations transmitted through the ground on residential buildings. *Geosciences* **2019**, *9*, 339. [CrossRef]
38. Khatami, S.M.; Naderpour, H.; Barros, R.C.; Jakubczyk-Gałczyńska, A.; Jankowski, R. Effective formula for impact damping ratio for simulation of earthquake-induced structural pounding. *Geosciences* **2019**, *9*, 347. [CrossRef]
39. Khatami, S.M.; Naderpour, H.; Barros, R.C.; Jakubczyk-Gałczyńska, A.; Jankowski, R. Determination of peak impact force for buildings exposed to structural pounding during earthquakes. *Geosciences* **2020**, *10*, 18. [CrossRef]
40. Kamgar, R.; Gholami, F.; Zarif Sanayei, H.R.; Heidarzadeh, H. Modified tuned liquid dampers for seismic protection of buildings considering soil-structure interaction effects. *Iran. J. Sci. Technol. Trans. Civ. Eng.* **2020**, *44*, 339–354. [CrossRef]
41. Kamgar, R.; Askari Dolatabad, Y.; Babadaei Samani, M.R. Seismic optimization of steel shear wall using shape memory alloy. *Int. J. Optim. Civ. Eng.* **2019**, *9*, 671–687.
42. Lee, S.; Lee, C. Prediction of shear strength of FRP-reinforced concrete flexural members without stirrups using artificial neural networks. *Eng. Struct.* **2014**, *61*, 99–112. [CrossRef]
43. Sobhani, J.; Najimi, M.; Pourkhorshidi, A.R.; Parhizkar, T. Prediction of the compressive strength of no-slump concrete: A comparative study of regression, neural network and ANFIS models. *Constr. Build. Mater.* **2010**, *24*, 709–718. [CrossRef]
44. Cheng, M.Y.; Cao, M.T. Evolutionary multivariate adaptive regression splines for estimating shear strength in reinforced-concrete deep beams. *Eng. Appl. Artif. Intell.* **2014**, *28*, 86–96. [CrossRef]
45. Behnood, A.; Olek, J.; Glinicki, M.A. Predicting modulus elasticity of recycled aggregate concrete using M5′ model tree algorithm. *Constr. Build. Mater.* **2015**, *94*, 137–147. [CrossRef]
46. Ebrahimpour Komleh, H.; Maghsoudi, A.A. Prediction of curvature ductility factor for FRP strengthened RHSC beams using ANFIS and regression models. *Comput. Concr.* **2015**, *16*, 399–414. [CrossRef]
47. Gu, Z.Q.; Oyadiji, S.O. Application of MR damper in structural control using ANFIS method. *Comput. Struct.* **2008**, *86*, 427–436. [CrossRef]
48. Amani, J.; Moeini, R. Prediction of shear strength of reinforced concrete beams using adaptive neuro-fuzzy inference system and artificial neural network. *Sci. Iran.* **2012**, *19*, 242–248. [CrossRef]
49. Mashrei, M.A.; Seracino, R.; Rahman, M. Application of artificial neural networks to predict the bond strength of FRP-to-concrete joints. *Constr. Build. Mater.* **2013**, *40*, 812–821. [CrossRef]
50. Mohammadhassani, M.; Nezamabadi-Pour, H.; Jumaat, M.; Jameel, M.; Hakim, S.J.S.; Zargar, M. Application of the ANFIS model in deflection prediction of concrete deep beam. *Struct. Eng. Mech.* **2013**, *45*, 319–332. [CrossRef]
51. Nehdi, M.; Nikopour, H. Genetic algorithm model for shear capacity of RC beams reinforced with externally bonded FRP. *Mater. Struct.* **2011**, *44*, 1249–1258. [CrossRef]
52. Li, Y.L.; Teng, J.G.; Zhao, X.L.; Singh Raman, R.K. Theoretical model for seawater and sea sand concrete-filled circular FRP tubular stub columns under axial compression. *Eng. Struct.* **2018**, *160*, 71–84. [CrossRef]
53. Li, P.; Wu, Y.F.; Gravina, R. Cyclic response of FRP-confined concrete with post-peak strain softening behavior. *Constr. Build. Mater.* **2016**, *123*, 814–828. [CrossRef]
54. Zhou, A.; Qin, R.; Chow, C.L.; Lau, D. Structural performance of FRP confined seawater concrete columns under chloride environment. *Compos. Struct.* **2019**, *216*, 12–19. [CrossRef]

55. Mansouri, I.; Ozbakkaloglu, T.; Kisi, O.; Xie, T. Predicting behavior of FRP-confined concrete using neuro-fuzzy, neural network, multivariate adaptive regression splines and M5 model tree techniques. *Mater. Struct.* **2016**, *49*, 4319–4334. [CrossRef]
56. Naderpour, H.; Kheyroddin, A.; Amiri, G.G. Prediction of FRP-confined compressive strength of concrete using artificial neural networks. *Compos. Struct.* **2010**, *92*, 2817–2829. [CrossRef]
57. Spoelstra, M.R.; Monti, G. FRP-confined concrete model. *J. Compos. Constr.* **1999**, *3*, 143–150. [CrossRef]
58. Toutanji, H. Stress-strain characteristics of concrete columns externally confined with advanced fiber composite sheets. *Mater. J.* **1999**, *96*, 397–404.
59. Samaan, M.; Mirmiran, A.; Shahawy, M. Model of concrete confined by fiber composites. *J. Struct. Eng.* **1998**, *124*, 1025–1031. [CrossRef]
60. Kono, S.; Inazumi, M.; Kaku, T. Evaluation of confining effects of CFRP sheets on reinforced concrete members. In Proceedings of the Second International Conference on Composites in Infrastructure National Science Foundation, Tucson, AZ, USA, 5–7 January 1998.
61. Karbhari, V.M.; Gao, Y. Composite jacketed concrete under uniaxial compression-Verification of simple design equations. *J. Mater. Civ. Eng.* **1997**, *9*, 185–193. [CrossRef]
62. Mander, J.B.; Priestley, M.J.; Park, R. Theoretical stress-strain model for confined concrete. *J. Struct. Eng.* **1988**, *114*, 1804–1826. [CrossRef]
63. Fardis, M.N.; Khalili, H. Concrete encased in fiberglass-reinforced plastic. *J. Am. Concr. Inst.* **1981**, *78*, 440–446.
64. Fardis, M.N.; Khalili, H.H. FRP-encased concrete as a structural material. *Mag. Concr. Res.* **1982**, *34*, 191–202. [CrossRef]
65. Ahmad, S.; Khaloot, A.; Irshaid, A. Behaviour of concrete spirally confined by fibreglass filaments. *Mag. Concr. Res.* **1991**, *43*, 143–148. [CrossRef]
66. Fam, A.Z.; Rizkalla, S.H. Concrete-filled FRP tubes for flexural and axial compression members. In Proceedings of the Third International Conference on Advanced Composite Materials in Bridges and Structures, Ottawa, ON, Canada, 15–18 August 2000.
67. Harmon, T.G.; Slattery, K.T. Advanced composite confinement of concrete. In Proceedings of the First International Conference on Advanced Composite Materials in Bridges and Structures, Sherbrooke, QC, Canada, 7–9 October 1992.
68. La Tegola, A.; Manni, O. Experimental investigation on concrete confined by fiber reinforced polymer and comparison with theoretical model. *Spec. Publ.* **1999**, *188*, 243–254.
69. Matthys, S.; Taerwe, L.; Audenaert, K. Tests on axially loaded concrete columns confined by fiber reinforced polymer sheet wrapping. *Spec. Publ.* **1999**, *188*, 217–228.
70. Micelli, F.; Myers, J.; Murthy, S. Effect of environmental cycles on concrete cylinders confined with FRP. In Proceedings of the CCC2001 International Conference on Composites in Construction, Porto, Portugal, 10–12 October 2001.
71. Mirmiran, A.; Shahawy, M. Behavior of concrete columns confined by fiber composites. *J. Struct. Eng.* **1997**, *123*, 583–590. [CrossRef]
72. Miyauchi, K. Estimation of strengthening effects with carbon fiber sheet for concrete column. In Proceedings of the 3rd International Symposium on Non-Metallic (FRP) Reinforcement for Concrete Structures, Sapporo, Japan, 14–16 October 1997.
73. Nanni, A.; Bradford, N.M. FRP jacketed concrete under uniaxial compression. *Constr. Build. Mater.* **1995**, *9*, 115–124. [CrossRef]
74. Picher, F.; Rochette, P.; Labossiére, P. Confinement of concrete cylinders with CFRP. In Proceedings of the International Conference on Composites in Infrastructure, Tuscon, AZ, USA, 15–17 January 1996.
75. Rochette, P.; Labossiere, P. Axial testing of rectangular column models confined with composites. *J. Compos. Constr.* **2000**, *4*, 129–136. [CrossRef]
76. Rousakis, T.; Tepfers, R. *Experimental Investigation of Concrete Cylinders Confined by Carbon FRP Sheets, under Monotonic and Cyclic Axial Compressive Load*; Chalmers University of Technology: Goteborg, Sweden, 2001.
77. Shahawy, M.; Mirmiran, A.; Beitelman, T. Tests and modeling of carbon-wrapped concrete columns. *Compos. Part B Eng.* **2000**, *31*, 471–480. [CrossRef]

78. Watanabe, K.; Nakamura, H.; Honda, Y.; Toyoshima, M. Confinement effect of FRP sheet on strength and ductility of concrete cylinders under uniaxial compression. In Proceedings of the Third International Symposium (FRPRCS-3) on Non-Metallic (FRP) Reinforcement for Concrete Structures, Sapporo, Japan, 14–16 October 1997; pp. 233–240.
79. Rojas, R. *Neural Networks: A Systematic Introduction*; Springer: Berlin/Heidelberg, Germany, 2013.
80. Dehkordi, A.N.; Kamali-Asl, A.; Wen, N.; Mikkelsen, T.; Chetty, I.J.; Bagher-Ebadian, H. DCE-MRI prediction of survival time for patients with glioblastoma multiforme: Using an adaptive neuro-fuzzy-based model and nested model selection technique. *NMR Biomed.* **2017**, *30*, e3739. [CrossRef]
81. Huang, X. *Study on ANN Noise Adaptability in Application of Industry Process Characteristics Mining*; Advanced Materials Research; Trans Tech Publications Ltd.: Stafa Zurich, Switzerland, 2012; Volume 462, pp. 635–640.

applied sciences

MDPI

Article

Temperature Impact on the Assessment of Reinforcement Corrosion Risk in Concrete by Galvanostatic Pulse Method

Wioletta Raczkiewicz * and Artur Wójcicki

Kielce University of Technology, Faculty of Civil Engineering and Architecture, Al. Tysiąclecia Państwa Polskiego 7, 25-314 Kielce, Poland; arturw@tu.kielce.pl

* Correspondence: wiolar@tu.kielce.pl; Tel.: +48-4134-24-582

Received: 10 December 2019; Accepted: 1 February 2020; Published: 6 February 2020

Abstract: The electrochemical galvanostatic pulse method (GPM) is used for the evaluation of the degree of corrosion risk of reinforcement in concrete. This non-destructive method enables determining the corrosion promoting conditions through the measurements of reinforcement stationary potential and concrete cover resistivity, and determining the probability of reinforcement corrosion in the tested areas. This method also allows for the estimation of the reinforcement corrosion activity and the prediction of the development of the corrosion process on the basis of corrosion current density measurements. The ambient temperature (and the temperature of the examined element) can significantly affect the values of the measured parameters due to electrochemical character of the processes as well as specific measurement technique. Differences in the obtained results can lead to a wrong interpretation of reinforcement corrosion risk degree in concrete. The article attempts to assess the effect of temperature on the measured parameters while using the galvanostatic pulse method. The GP-5000 GalvaPulse^{TM} set was used. The results of this study confirmed the impact of temperature changes on the values of three measured parameters (reinforcement stationary potential, concrete cover resistivity, and corrosion current density) and contributed to catching the trend of these changes.

Keywords: reinforced concrete diagnostics; non-destructive method; galvanostatic pulse method; reinforcement corrosion; temperature impact

---

1. Introduction

Durability is one of the main requirements for building structures [1–5]. In the steel and reinforced concrete structures, the steel corrosion is a direct factor that affects durability of concrete [2–7]. While in steel structures, corrosion protective coatings are usually visible, and the corrosion process itself is relatively easy to notice and evaluate, in reinforced concrete structures, the corrosion of the reinforcement will be undetected on the surface of concrete for many years, thus causing significant destruction and making corrosion assessment more difficult. However, there is a group of non-destructive methods that allow, in probabilistic terms, diagnosing the process of reinforcement corrosion in a structural element [6–12]. These are electrochemical methods that utilize the physico-chemical properties of concrete and steel. Small gel pores, larger capillary pores, and macro-pores form the porous structure of concrete, which form a system of interconnected channels that are filled with ion-carrying liquid. Steel reinforcing bars, on the other hand, are electron carriers. The flow of electrons between the existing local anode and cathode microcells (resulting from microdefects in steel) on the bar surface and the ion flow in the liquid filling the pores of concrete produce a kind of galvanic cell. The alkaline liquid in the pores is an electrolyte and the reinforcing bar is an electrode [6,13–15]. Reinforcement corrosion begins in favourable conditions, when the passive layer protecting the bar surface is damaged (usually as a result of concrete carbonation or action of chlorides), the pores are filled with liquid, and oxygen is available. The iron oxidation process occurs at the anode:

$$2Fe \rightarrow 2Fe^{2+} + 4e^{-}$$

and the reduction process occurs at the cathode:

$$2H_2O + O_2 + 4e^{-} \rightarrow 4OH^{-}.$$

Figure 1 schematically this.

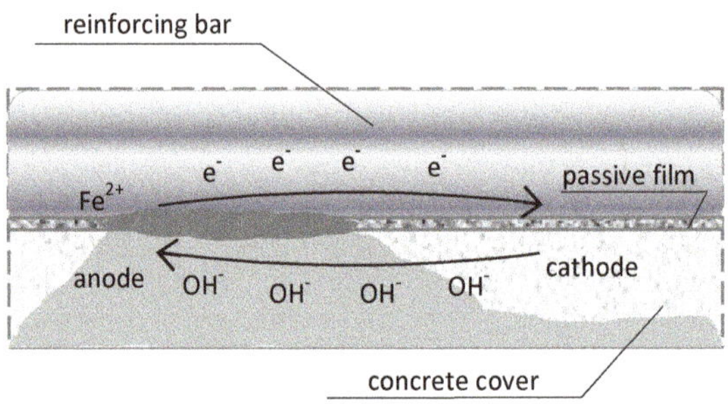

Figure 1. Reinforcement corrosion process in concrete.

In the course of the corrosion process, iron ions react with OH^{-}; ions, forming iron hydroxide:

$$Fe^{2+} + 2OH^{-} \rightarrow Fe(OH)_2$$

and then in the presence of oxygen:

$$2Fe(OH)_2 + 1/2O_2 \rightarrow 2FeOOH + H_2O.$$

As a result of the process, iron hydroxide is formed $Fe(OH)_3$ and iron oxide Fe_2O_3 are produced. The corrosion product, rust, is a mixture of $Fe(OH)_3$, Fe_2O_3, Fe_2O_3, H_2O, and magnetite [2,6,13,14]. The degree of reinforcing bar corrosion can be determined from Equation (1) by calculating the corrosion loss that is based on Faraday's law:

$$\Delta m = k \times i \times t \quad (1)$$

where:

Δm–loss of weight [g],
K–electrochemical equivalent [g/A × s]
i–corrosive current strength [A], and
t–current flow time [s].

The corrosive current can be determined from Equation (2):

$$i = \frac{\Delta E}{R_K + R_A + R_O + R_Z} \quad (2)$$

where:

ΔE–anode and cathode potential difference,
R_K, R_A–cathode and anode polarization resistance, and
R_O, R_Z–internal and external circuit resistance.

Equation (2) indicates that the moisture level-dependent internal resistance that inhibits ion transport is important. The diffusion is practically stopped at the moisture content below 75%.

The temperature of the tested element also affects the ion diffusion coefficient [2,13]. The diffusion coefficient increases with the increase in temperature (the increase rate depends on the concrete composition), as shown by the Arrhenius equation:

$$D = D_0 e^{-\frac{E}{KT}} \tag{3}$$

where:

D–diffusion coefficient at temperature T [cm^2/s],

D_0–diffusion coefficient at temperature T = 0,

E–activation energy of the diffusion process [kJ/mol] (dependent on w/c and type of cement),

R–gas constant (R = 8.3144598(48) [J/mol × K]), and

T–absolute temperature [K].

As a result of the electrochemical process of reinforcement corrosion in concrete, certain electrical quantities change. Stationary potential, concrete resistivity, or corrosion current density can be measured and used for the estimation of the extent of reinforcement corrosion. Measurements are made while using specialized devices and the results are analyzed against the criteria values pre-determined for a given device [6,12,16–32]

The galvanostatic pulse method (GPM) is one of the methods used for this purpose [21–32]. It is a non-destructive polarization technique that allows for measurements of the reinforcement stationary potential on the concrete surface (E_{st}), concrete cover resistivity (Θ), and corrosion current density (i_{cor}). The measurements of the reinforcement stationary potential and the concrete cover resistivity are called basic measurements. It should be noted that the values of the reinforcement stationary potential and the concrete cover resistivity are only estimates of the probability of reinforcement corrosion risk in the investigated areas, and the measurement results are not meaningful. Corrosive current density data, on the other hand, allow for the estimation of corrosion activity in the reinforcement and prediction of its rate over time. The dynamic equilibrium on the electrode (reinforcing bar) immersed in the electrolyte (alkaline liquid filling the pores) has to be disturbed (polarization of the reinforcement) to conduct the measurements. In the case of the galvanostatic pulse method, a current of certain intensity generates such a disturbance.

External factors, including electrolyte temperature, have some influence on the intensity of the process due to the fact that ions flow as a result of diffusion in the pore liquid, and thus also on the measured electrical values. As reported by Neville [2], the diffusion coefficient (D) increases as the electrolyte temperature increases (3), which is directly related to the ion flow rate in the electrolyte and the measurement of this flow. In addition, the increase in temperature accelerates depolarization, which is important when galvanostatic pulse measurements are used. Although the galvanostatic pulse method has been used for several years [21–31], few authors have evaluated the impact of temperature on the measurement results [20,23,25,28,30,31]. Most research has been devoted to the impact of environmental conditions, i.e., both temperature and moisture content. Little or no information regarding the possible correlation between the temperature of the tested element and the obtained measurement results can be found in the literature, which suggests that it is not taken into account.

The authors' previous research [21,25,28] prompted the thesis that the temperature of the tested element has a significant impact on the values of the measured parameters, which is also indicated by the physico-chemical nature of the reinforcement corrosion process in concrete. The results of measurements that were made on the same laboratory specimen, both outdoors in winter (in temperature of about 0 °C) and indoors at room temperature (about 23 °C), varied noticeably [25,28]. Similar observations of differences in the corrosion rates between September 2000 and April 2001 were reported in [23]. For this reason, tests were carried out on a larger number of samples under gradually changed thermal conditions. The purpose of the work was to determine an influence of the temperature of the tested element on the measurement results of three parameters: the corrosion current density, the reinforcement stationary potential, and the concrete cover resistivity. The influence of temperature on the results of potential and resistivity measurements may result in incorrect marking of areas for advanced measurements and, as a consequence, disturb corrosion risk assessment, although measurements of stationary reinforcement potential and concrete cover resistivity are less important than measurements of corrosive current density in general GPM tests. Analysis of the values of these parameters allows for the estimation of actual steel reinforcement corrosion degree.

2. Measuring Device

The GP-5000 GalvaPulse™ set is one of the devices designed to measure stationary reinforcement potential and concrete cover resistivity (basic measurements) or, additionally, corrosion current density (advanced measurements) whileusing the polarization method [32]. The half-cell potential is measured to an accuracy of ± 5 mV with the Ag/AgCl electrode. The electrical resistance is estimated to be measured with an accuracy of ± 5% [32]. The main elements of the set include the control and recording device (PSION minicomputer), silver-chloride reference electrode (Ag/AgCl), and calibration device (Figure 2).

Figure 2. The GP-5000 GalvaPulse™ set.

The GP-5000 GalvaPulse™ can be used in both laboratory and field conditions. Measurements are made on the surface of the element at the points evenly distributed above the reinforcement. This allows for creating graphical maps of the distribution of measured parameters and facilitates further comprehensive analysis of the results, especially for large-size elements. For the measurements, the reinforcing bar (local hole is required) is connected with the calibrated control and recording device (PSION minicomputer) and with the reference electrode.

Before the measurements, the location of the tested bar is determined and its continuity along the tested section is checked. Where the electrode is applied, the concrete surface must be properly cleaned and wetted. The recommended concrete resistivity should be no more than 50 kΩ·cm. Changes in the moisture content in concrete result in resistivity changes, which, at a given constant current strength, could lead to incorrect measurements of the reinforcement stationary potential. Therefore, the moisture concentration on the tested surface should be maintained as constant throughout the measurement period. Figure 3 shows the connection scheme of the GP-5000 GalvaPulse™ set with the tested reinforcement.

Figure 3. The connection scheme of the GP-5000 GalvaPulse™ set with the tested reinforcement.

The measuring device must be properly calibrated by entering the coordinates of the measurement points on the test surface, pulse duration (5 ÷ 20 s), current strength (5 ÷ 400 μA; a higher value is expected for bigger quantity of corrosion products), reinforcing bar parameters (diameter, length, surface area), and by activating (if necessary) the ring that limits the electrode area of operation.

Information on the criteria for the interpretation of the test results is attached to the apparatus (Table 1). It is possible to infer about the probability of reinforcement corrosion in the examined area, and the corrosion process rate can be estimated based on the value of the corrosion current density, depending on the obtained values of the reinforcement stationary potential and the concrete cover resistivity. Table 1 summarizes the appropriate criteria for measurements performed while using a GP-5000 GalvaPulse™ set. The reference values should not be compared with the measurements that were obtained with other devices [6,32].

Table 1. Criteria for assessing the reinforcement corrosion risk level [32].

Criteria for Assessing the Degree of Reinforcement Corrosion Risk by use a GP-5000 GalvaPulse™ Set			
	Reinforcement Corrosion Activity, i_{cor}[μA/cm^2]		**Corrosion Rate, [mm/year]**
Corrosion current density	<0.5	not forecasted corrosion activity	<0.006
	0.5 ÷ 2.0	irrelevant corrosion activity	0.006 ÷ 0.023
	2.0 ÷ 5.0	low corrosion activity	0.023 ÷ 0.058
	5.0 ÷ 15.0	moderate corrosion activity	0.058 ÷ 0.174
	>15.0	high corrosion activity	>0.174
Reinforcement stationary potential, E_{st} [mV] vs. Ag/AgCl	<−350	95% of corrosion probability	
	−350 ÷ −200	50% of corrosion probability	
	>−200	5% of corrosion probability	
Concrete cover resistivity, Θ [kΩ·cm]	≤10	high corrosion probability	
	10 ÷ 20	medium corrosion probability	
	≥20	small corrosion probability	

3. Research Methodology and Material

Seven rectangular specimens with dimensions 210 × 228 × 100 mm were cast. The specimens were made from the same concrete mixture (C40/45) under identical laboratory conditions. The following quantities of constituents were assumed for 1 m^3 of the mixture: cement (CEM I 52.5) – 390 kg, sand – 660 kg, aggregate 2 ÷ 8 – 617 kg, aggregate 8 ÷ 16 – 694 kg, water – 155 L, plasticizer (1.84 g), and air-entraining admixture (0.47 g). Two parallel ribbed bars with a diameter of 8 mm that were made of BST 500 steel were placed in each specimen 70 mm from the side edges and 25 mm from the upper specimen surface – the cover (Figure 4). The specimens were stored at a temperature of 20 °C ± 2 °C, and relative humidity of 50% ± 5%.

Figure 4. The test specimen.

An orthogonal grid of four measuring points was established for measurement purposes, being spaced evenly every 70 mm on each specimen (two points above each bar) in which reinforcement stationary potential, concrete cover resistivity and corrosion current density were measured (Figure 5). The measurements were carried out in accordance with the requirements of the GP-5000 GalvaPulse^{TM} use. The obtained results were archived in the created database.

Figure 5. Photo of measurements carried out on one of the tested specimens.

The test temperatures were varied and they included 1 °C, 5 °C, 8 °C, 10 °C, 15 °C, 18 °C, and 21 °C. The choice of temperature range resulted from the intention to reflect the real conditions in which in-situ research is usually carried out. Measurements were not possible at minus temperatures due to the specifics of the measurement technique, i.e., the use of the electrolysis process and the need for strong hydration of the concrete surface in the studied area. The temperature of specimen surface was measured every time at three points while using a non-contact infrared thermometer with a range of −50 °C ÷ 380 °C and tolerance of +/−0.5 °C.

4. Research Results and the Analysis

The data sets of the measured parameters (reinforcement stationary potential-Est, concrete cover resistivity-Θ, and corrosion current density-i_{cor}) as a function of temperature were produced, as shown in Figure 6 (measured values) and Figure 7 (relative values referred to the values measured at T = 1 °C). The graphs show the relation between the obtained results and the specimen temperature.

Analysis of the values of parameters determining the probability of reinforcement corrosion in the examined region, i.e., reinforcement stationary potential (Est) and concrete cover resistivity (Θ), showed that the obtained values were in the ranges: $E_{st} = -212 \div 295$ mV and $\Theta = 5.3 \div 57.5$ kΩ·cm. As described earlier, the measurements were made on the same specimens at the same measuring points, but at varied temperature values. While the values of reinforcement stationary potential at almost all assumed temperatures were less than −200 mV, which indicated the same 5% probability of reinforcement corrosion in the tested region (although with a wide spread of results), the values of concrete cover resistivity exhibited relatively high variability and, thus, led to significantly different conclusions about the corrosion probability-from low to high. Thus the temperature of the concrete cover has a large impact on its resistivity. The results of the third measured parameter, i.e., corrosion current density, were in the range of irrelevant corrosion activity, $i_{cor} = 0.18 \div 1.0$ μA/cm^2, which corresponded to the real condition of the tested specimens.

In addition, at a given measurement point for a given parameter, the changes in its value versus temperature were quite similar. This indicates a constant (though not equally significant) effect of specimen temperature on the obtained values of the measured parameter. The values were averaged in order to discover the global trend of changes in the value of each parameter, as shown in Figure 6. The plot of average values is marked with a red thick line, whereas the black thin line denotes the approximating function in the form of a three-degree polynomial.

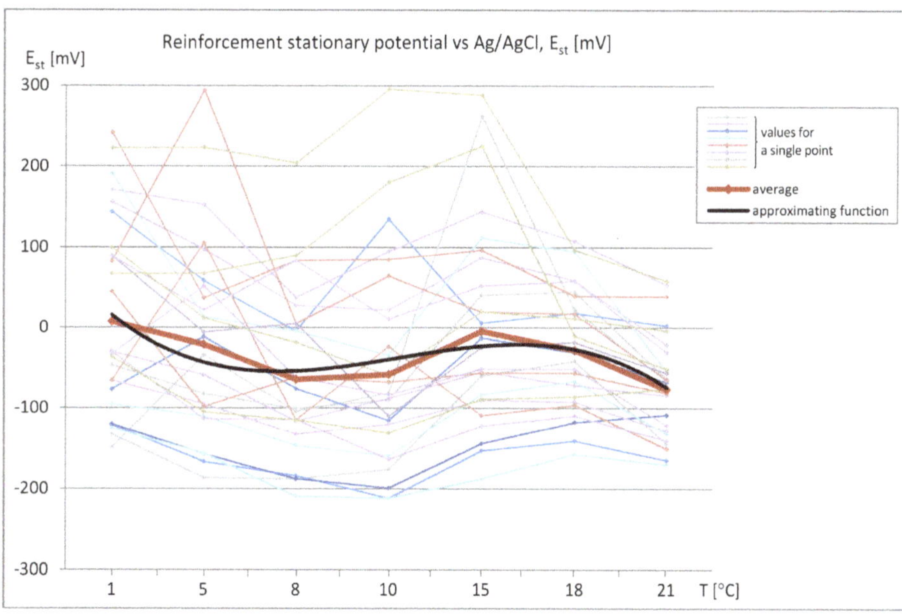

(a)

Figure 6. *Cont.*

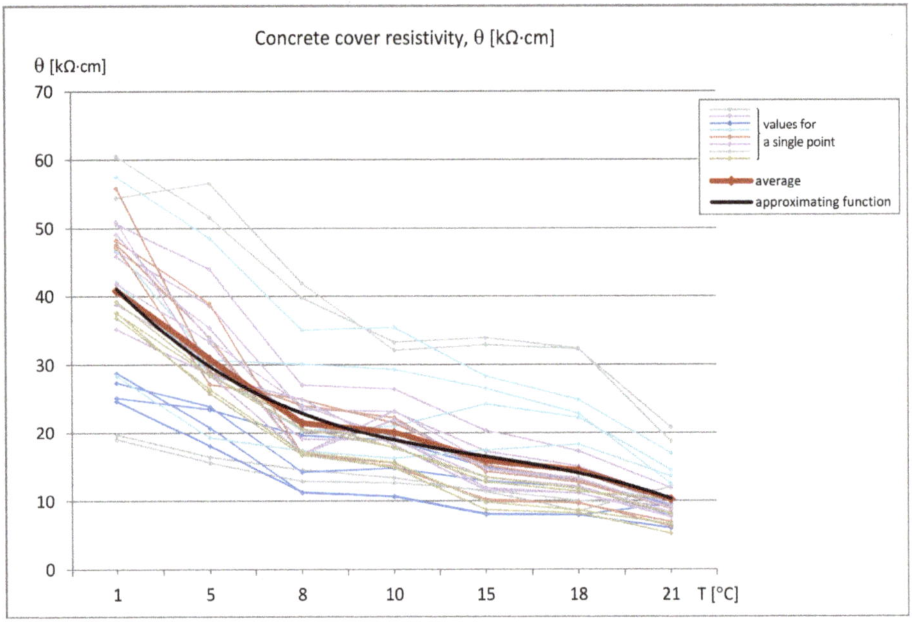

(b)

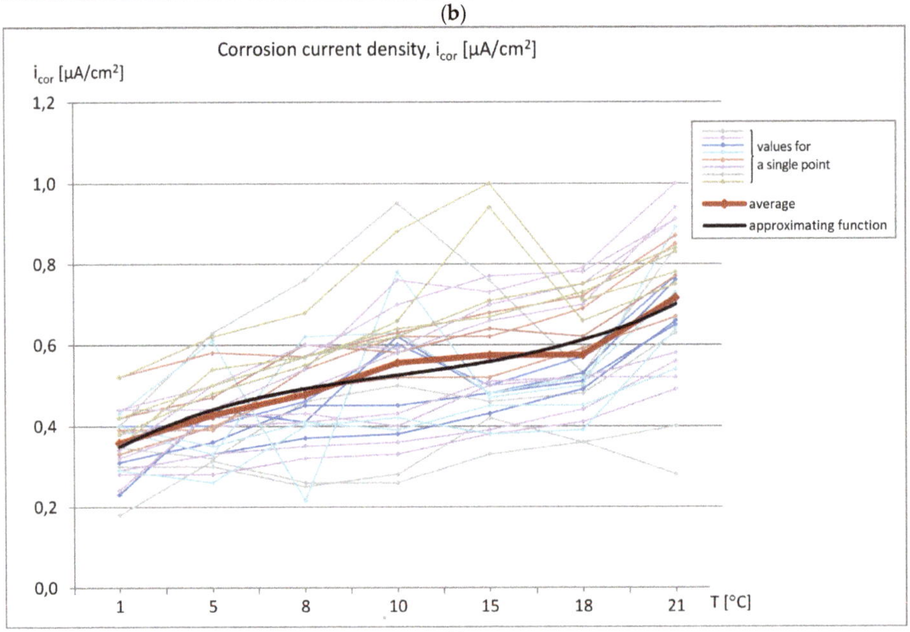

(c)

Figure 6. Parameter values measured on specimens at various temperatures; (**a**) reinforcement stationary potential, (**b**) concrete cover resistivity, (**c**) corrosion current density.

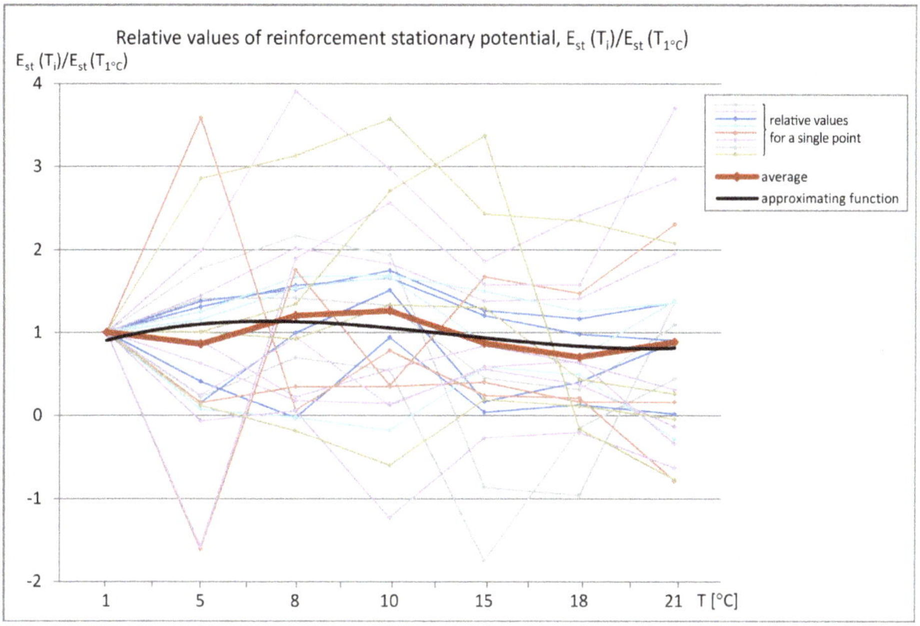

(a)

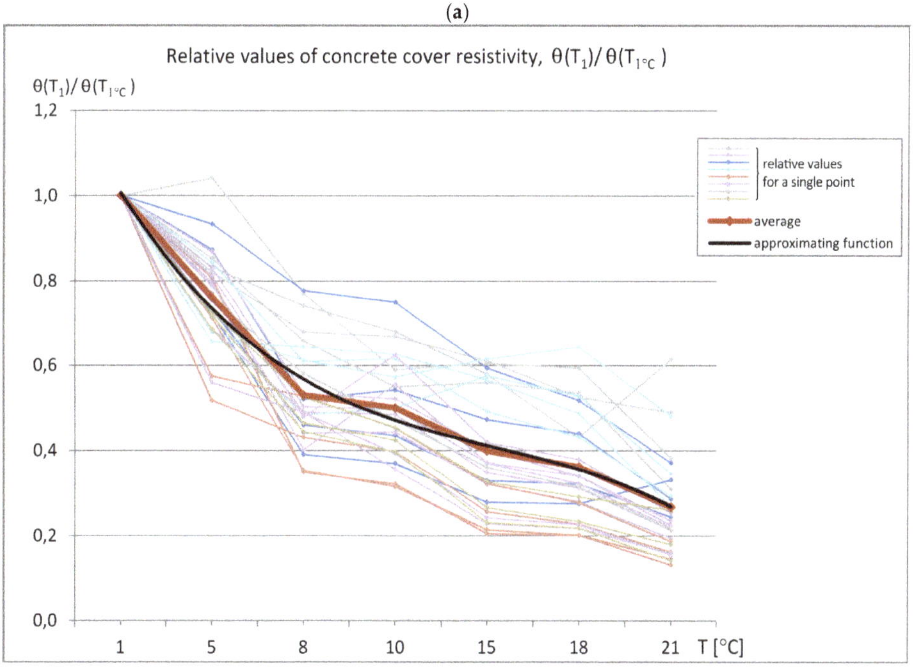

(b)

Figure 7. *Cont.*

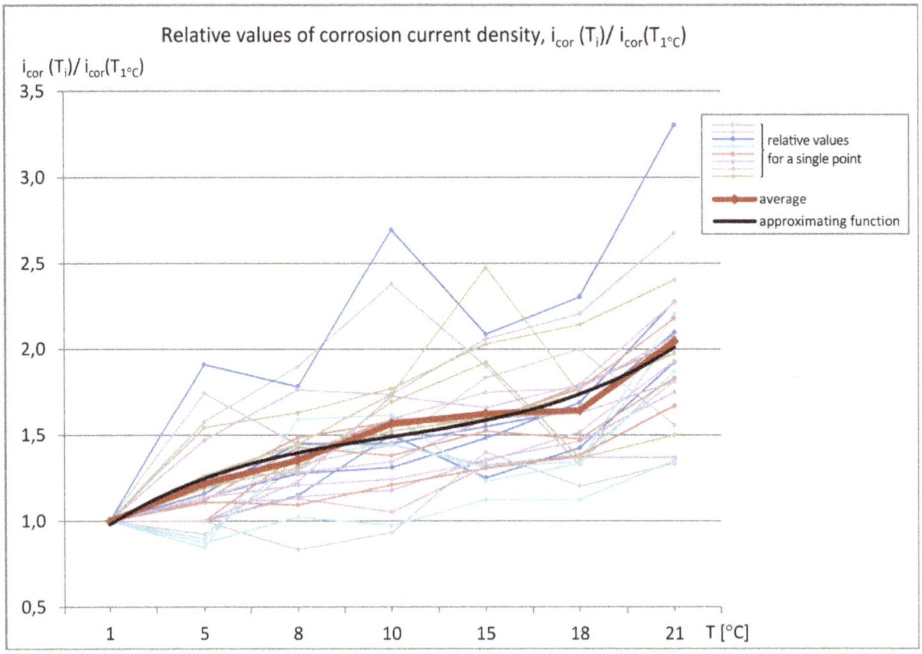

(**c**)

Figure 7. Relative values of measured parameters on samples at different temperatures; (**a**) reinforcement stationary potential, (**b**) concrete cover resistivity, (**c**) corrosion current density.

The average value of the reinforcement stationary potential noticeably dropped with the increase in temperature of the tested specimen. The plot of the absolute value changes ranged from $E_{st} \approx 0$ mV at T = 1 °C to $E_{st} \approx -77$ mV at T = 21 °C. These changes are not strictly monotonic and they are in the range of insignificant values (Figure 6a).

The changes in the concrete cover resistivity values were more explicit and unidirectional. The average value decreased from $\Theta \approx 41$ kΩ·cm at T = 1 °C to $\Theta \approx 10$ kΩ·cm at T = 21 °C. The graph does not differ significantly from the linear one (Figure 6b).

Changes in the average corrosion current density values were also proportional to changes in specimen temperature and ranged from $i_{cor} \approx 0.36$ μA/cm^2 at T = 1 °C to $i_{cor} \approx 0.72$ μA/cm^2 at T = 21 °C.

To assess changes in the values of the measured parameters (E_{st}, i_{cor}, Θ) obtained at different temperatures, the plots of relative values were prepared. All of the values were referred to those that were obtained at T = 1 °C chosen in this case as the reference temperature. This approach gives the possibility to correct the results in the case of need to compare them with the results that were obtained during measurements performed in different conditions, e.g., in different seasons.

The observed relatively small changes in the average relative values of the reinforcement stationary potential E_{st} are visible in Figure 7a. The positive and negative values of this parameter disturb general analysis of the relative values. It should be remembered that, in in-situ studies, only negative values represent a reliable and significant range of results. Values that are greater than $E_{st} = -200$ mV indicate very low corrosion probability. Therefore, the trend of changes as a function of temperature was verified on specimens with advanced reinforcement corrosion and presented in a separate publication.

The average value of the concrete cover resistivity (Θ) at higher temperatures clearly decreases down to 25% of the value at T = 1 °C. As the changes are significant, the inference about the corrosion risk of the tested element will be different.

The most important factor enabling the determination of the reinforcement corrosion activity, i.e., the corrosion current density (i_{cor}) changed by ~100% at the temperature T = 21 °C in relation to the reference temperature (i.e., T = 1 °C). While the absolute values, regardless of the temperature, were insignificant and indicated low reinforcement corrosion activity, the observed relative changes of the i_{cor} values in relation to those that were measured at the reference temperature are considerable. Therefore, the prediction of the reinforcement corrosion activity requires taking into account the temperature of the element being tested. Otherwise, the measurement will be inaccurate.

There were extreme differences in the relative values of the measured parameters at individual points in relation to the average values at these points: the reinforcement stationary potential - Est ± 200 ÷ 500 %, the concrete cover resistivity - Θ ± 40 ÷ 85%, and the corrosion current density - i_{cor} ± 25 ÷ 300%.

5. Conclusions

The tests confirmed the effect of temperature on the assessment of reinforcement corrosion risk in concrete by the galvanostatic pulse method. The differences in the values of the reinforcement stationary potential, concrete cover resistivity, and corrosion current density, measured on the same specimens but at different temperatures, in some cases amounted to several dozen percent. This means that measurements of real structural elements made while using the GPM at different temperatures (seasons) can lead to incorrect estimation of the probability of reinforcement corrosion in the examined area and the incorrect assessment of its corrosion activity over time. Therefore, it is advisable to specify the appropriate temperature correction factors for measurements made while using the GPM. The authors' intention is to estimate the value of the necessary correction factors.

Author Contributions: Conceptualization, W.R. and A.W.; methodology and tests, W.R. and A.W.; formal analysis, W.R. and A.W.; data interpretation, W.R. and A.W.; writing—Original draft preparation, W.R. and A.W.; editing, W.R.; funding acquisition, W.R. and A.W. All authors have read and agreed to the published version of the manuscript.

Funding: This research was funded by grant number 02.0.06.00/2.01.01.01. 0007; MNSP. BKWB. 16.001 "Analysis of limit states, durability and diagnostics of structures and methods and tools for quality assurance in construction" [Kielce University of Technology, Kielce, Poland].

Conflicts of Interest: The authors declare no conflict of interest.

References

1. *EN 1992-1-1:2008 Eurocode 2. Design of Concrete Structures. Part 1-1: General Standard and Standards for Buildings;* Polish Committee for Standardization: Warsaw, Poland, 2008.
2. Neville, A.M. *Properties of Concrete*, 5th ed.; Pearson: London, UK, 2011.
3. Tang, S.W.; Yao, Y.; Andrade, C.; Li, Z.J. Recent durability studies on concrete structure. *Cem. Concr. Res.* **2015**, *78 Pt A*, 143–154. [CrossRef]
4. Baltazar-Zamora, M.A.; Bastidas, D.; Santiago-Hurtado, G.; Mendoza-Rangel, J.M.; Gaona-Tiburcio, C.; Bastidas, J.M.; Almeraya-Calderón, F. Effect of Silica Fume and Fly Ash Admixtures on the Corrosion Behavior of AISI 304 Embedded in Concrete Exposed in 3.5% NaCl Solution. *Materials* **2019**, *12*, 4007. [CrossRef] [PubMed]
5. Gjørv, O.E. Durability of Concrete Structures. *Arab. J. Sci. Eng.* **2011**, *36*, 151–172. [CrossRef]
6. Drobiec, Ł.; Jasiński, R.; Piekarczyk, A. *Diagnosis of Reinforced Concrete Structures. Methodology, Field and Laboratory Testing of Concrete and Steel*, 1st ed.; PWN: Warsaw, Poland, 2010.
7. Luo, D.; Li, Y.; Li, J.; Lim, K.S.; Nazal, N.A.M.; Ahmad, H. A Recent Progress of Steel Bar Corrosion Diagnostic Techniques in RC Structures. *Sensors* **2018**, *19*, 34. [CrossRef] [PubMed]
8. Jaśniok, M.; Jaśniok, T. Measurements on corrosion rate of reinforcing steel under various environmental conditions, using an insulator to delimit the polarized area. *Procedia Eng.* **2017**, *193*, 431–438. [CrossRef]
9. Verma, S.K.; Bhadauria, S.S.; Akhtar, S. Monitoring Corrosion of Steel Bars in Reinforced Concrete Structures. *Sci. World J.* **2014**, *2014*. [CrossRef] [PubMed]

10. Raczkiewicz, W.; Kossakowski, P.G. Electrochemical Diagnostics of Sprayed Fiber-Reinforced Concrete Corrosion. *Appl. Sci.* **2019**, *9*, 3763. [CrossRef]
11. Schabowicz, K. Non-destructive testing of materials in civil engineering. *Materials* **2019**, *12*, 323. [CrossRef] [PubMed]
12. Hoła, J.; Schabowicz, K. State-of-the-art non-destructive methods for diagnostic testing of building structures—anticipated development trends. *Arch. Civ. Mech. Eng.* **2010**, *10*, 5–18. [CrossRef]
13. Stankiewicz, H. (Ed.) *Corrosion Protection Technique. Sec. XI: Corrosion and Resistance of Mineral Materials*; PWSZ: Warsaw, Poland, 1973.
14. Kurdowski, W. *Cement and Concrete Chemistry. Industrial Chemistry and Chemical Engineering*; Springer: Berlin/Heidelberg, Germany, 2014.
15. Bertolini, L.; Elsener, B.; Pedeferri, P.; Polder, R. *Corrosion of Steel in Concrete*, 2nd ed.; Wiley VCH: Weinheim, Germany, 2004.
16. Chess, P.; Green, W. *Durability of Reinforced Concrete Structures*, 1st ed.; CRC Press: Boca Raton, FL, USA, 2019.
17. Brodnan, M.; Kotes, P.; Vanerek, J.; Drochytka, R. Corrosion Determination of Reinforcement Using The Electrical Resistance Method. *MaterialiinTehnologije* **2017**, *51*, 85–93. [CrossRef]
18. Brodnan, M.; Koteš, P.; Bahleda, F.; Šebök, M.; Kučera, M.; Kubissa, W. Using non-destructive methods for measurement of reinforcement corrosion in practice. *Prot. Against Corros.* **2017**, *21*, 55–58.
19. Morris, W.; Vico, A.; Vazquez, M.; de Sanchez, S.R. Corrosion of reinforcing steel evaluated by means of concrete resistivity measurements. *Corros. Sci.* **2002**, *44*, 81–99. [CrossRef]
20. Østvik, J.-M. Thermal Aspects of Corrosion of Steel in Concrete: Effect of Low Temperature on the Resistivity and the Cathodic Reaction Rate. Ph.D. Thesis, Norwegian University of Science and Technology, Trondheim, Norway, 2005.
21. Raczkiewicz, W.; Wójcicki, A.; Grzmil, W.; Zapala-Slaweta, J. Impact of Environment Conditions on the Degradation Process of Selected Reinforced Concrete Elements. IOP Conference Serie—Materials Science and Engineering. In Proceedings of the 3rd World Multidisciplinary Civil Engineering, Architecture, Urban Planning Symposium (WMCAUS 2018), Prague, Czech RepublicPrague, Czech Republic, 18–22 June 2018; IOP: London, UK, 2019; Volume 471. [CrossRef]
22. Bäßler, R.; Burkert, A.; Frølund, T.; Klinghofer, O. Usage of GPM-Portable equipment for determination of corrosion stage of concrete structures. *Corrosion* **2003**, 03388.
23. Frølund, T.; Jensen, F.M.; Bäßler, R. *Smart Structures: Determination of Reinforcement Corrosion Rate by Means of the GPT*; IABMAS Congress: Barcelona, Spain, 2002.
24. Sørense, H.E.; Frølund, T. Monitoring of reinforcement corrosion in marine concrete structures by the GPM. In Proceedings of the International Conference on Concrete in Marin Environments, Hanoi, Vietnam, October 2012; pp. 213–220. [CrossRef]
25. Raczkiewicz, W.; Wójcicki, A. Selected aspects of forecasting the level of reinforcing steel corrosion in concrete by electrochemical method. *Weld. Rev.* **2017**, *89*, 28–33.
26. Elsener, B.; Klinghoffer, O.; Frolund, T.; Rislund, E.; Schiegg, Y.; Böhni, H. Assessment of reinforcement corrosion by means of galvanostatic pulse technique. In Proceedings of the International Conference—Repair of Concrete Structures, Svolvaer, Norway, 28–30 May 1997.
27. Frølund, T.; Klinghofer, O.; Poulsen, E. *Rebar Corrosion Rate Measurements for Service Life Estimates*; ACI Fall Convention: Toronto, ON, Canada, 2000.
28. Raczkiewicz, W.; Wójcicki, A. Evaluation of effectiveness of concrete coat as a steel bars protection in the structure—galvanostatic pulse method. In Proceedings of the 26th International Conference On Metallurgy And Materials (METAL), Brno, Czech Republic, 24–26 May 2017.
29. Raczkiewicz, W.; Grzmil, W. Assessment of the impact of cement type on the process of concrete carbonation and reinforcement corrosion in reinforced concrete specimens. *Cem. Lime Concr.* **2017**, *4*, 311–319.
30. Vedalakshmi, R.; Balamurugan, L.; Saraswathy, V.; Kim, S.-H.; Ann, K.Y. Reliability of Galvanostatic Pulse Technique in assessing the corrosion rate of rebar in concrete structures: Laboratory vs field studies. *Ksce J. Civ. Eng.* **2010**, *14*, 867–877. [CrossRef]

31. Allampallewar, S.B.; Srividya, A. Corrosion performance of reinforced concrete member along Indian coasts: Effect of temperature & relative humidity. *Int. J. Perform.* **2008**, *4*, 285–292.
32. Available online: http://www.germann.org/TestSystems/GalvaPulse/GalvaPulse.pdf (accessed on 28 September 2019).

MDPI
St. Alban-Anlage 66
4052 Basel
Switzerland
Tel. +41 61 683 77 34
Fax +41 61 302 89 18
www.mdpi.com

Applied Sciences Editorial Office
E-mail: applsci@mdpi.com
www.mdpi.com/journal/applsci

www.ingramcontent.com/pod-product-compliance
Ingram Content Group UK Ltd.
Pitfield, Milton Keynes, MK11 3LW, UK
UKHW060124010826
14089UKWH00006B/137